Logical and Computational Aspects of Model-Based Reasoning

APPLIED LOGIC SERIES

VOLUME 25

Managing Editor

Dov M. Gabbay, *Department of Computer Science, King's College, London, U.K.*

Co-Editor

Jon Barwise†

Editorial Assistant

Jane Spurr, *Department of Computer Science, King's College, London, U.K.*

SCOPE OF THE SERIES
Logic is applied in an increasingly wide variety of disciplines, from the traditional subjects of philosophy and mathematics to the more recent disciplines of cognitive science, computer science, artificial intelligence, and linguistics, leading to new vigor in this ancient subject. Kluwer, through its Applied Logic Series, seeks to provide a home for outstanding books and research monographs in applied logic, and in doing so demonstrates the underlying unity and applicability of logic.

The titles published in this series are listed at the end of this volume.

Logical and Computational Aspects of Model-Based Reasoning

Edited by

LORENZO MAGNANI

University of Pavia, Pavia, Italy

NANCY J. NERSESSIAN

Georgia Institute of Technology,
Atlanta, Georgia, U.S.A.

and

CLAUDIO PIZZI

University of Siena, Siena, Italy

SPRINGER SCIENCE+BUSINESS MEDIA, B.V.

A C.I.P. Catalogue record for this book is available from the Library of Congress.

ISBN 978-1-4020-0791-0   ISBN 978-94-010-0550-0 (eBook)
DOI 10.1007/978-94-010-0550-0

Printed on acid-free paper

Mathematical reasoning consists in constructing a diagram according to a general precept, in observing certain relations between parts of that diagram not explicitly required by the precept, showing that these relations will hold for all such diagrams, and in formulating this conclusion in general terms. All valid necessary reasoning is in fact thus diagrammatic.

Charles Sanders Peirce, *Collected Papers*, 1.54

Series Editor's Preface

Information technology has been, in recent years, under increasing commercial pressure to provide devices and systems which help/replace the human in his daily activity. This pressure requires the use of logic as the underlying foundational workhorse of the area. New logics were developed as the need arose and new foci and balance has evolved within logic itself. One aspect of these new trends in logic is the rising importance of model based reasoning. Logics have become more and more tailored to applications and their reasoning has become more and more application dependent. In fact, some years ago, I myself coined the phrase "direct deductive reasoning in application areas", advocating the methodology of model-based reasoning in the strongest possible terms. Certainly my discipline of Labelled Deductive Systems allows to bring "pieces" of the application areas as "labels" into the logic.

I therefore heartily welcome this important book to Volume 25 of the Applied Logic Series and see it as an important contribution in our overall coverage of applied logic.

Dov M. Gabbay, London

Preface

The volume is based on the papers that were presented at the International Conference *Model-Based Reasoning: Scientific Discovery, Technological Innovation, Values* (MBR'01), held at the Collegio Ghislieri, University of Pavia, Pavia, Italy, in May 2001. The previous volume *Model-Based Reasoning in Scientific Discovery*, edited by L. Magnani, N.J. Nersessian, and P. Thagard (Kluwer Academic/Plenum Publishers, New York, 1999; Chinese edition, China Science and Technology Press, Beijing, 2000), was based on the papers presented at the first "model-based reasoning" international conference, held at the same place in December 1998.

The presentations given at the Conference explored how scientific thinking uses models and exploratory reasoning to produce creative changes in theories and concepts. Some addressed the problem of model-based reasoning in ethics, especially pertaining to science and technology, and stressed some aspects of model-based reasoning in technological innovation.

The study of diagnostic, visual, spatial, analogical, and temporal reasoning has demonstrated that there are many ways of performing intelligent and creative reasoning that cannot be described with the help only of traditional notions of reasoning such as classical logic. Understanding the contribution of modeling practices to discovery and conceptual change in science requires expanding scientific reasoning to include complex forms of creative reasoning that are not always successful and can lead to incorrect solutions. The study of these heuristic ways of reasoning is situated at the crossroads of philosophy, artificial intelligence, cognitive psychology, and logic; that is, at the heart of cognitive science.

There are several key ingredients common to the various forms of model-based reasoning. The term "model" comprises both internal and external representations. The models are intended as interpretations of target physical systems, processes, phenomena, or situations. The models are retrieved or constructed on the basis of potentially satisfying salient constraints of the target domain. Moreover, in the modeling process, various forms of abstraction are used. Evaluation and adaptation take place in light of structural, causal, and/or functional constraints. Model simulation can be used to produce new states and enable evaluation of behaviors and other factors.

The various contributions of the book are written by interdisciplinary researchers who are active in the area of creative reasoning in science and technology, and are *logically* and *computationally* oriented: the most recent results and achievements about the topics above are illustrated in detail in the papers.

In recent years novel analyses of logical models of model-based reasoning have been undertaken. Many of these are represented in this volume. For example, the recent attention given to diagrammatic reasoning (in general and especially in the case of geometrical aspects) has promoted the creation of many new logical so-called "heterogeneous" models (G. Allwein and N. Swoboda). It is possible to translate diagrams into a formal predicate logic, so exploiting linear description language and rules of inferences, and consequently to use deductive methods to make model-based inferences (L. Pineda; A. Shimojima). Moreover, adaptive logic is able to construct models of metaphorical (J. D'Hanis), abductive (J. Meheus, L. Verhoeven, M. Van Dyck, and D. Provijn), and explanatory diagnostic reasoning (D. Provijn and E. Weber). New perspectives belonging to the tradition of nonmonotonic and defeasible logic oriented to the clarification of logical foundations of abductive reasoning are being developed (I.C. Burger and J. Heidema; A. Nepomuceno-Fernández). Additionally, problems related to theorem proving in modeling the reasoning capabilities of human mathematicians are being addressed (S. Choi).

Moreover, very interesting artificial intelligence computer programs and computational frameworks and architectures have been built, that represent various model-based reasoning performances, for example, in scientific discovery. Several of the papers in this volume aim to increase "computational" knowledge about the role of model-based reasoning in various tasks: computational methods and representation for models for discovering communicable knowledge in two scientific domains (P. Langley); PRET, a computer program that automatically builds the models which the engineers need in the case of a physical system balances accuracy and parsimony (R. Stolle, M. Easley, and E. Bradley); the integration of knowledge-based modeling or modeling from first principles, with data-driven modeling of dynamic systems in the field of automated equation discovery (S. Dzeroski and L. Todorovski); consistency-based diagnosis as a kind of model-based approach to diagnostic reasoning (B. Górny and A. Ligeza); modeling in planning and prognosis to provide a general architecture for building normative "objective models" (E. Finkeissen), and, finally, the design of modeling as a process in which human intervention is considered in three different levels, namely, individual, working group, and social level, where a structural framework for computational models in terms of a mathematical language is proposed (G.M. da Nóbrega, P. Malbos, and J. Sallantin).

The conference, and thus indirectly this book, was made possible through the generous financial support of the MURST (Italian Ministry of the University), University of Pavia, CARIPLO (Cassa di Risparmio delle Provincie Lombarde) and of Georgia Institute of Technology. Their support is gratefully acknowledged.

The editors express their appreciation to the other co-chair of the conference K. Knoespel (Georgia Institute of Technology, Atlanta, GA, USA), and to the members of the Scientific Committee for their suggestions and assistance: A. Bostrom, Georgia Institute of Technology, Atlanta, GA, USA; E. Gagliasso, University of Rome "La Sapienza", Rome, Italy; D. Gentner, Northwestern University, Evanston, USA; R. Giere, University of Minnesota, Minneapolis, MN, USA; M.L. Johnson, University of Oregon, Eugene, OR, USA; P. Langley, Stanford University, Stanford, CA, USA; B. Norton, Georgia Institute of Technology, Atlanta, GA, USA; Mario Stefanelli, University of Pavia, Pavia, Italy; P. Thagard, University of Waterloo, Waterloo, Ontario, Canada; Ryan D. Tweney, Bowling Green State University, Bowling Green, OH, USA; S. Vosniadou, National and Capodistrian University of Athens, Athens, Greece.

Special thanks to the members of the Local Organizing Committee R. Dossena, E. Gandini, M. Piazza, and S. Pernice, for their contribution in organizing the conference, to R. Dossena for his contribution in the preparation of this volume, and to the copy-editors L. d'Arrigo and M. Piazza. The preparation of the volume would not have been possible without the contribution of resources and facilities of the Computational Philosophy Laboratory and of the Department of Philosophy, University of Pavia. Also special thanks to Dov M. Gabbay for having accepted this book in the "Applied Logic Series" he edits (formerly with the late Jon Barwise) for KluwerAcademic Publishers.

The more philosophical, epistemological, and cognitive oriented papers deriving from the presentations given at the Conference have been published in the book L. Magnani and N.J. Nersessian, eds., *Model-Based Reasoning: Science, Technology, Values*, Kluwer Academic/Plenum Publishers, New York, 2002. The remaining selected papers will be published in five Special Issues of Journals: in *Foundations of Science*, Abductive Reasoning in Science; in *Foundations of Science*, Model-Based Reasoning: Visual, Analogical, Simulative; in *Mind and Society*, Scientific Discovery: Model-Based Reasoning; in *Mind and Society*: Commonsense and Scientific Reasoning, all edited by L. Magnani and N.J. Nersessian; in *Philosophica*, Diagrams and the Anthropology of Space, edited by K. Knoespel.

Lorenzo Magnani, Pavia, Italy
Nancy J. Nersessian, Atlanta, GA, USA
Claudio Pizzi, Siena, Italy

Contents

LOGICAL ASPECTS OF
MODEL-BASED REASONING

A CASE STUDY OF THE DESIGN AND IMPLEMENTATION OF HETEROGENEOUS REASONING SYSTEMS

Nik Swoboda

Indiana University and ATR Media Integration and Communications Laboratory,
Department of Computer Science, Bloomington, IN 47405, USA
nswoboda@cs.indiana.edu

Gerard Allwein

Indiana University, Department of Computer Science, Bloomington, IN 47405, USA
gtall@cs.indiana.edu

Abstract   In recent years we have witnessed a growing interest in heterogeneous reasoning systems. A heterogeneous reasoning system incorporates representations from a number of different representation systems, in our case a sentential and a diagrammatic system. The advantage of heterogeneous systems is that they allow a reasoner to bridge the gaps among various formalisms and construct threads of proof which cross the boundaries of the systems of representation. In doing this, these heterogeneous systems allow the reasoner to take advantage of each component system's ability to express information in that component's area of expertise. The purpose of this paper is twofold: to propose a general theoretical framework, inspired by Barwise and Seligman's work in *Information Theory* [Barwise and Seligman, 1997], for the design of heterogeneous reasoning systems and to use this framework as the basis of an implementation of a First Order Logic and Euler/Venn reasoning system.

1.   Background

Recently there has been a significant amount of interest in a new and growing area of research referred to as diagrammatic or visual reasoning. Researchers in this area are interested in how "picture-like" representa-

L. Magnani, N.J. Nersessian, and C. Pizzi (eds.),
Logical and Computational Aspects of Model-Based Reasoning, 3–20.

tions such as diagrams and maps can play a role in reasoning processes. This field has drawn interest from scientists from many different backgrounds. Some of these include psychologists interested in the role that diagrams play in cognition, computer scientists interested in how diagrams can play a role in computation, and semanticists interested in the notion of meaning that these representations employ, just to name a few. In this paper the problems associated with the design of a heterogeneous reasoning system will be approached from the point of view of logicians interested in developing a theoretical framework for this kind of reasoning. The ultimate goal of this work is to apply that framework to the design of a First Order Logic (FOL) and Euler/Venn reasoning system, and in doing so provide a system that is robust enough to serve as the basis of an implementation.

Hyperproof was the first case study of a computer aided reasoning system that makes extensive use of both diagrams and sentences [Barwise and Etchemendy, 1995; Barwise and Etchemendy, 1998; Barwise and Etchemendy, 1994]. *Hyperproof* allows the user to reason with sentences of FOL and diagrams representing a blocks world. Currently, work is under way to expand the *Hyperproof* framework to allow the user to reason with any number of sentential and diagrammatic systems; this new system will be called *Openproof*. The work done to implement the system presented here is based upon this new *Openproof* framework and is the first heterogeneous system built with this framework.

2.   Overview

Designing a reasoning system requires one to formalize the syntax, semantics and notion of inference for that system. Many such formalisms exist today, a number of which are specially designed for reasoning in a very specific domain. These specialized systems, such as those for using wiring diagrams and architectural blueprints, have shown their usefulness when reasoning within their respective domains. The disadvantage of using these specialized systems is that, when one wishes to reason about more than one of these domains together, one must either use a general purpose language or design a new system which is *conceptually* the union of the individual systems. By electing to employ a general purpose language, we lose the advantages gained by using a specialized reasoning system; hence designing a new system is in many cases preferable. However, this can be a difficult and time consuming task. Thus what will be presented here is a case study of a heterogeneous reasoning system which could be thought of as a starting point for a general

framework, based in Barwise/Seligman's *Information Theory* [Barwise
and Seligman, 1997], which will more easily allow the coupling of various
systems of reasoning into one unit, a heterogeneous reasoning system.

We start with two separate reasoning systems, which for our purposes
here will consist of a FOL reasoning system and an Euler/Venn reasoning
system. Both of these systems, separately, are well understood. For
the basis of the sentential system we will use a standard formalization
of a FOL reasoning system, for details of one such formalism consult
Barwise and Etchemendy's system described in [Barwise et al., 1999].
For the Euler/Venn system we take a slightly modified version of Shin
and Hammer's work [Shin, 1996; Hammer, 1995] which is described in
detail in [Swoboda, 2002]. It should be noted that although FOL can
be thought of as a general purpose reasoning language, it was not chosen
(and will not be used) for this reason. It was chosen merely as a common
example of a sentence-based reasoning system.

Since we begin with the assumption that each of our system's representations carries information regarding portions of the same domain,
our next step in modeling the heterogeneous system will be to develop a
common domain for all the representations with which we plan to reason. In the case of our FOL and Euler/Venn system, we can take as the
common domain a variation of standard FOL models having only unary
predicates with the standard notion of truth.[1] We will use the modified
version of Hammer's system as the basis of truth in the diagrammatic
system. See Figure 1 for an illustration of this general framework.

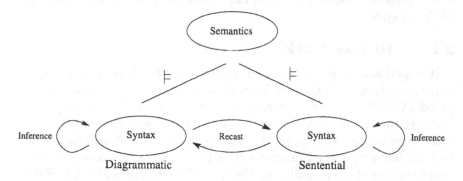

Figure 1.   A heterogeneous system.

---

[1]Though the common domain of these two systems will consist of only models having unary
predicates, the homogeneous FOL system will allow reasoning with n-ary relations.

Once we have a common domain, we are left with defining rules of inference that allow the exchange of information between the various representations. We will refer to these rules generically as the **Recast Rule**, but will separately define the use of the rule for each of the pairs of representations in the system. For each pair there will also be a separate instance of the rule for the exchange of information in both directions. For example, in the case of our FOL and Euler/Venn system we will have two ways to use this rule, one for the extraction of information from a diagram to be expressed in sentential form and another that allows the extraction of information from a formula to be incorporated into a diagram. These rules of inference will be defined by introducing a fourth collection of mathematical structures, which we will call *information types*, into the middle of Figure 1, and then by defining relations between the semantics, the two systems of syntax, and this new collection. For any given heterogeneous systemheterogeneous systems there may be many different collections of information types and relations that could be selected, each resulting in different kinds of recasting rules. In the following case study we will advocate a notion of recasting that is based upon the explicit information content of a representation and one which is symmetric, i.e., recasting from one kind of representation to a second kind of representation will be defined using the same principles as recasting from the second kind of representation to the first.

Once this general framework is presented, along with its application to the design of a FOL and Euler/Venn reasoning system, we will then briefly present some details of the implementation of this FOL and Euler/Venn system.

2.1   Related work

It should also be noted that this system is not the first attempt at an implementation of a FOL and Venn Reasoning system. A basic system called JVENN has been presented in [Sawamura and Kiyozuka, 2000]. However that implementation only allows Venn diagrams containing at most three curves (and no Euler type diagrams in which one curve can be completely contained in another). Also the system has only a very simple notion of rules similar to *Hyperproof's* Observe and Apply Rules and does not allow the construction of heterogeneous proofs.

3.   Some preliminary definitions

Before beginning, it is useful to give a brief introduction to Euler/Venn diagrams and to explain a number of terms that will be used in the following sections. While this treatment was inspired by and is quite similar to that found in [Hammer, 1995], there are a number of important differences that should be noted, the most important of which include the fact that the grammar informally presented here adds more well-formed diagrams. See [Swoboda, 2002] for more formal treatment of these diagrams.

An Euler/Venn diagram consists of the following syntactic features:

1 Rectangles - Each rectangle denotes the domain of discourse to be represented by the diagram.

2 Closed Curves - A countably infinite set $C_1, C_2, C_3, \ldots$ of uniquely labeled closed curves. Each closed curve must not intersect itself. Each curve is taken to represent the set which corresponds to its label.

3 Shading - The shading of any part of the diagram denotes that the set represented by that part of the diagram is empty.

4 Constant Symbols - A countably infinite set of individual constant symbols, each of which will be used to indicate the existence of members in the set represented by the constant symbol's section of the diagram.

5 Lines - Lines will be used to connect matching individual constant symbols in different parts of the diagram to illustrate the uncertainty of which set contains the object referred to by that constant. A series of constant symbols connected by lines will be referred to as a *constant sequence*.

An example Euler/Venn diagram can be found in Figure 2. This diagram contains a small amount of information regarding Japanese sake. From the diagram, one can see that Nigori and Ginjo are two different kinds of sake, and that no sake is both a Nigori and a Ginjo due to the shading of the intersection of the two curves. One can also see that Umeshu is either a sake which is neither a Ginjo nor a Nigori, or a Ginjo due to the placement of the two constant symbols connected by a line.

A *region* is any area of a diagram completely enclosed by lines of that diagram. Regions are closed under union, intersection, and complement; thus a region may contain disconnected parts. Any region of the diagram completely enclosed by a closed curve is referred to as a *basic region*. A *minimal region* is any region which is not crossed by any of the lines of that diagram. A *complete Euler/Venn diagram*, a diagram in which all possible intersections of curves exist, that has n basic regions will contain 2^n minimal regions. Other diagrams with Euler-like features, expressing set containment, will contain fewer minimal regions. This

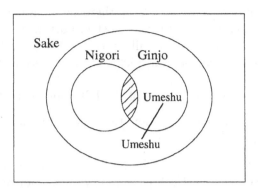

Figure 2.   An example Euler/Venn diagram.

lack of certain minimal regions in these diagrams is taken to mean that
the set the missing region represents is empty. Since this "lack of a
region" is meaningful, we will refer to these regions as *missing regions*.
For example, in the left diagram of Figure 3 the region denoting the
intersection of C and the complement of B is missing.

Lastly we will look at an informal definition of the notion of a well-
formed Euler/Venn diagram. Any diagram only consisting of a rectangle
is a well-formed Euler/Venn diagram. If a diagram is well-formed, that
diagram with the following modifications is also well-formed:

1 The addition of any closed curve C with a unique label N completely within the rectangle of the diagram so that the regions
   intersected by C are split into at most two new regions.[2]

2 The shading of any region.

3 The addition of an individual constant symbol that is not already
   in the diagram to any region.

4 The addition of a constant symbol that is already in the diagram
   to any region not containing that constant symbol along with a
   line connecting this new constant symbol to an existing symbol.

Some examples of well-formed and not well-formed Euler/Venn diagrams can be found in Figure 3.

---

[2]This stipulation, while more general than that used by Hammer [Hammer, 1995], is still
not as general as one might like. But it can be proved that any Euler/Venn diagram can be
expressed under this restriction.

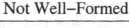

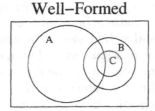 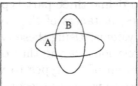

Figure 3. Examples of well-formed and not well-formed Euler/Venn diagrams.

4. A framework for heterogeneous reasoning

When working with a heterogeneous reasoning system, one of the underlying assumptions is that all of the representations carry information about some common domain. In other words, though each kind of representation can represent different kinds of information in different ways, there is a relationship between the information contained in the various representations. If this were not the case, the system as a whole would be of little use. If there were no connection between the information expressed by the various parts, there would be no point in having combined them into a conglomerate reasoning system.

Thus, we begin with a reasoning system consisting of a common domain, in the form of a collection of mathematical structures, two collections of well-formed representations, and notions of truth relating these representations to the domain.[3] One might think that this alone would suffice for the definition of notions of recasting between these two kinds of representations. For example, we could say that you can recast one representation into a second when all the domain instances in which the first was true were instances in which the second representation was true as well. However, this notion of recasting would be too reminiscent of a notion of logical consequence, e.g., you could recast an inconsistent representation into any representation, and recast any representation into a logically true representation. Unfortunately such a relation does not seem to capture our goal of modeling the explicit information content of a representation.

Thus, to accomplish our goal of being able to define many different kinds of recast relations, including those based upon the explicit information content of a representation, we will introduce another collection

---

[3]Simply put, a diagram is true in a domain instance if the information represented by the diagram is contained in the state of the world represented by the domain instance.

of structures, information types, into our framework. Once introduced, we can then define three part type-token relations between these information types, instances of the domain, and well-formed representation from both systems. Using these type-token relations, we will then define our recast relations, saying that one representation can be recast into a second when all of the types to which the first representation belongs are also types to which the second representation belongs. In general, it is advantageous to carefully select this collection of information types to precisely capture the information content of each of the system's representations. The more tailored the information types are to the kind of information that the system's representations can carry, the closer the resulting notions of recasting will be to the explicit information content of those representations. Note that in defining a notion of recasting there is no interlingua, no "combined" syntax, being used. We can reason with the systems together solely in terms of their connections to the common domain being represented.

5. Defining the notion of recasting for Euler/Venn and FOL

Before giving a summary of how the instances of the Recast Rule will be defined, it is necessary to embark on a short discussion of some of the philosophical base for these rules.

Our first step will be to define a notion of recasting information from Euler/Venn diagrams to FOL. The basic goal for the definition of this rule will be that only things that the reasoner can *see* to be the case from the diagram should be able to be recast from that diagram. We will refer to this special kind of recasting as *observation*. However, as you can imagine, what one can see to be the case from the diagram is a matter open for debate. There is a range of attitudes, ranging from claiming that anything that follows from the diagram should be capable of being observed, to arguing that only the basic syntactic features of the diagram can be observed (for example, that there is an icon labeled "Bob" in the lower left hand corner of the diagram). Both of these extremes are undesirable. At one extreme, we have a point of view that weakens the connection between observation and visual activity and results in the adoption of a notion that is closer to logical consequence than seeing. At the other extreme, we would be denying the fact that the diagram has some interpretation, and carries information in virtue of this interpretation. A notion of observation that is between these two extremes will be presented, one that should preserve the strong connection between observation and seeing while keeping in mind the interpretation of the diagram.

Dretske's definition of *secondary seeing that* [Dretske, 1969] was found to capture this intuition for observation. Put simply, for Dretske an agent can see that b is P when there is a c that can be non-epistemically seen, and in virtue of the state of c it must be the case that b is P. Thus using Dretske's notion of seeing that as a guide, we will say that a sentence can be observed from the diagram when there exist one or more syntactic features of the diagram whose interpretation is evidence for that sentence. Likewise, we will say that a sentence cannot be observed from a diagram when there exist one or more syntactic features that cause that sentence to fail.

Unfortunately, we are not out of the woods yet; the task of defining the notion of observation for these diagrams is still difficult. Part of these difficulties arise from an Euler/Venn diagram's unique ability to express:

1 Negative information - By shading a region of the diagram, one is able to express that the set denoted by that region is empty. When using traditional FOL models, this causes problems since it is not possible to explicitly represent in these models the fact that the intersection of the extensions of two predicates is empty.

2 Inconsistent information - By placing all the links of a constant sequence in shaded regions of the diagram, one is able to construct a diagram which has no FOL models. This is problematic because we would still like to be able to make observations from these diagrams and exclude the observation of inconsistent sentences.

3 Disjunctive information - By constructing constant sequences, one is able to express uncertainty regarding which set contains the object indicated by the constant symbol. This complicates things because, when working with traditional FOL models, it is not possible to have uncertainty regarding the assignment of objects to the extensions of predicates.

4 Empty domain - By shading the entire box of an Euler/Venn diagram, one is able to express that domain of the diagram is empty. This is problematic because this violates a basic requirement on traditional FOL models, that their domains be non-empty.

Examples of diagrams containing these sorts of information can be found in Figure 4.

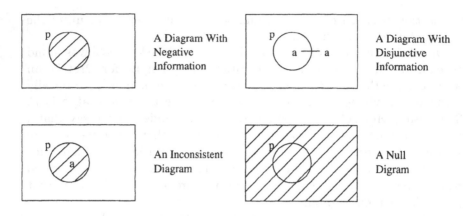

Figure 4.   Some interesting Euler/Venn diagrams.

Two different but essentially equivalent notions of observation from Euler/Venn diagrams to FOL will be presented. First, a strong notion of observation will be defined for a special fragment of monadic first order logic (MFOL). This strong notion will be defined recursively on the structure of the formulas in this fragment of MFOL and features of the Euler/Venn diagram. The main virtues of this definition are that it is simple and intuitive. Next, a more semantic notion of observation using a generalization of partial structures will be presented. This second notion will be defined for all formulas of MFOL, not just a fragment, and will more precisely characterize the information content of the diagram.

5.1   A strong notion of observation from Euler/Venn to FOL

By carefully choosing a fragment of MFOL so as to have only formulas that represent information very similar to the information contained in the Euler/Venn diagrams themselves, a recursive definition for observation from Euler/Venn to FOL can be presented. This fragment will be referred to as Euler/Venn Observational Formulas (**EVOF**). In this fragment there are no embedded quantifiers, and only predicate expressions containing the quantified variable can occur in the scope of a quantifier. Furthermore a number of generalized quantifiers will be used in place of the traditional universal and existential quantifiers of FOL. These new quantifiers include: Nx $\varphi(x)$ which will be read as "there is no x such that $\varphi(x)$", All x $(\varphi_1(x), \varphi_2(x))$ which will be read as "all x such that

$\varphi_1(x)$ then $\varphi_2(x)$", and Some x $(\varphi_1(x), \neg\varphi_2(x))$ which will be read as "there is some x such that $\varphi_1(x)$ and not $\varphi_2(x)$".

It should also be noted that the strong observation relation that we are proposing does not preserve logical equivalence. In other words, we will only be able to observe formulas which correspond to the interpretation of syntactic features of the diagram and we will not be able to observe logical consequences of those formulas which do not correspond to the interpretation of features of the diagram. For example, from the left diagram of Figure 3, a reasoner will be able to observe that All x $(C(x), B(x))$, while Nx $(C(x) \wedge \neg B(x))$ cannot be observed that from that same diagram.

The basic idea behind formulas of **EVOF** is that unquantified formulas make a statement similar in meaning to the placement of constant symbols in the diagram, and that quantified formulas make a statement similar in meaning to the shading of regions of the diagram, the containment of curves in the diagram, or the existence of constant symbols in the diagram. In other words, every formula in **EVOF** can be thought of as making a statement resembling the meaning of various features of an Euler/Venn diagram.

Once a partial function associating unquantified formulas of **EVOF** to regions of a diagram is defined, we can define the relation of strong observation for Euler/Venn diagrams recursively on the structure of formulas of **EVOF**. Then it can be shown that if a formula can be observed from the diagram, then it is a logical consequence of that diagram, and that strong observation is decidable. Lastly, it can be shown that a diagram is the logical consequence of all the formulas observable from it. From this result, we get as a corollary that any logical consequence of a diagram is a consequence of formulas that can be observed from that diagram. Further details can be found in [Swoboda, 2001].

5.2   Observation using Euler/Venn information types

When thinking about a more semantic way of defining the notion of observation, one's first inclination is to use FOL models as the basis of the definition. Unfortunately, this is not possible due to the problematic diagrams mentioned above. An example of one such problem is that traditional FOL models can not contain uncertainty regarding whether or not an object is contained in the extension of a predicate. Once this is realized, one might be inclined to use a traditional collection of FOL models, but one can not explicitly express inconsistent information using these models either. Yet another approach would be to use a collection of partial models. Again problems arise. Partial models do not allow the expression of the kind of negative information contained in these diagrams, e.g., that the intersection of the extensions of a number of predicates is empty. After realizing these difficulties, the conclusion was reached that a new approach was needed. This new approach will involve the use of what will be called *Euler/Venn information types* which are a generalization of partial structures.

An Euler/Venn information type consists of four parts:

- A non-empty domain.

- A function assigning disjoint subsets of the domain to each potential minimal region of a diagram.

- A function assigning either $\emptyset$ or $\bullet$ to each potential minimal region of a diagram. When a region is assigned $\bullet$, this indicates that the set represented by the region in question is definitely empty.

- A partial function assigning members of the domain to constant symbols.

Once defined, these information types will serve a central role in the definition of observation. By defining relations between these types and well-formed Euler/Venn diagrams, Hammer models (a specialized model for MFOL used as the basis of reasoning with Euler/Venn diagrams), and formulas of MFOL, we then have a way of reasoning about these various parts together. For example, one can define the notion of observation between Euler/Venn diagrams and formulas of MFOL in terms of the relations that each of these have with Euler/Venn information Types. An illustration of this system can be found in Figure 5. Note that, although the terminology used here is somewhat different from that used by Barwise and Seligman, this framework is inspired by their recent work on Information Flow [Barwise and Seligman, 1997].

After this second notion of observation is defined, it can be shown that, once the distinction between the generalized quantifiers of **EVOF** is blurred in the definition of the relation of strong observation, the Euler/Venn information type-based notion of observation can be characterized by the strong notion of observation. With this result, the second notion of observation can then be shown to be decidable. A more formal treatment of this material can be found in [Swoboda, 2001].

5.3   Recasting from FOL to Euler/Venn

In the previous sections, a special kind of recasting, which we referred to as "observation", was discussed that allowed the extraction of

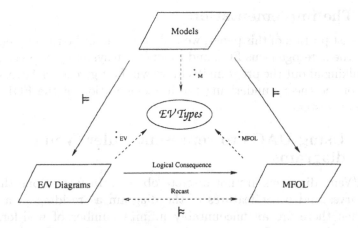

Figure 5.   The use of Euler/Venn Information Types to define Observation.

information from an Euler/Venn diagram to be expressed in FOL. We will now briefly look at the opposite of this, the use of recasting in the construction of an Euler/Venn diagram from information in sentential form. Though the notion of the explicit information content of a diagram based upon the existence of visual features in the diagram required a good deal of motivation (which we provide in the form of the strong notion of observation), it seems even harder to think about the explicit information content of a formula of FOL. To define a notion of recasting from FOL to Euler/Venn diagrams, we will attempt to mirror the work done to define observation from Euler/Venn diagrams to formulas of FOL, but we will use a slightly different approach in the motivating notion of observation. To do this, we will use Shin's notion of representing facts to make explicit the features of a diagram and define when a formula of **EVOF** supports a representing fact. Using these definitions we will then say that a diagram can be strongly observed from a formula if all of the diagram's representing facts are supported by the formula. For a more general notion of observation from formulas of MFOL to Euler/Venn diagrams, we will again use the collection of information types just presented. In fact, we will define this kind of recasting in a manner completely analogous to that discussed in the last section. Once this is done, the close relationship between the notion of strong observation and the information type-based notion of observation can be proved.

6.   The implementation

In the last portion of this paper, we will briefly describe the implementation of the heterogeneous FOL and Euler/Venn system just presented. Before talking about the program itself, we will first give a brief overview of some of the theory underlying the implementation of the FOL and Euler/Venn system.

6.1   Using DAGs to represent Euler/Venn diagrams

Euler/Venn diagrams are not discrete objects. In fact, if one thinks of the curves and other features of the diagram as residing in a real plane, then there are an uncountably infinite number of well-formed Euler/Venn diagrams. Fortunately, the information that these diagrams carry is not dependent on the continuous nature of the diagram. One can take any Euler/Venn diagram and re-arrange its curves in many different ways to produce logically equivalent diagrams. The first step in the implementation of the heterogeneous system just presented will be to design a discrete way of representing Euler/Venn diagrams which is rich enough to capture all of the diagram's information. In this project, directed acyclic graphs (DAGs) will serve the role of representing the Euler/Venn diagrams with which we plan to reason.

These DAGs consist of nodes representing regions of the diagram. Edges connecting these nodes represent the cover relation between regions of the diagram.[4] Nodes of the DAG also contain annotations which indicate the existence of shading and constant sequence links in the regions of the diagram.

Each DAG will have one root node to represent the diagram's rectangle, the domain of discourse. Below this node will be nodes representing some of the basic regions of the diagram and nodes representing the complement of some the basic regions of the diagram. Below these will be more nodes representing regions which can be thought of as the intersection of basic regions and complements of basic regions. Hence, leaf nodes of the DAG, nodes having no children, will correspond to each of the diagram's minimal regions.

Here are a few sample Euler/Venn diagrams and their corresponding DAGs:

1 A diagram with one curve.

---

[4] "*A covers B*" iff $B \subsetneq A$ and there is no region C such that $B \subsetneq C$ and $C \subsetneq A$.

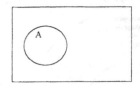

2 A Venn diagram with two curves.

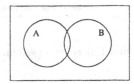

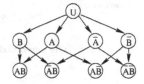

3 An Euler diagram representing set containment.

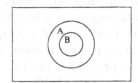

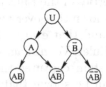

To show that the DAG system is rich enough to represent Euler/Venn diagrams, we can think of this system of DAGs as a second reasoning system. The notion of a well-formed DAG is presented as well as notions of inference and truth for the system. The notions of inference and truth are designed so as to mirror the corresponding notions of the Euler/Venn system. An illustration of the combined DAG and Euler/Venn system can be found in Figure 6.

We will then establish relationships between the two systems (the inner and outer triangles of Figure 6) and thereby show that the DAG system "captures the essential properties"[5] of the Euler/Venn system. First, inductive and non-inductive methods for translating Euler/Venn diagrams to DAGs are presented. The inductive algorithm will allow one to add a curve to an existing DAG, and the non-inductive algorithm will

---

[5]One system *captures the essential properties* of another system if there is a translation or mapping between them that preserves deductive and semantic relations.

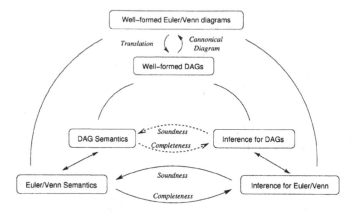

Figure 6.   The combined DAG and Euler/Venn system.

allow the translation of any existing Euler/Venn diagram into a DAG.
Once this is done, it can be shown that the DAG system in itself is both
sound and complete, and finally that the DAG system captures the essential properties of the Euler/Venn system. In order to show this result,
it is first be shown that every Euler/Venn diagram can be translated into
some DAG, and that every DAG is the translation of a unique class of
isomorphic Euler/Venn diagrams. Then to prove the main result, that
the DAG system captures the essential properties of the Euler/Venn
system, it can be shown that, if given a set of Euler/Venn diagrams V
and a single Euler/Venn diagram V along with their respective translations into DAGs, namely $T(V)$ and $T(\{V\})$, V can be inferred from V iff
$T(\{V\})$ can be inferred from $T(V)$ and that V is a logical consequence
of V iff $T(\{V\})$ is a logical consequence of $T(V)$. Further details of these
proofs and the DAG system can be found in [Swoboda, 2002].

6.2   The program

Screen shots from the implementation of the heterogeneous FOL and
Euler/Venn system can be seen in Figure 7. On the left hand side of
the figure you will find the proof window, in which formulas of FOL and
icons representing diagrams are used to construct Fitch style proofs.
On the right of the figure is a sample Euler/Venn diagram being used
as the premise of the yet to be completed proof. The program allows
the user to reason with any number of Euler/Venn diagrams as well as
formulas of FOL. The user can employ all the homogeneous FOL and
homogeneous Euler/Venn rules of inference in any proof. Also, the user
can recast information from any diagram into the form of a formula of

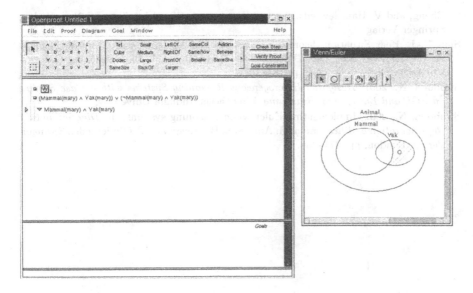

Figure 7.   Screen shots from the implementation.

FOL and to recast any formula of FOL to a Euler/Venn diagram. Lastly, once the user has written a proof candidate, the user can then check the correctness of all inferences made in that proof candidate.

References

Barwise, J. and Etchemendy, J., 1994, *Hyperproof*, CSLI Publications, Stanford.

Barwise, J. and Etchemendy, J., 1995, Heterogeneous logic, in: *Heterogeneous logic*, J.I. Glasgow, N.H. Narayanan, and B. Chandrasekaran, eds., AAAI Press and MIT Press, Menlo Park, pp. 209–232.

Barwise, J. and Etchemendy, J., 1998, Computers, visualization, and the nature of reasoning, in: *The Digital Phoenix: How Computers are Changing Philosophy*, T.W. Bynum and J.H. Moor, eds., Blackwell Publishers, Oxford.

Barwise, J., Etchemendy, J., Allwein, G., Barker-Plummer, D., and Liu, A., 1999, *Language Proof and Logic*, CSLI and Seven Bridges Press, Stanford.

Barwise, J. and Seligman, J., 1997, *Information Flow - The Logic of Distributed Systems*, Number 44 in Cambridge Tracts in Theoretical Computer Science, Cambridge University Press, Cambridge.

Dretske, F.I., 1969, *Seeing and Knowing*, Chicago University Press, Chicago.

Hammer, E.M., 1995, *Logic and Visual Information*, CSLI and FOLLI, Stanford.

Sawamura, H. and Kiyozuka, K., 2000, Jvenn: A visual reasoning system with diagrams and sentences. in: *Theory and Application of Diagrams*, M. Anderson, P.

Cheng, and V. Haarslev, eds., Lecture Notes in Artificial Intelligence, New York, Springer-Verlag.

Shin, S.-J., 1996, Situation-theoretic account of valid reasoning with Venn diagrams, in: *Logical Reasoning with Diagrams*, G. Allwein and J. Barwise, eds., Oxford University Press, New York, pp. 81–108.

Swoboda, N., 2001, *Designing Heterogeneous Reasoning Systems with a Case Study on FOL and Euler/Venn Reasoning*, PhD thesis, Indiana University.

Swoboda, N., 2002, Implementing Euler/Venn reasoning systems, in: *Diagrammatic Representation and Reasoning*, M. Anderson, B. Meyer, and P. Olivier, eds., Springer-Verlag, London, pp. 371–386.

A LOGICAL APPROACH TO THE ANALYSIS OF METAPHORS

Isabel D'Hanis

Centre for Logic and Philosophy of Science, Ghent University,
Blandijnberg 2, 9000 Ghent, Belgium
isabel.dhanis@rug.ac.be

Abstract   In this paper, I will present an adaptive logic that grasps the way we analyze metaphors. Metaphors are powerful tools to generate new scientific ideas. Therefore, it is important to have a good theory on what metaphors are and how they function. The first question we have to answer when we want to develop such a theory is obviously "what metaphors are". Philosophy of language can offer some interesting ideas but most views do not allow for a cognitive function of metaphors. One of the sparse views that does allow for it is interactionism. The basic version, however, has some serious shortcomings that need solving when we want to use this theory. First of all the terminology is too vague. Furthermore, the description of the reasoning process we use when we analyze a metaphor, only works for very simple examples. The logic I will present, **ALM**, is based on a broadened version of this view. A logical approach of metaphors allows us to gain a profound insight in the way we analyze metaphors. The analysis of metaphors is a dynamical reasoning process. When we want to capture this process in a logical system, we need a logic that is capable of grasping that specific type of dynamics. An adaptive logic seems to be the best choice. Therefore, I shall present an adaptive logic that grasps the analysis of metaphors.

1.   Introduction

Metaphors are powerful instruments to generate new ideas. This is why they play an important role in scientific development. A famous example is the clock metaphor in psychology, studied by McReynolds in [McReynolds, 1980] (see also the next section). Another remarkable example is the role of the metaphor

(1) The ocean is a conveyor belt.

L. Magnani, N.J. Nersessian, and C. Pizzi (eds.),
Logical and Computational Aspects of Model-Based Reasoning, 21–37.
© 2002 Kluwer Academic Publishers. Printed in the Netherlands.

played in oceanography. Brüning and Lohmann describe in [Brüning and Lohmann, 1999] how this metaphor was used to construct an explanation for certain temperature changes.

However, when we compare research results of philosophers and historians of science concerning the use of metaphors in scientific inquiry, there seems to be a lot of confusion about a fundamental question as what metaphors are. Different names are used to refer to the same phenomena, or the same terms are used to refer to different phenomena. Especially the confusion between metaphors and analogies causes many problems. If we want to study the function of metaphors in scientific development, we have to start from a theory on what metaphors are: understanding their function presupposes that we have a grip on their nature. Only in this way, we can distinguish them from other forms of model based reasoning, such as analogies and similarity reasoning, and overcome the present confusion.

In the next part of this paper, I will discuss two different metaphor theories, namely the comparison view on metaphors and the interactionist view, and I will show that interactionism is the best choice. In the third part, I will tackle some problems with the basic version of interactionism. I will explore some possible solutions and show that a logical approach of metaphors is an inherent part of a metaphor theory. In the fourth part, I will present an adaptive logic for metaphors. This logic differs from other logical approaches in the sense that it is very close to actual human reasoning and also in the sense that *the logic itself* is capable of capturing a large part of the metaphorical analysis. Therefore, it can overcome some of the fundamental problems of other logical approaches.

2. The interactionist view as a basis for a metaphor theory

Philosophers of language did a lot of research on metaphors. Unlike philosophers and historians of science, they concentrated on the question *what metaphors are*. However, most philosophers of language did not take into account that metaphors can have a *cognitive function*. They can play a role in *reasoning processes* and not only on a language level. Therefore, these philosophers cannot account for the innovative power of metaphors. There is nevertheless a theory that allows for a cognitive function: interactionism. The merits of this metaphor theory become clear when we confront it with the comparison view, a view on metaphors that is very popular among philosophers of language.

Miller shows in [Miller, 1993] what a comparison view amounts to. Let us consider the metaphor

(2) Man is a wolf.

Miller states that, while analysing this metaphor, we consider the properties of wolves and the properties of men. What these two sets have in common is the *meaning* of the metaphor. According to Miller, analysing a metaphor is constructing a *comparison* between the two elements the metaphor consists of. For metaphor (2), this means that we interpret the metaphor as: men are like wolves – in being cruel for example. Here, cruel is one of the properties that men and wolves have in common.

The problem with this view is that everything is *already known* about the two parts the metaphor consists of. Understanding a metaphor requires only the analysis of the information that is *already given*. In cases of scientific innovation, however, a metaphor may cause the person who hears a metaphor to change his or her ideas on what the metaphor is about.

(3) The human mind is a clock

was such a powerful metaphor, because it caused people to apply certain properties of clocks to the human mind. According to the comparison view, people knew already from the beginning that the mind had certain properties of clocks. They only had to analyze the properties of minds and the properties of clocks and decide what they had in common. This scenario seems plausible in cases where the metaphor is already known, but it cannot explain the conceptual innovations this metaphor caused (see also [McReynolds, 1980]). One of the merits of interactionism is that it can explain scientific innovations like this.

The interactionist view was developed by Max Black in [Black, 1962] and revised in [Black, 1993]. According to Black, a metaphor consists of two parts, a primary subject and a secondary subject. The *primary subject* is the subject the sentence expression is about – in metaphor (2), "man" is the primary subject. In the same metaphor, the *secondary subject* is "wolf". It is typical for the secondary subject that it is not used in its literal meaning, but in a metaphorical or "non-literal" meaning. The secondary subject thus *modifies* the primary subject. On Black's account, we associate a "system of associated commonplaces" with the secondary subject. This system consists of things we immediately associate with wolves. The elements (implications in Black's terminology) of the "system of associated commonplaces" of the secondary subject are *projected* on or *transferred* to the primary subject. The result is that the primary subject enlarges, gets reorganized or even changes. This metaphor theory can therefore account for the role metaphor (3) played in the development of psychology. It explains how the metaphor enabled people to attribute certain properties to the human mind, they did not attribute to it before.

Let us briefly summarize this section. According to the comparison view, the analysis of a metaphor consists in deciding what properties the two parts of the metaphor have in common. Once this decision is made, we reconstruct a comparison stating that one thing is like another in having this or that property. According to the interactionist view, we distinguish two parts of a metaphor (a primary and a secondary subject), and we consider a "system of associated commonplaces" associated with the secondary subject. We then analyze a metaphor by transferring the elements of the "system of associated commonplaces" to the primary subject and by attributing the elements to the primary subject. The result is that the primary subject is reorganized, elaborated or even changed. This way, interactionism can account for a cognitive function of metaphors.

3.   Some problems with the basic formulation of the interactionist view

Black's approach seems very promising. There are nevertheless some problems with the original view. A first group of problems has to do with the *vagueness* of the terminology Black used. What exactly is a "system of associated commonplaces"? How do we have to interpret the idea of *projection* or *transfer*? What exactly happens when we attribute the elements of the "system of associated commonplaces" of the secondary subject to the primary subject? In what way does this transfer modify the primary subject? If we want to develop a metaphor theory, we need to find an answer to these questions. This theory can provide us with a profound insight in the analysis of metaphors and can point us also to an answer for the question how metaphors function.

A second important problem is the *relevance problem*. The way Black states it, every implication of the "system of associated commonplaces" is projected on the primary subject. This means that in the case of metaphor (2), also implications as "howl at the moon" are projected on the primary subject, which provokes us to derive absurd conclusions as "men howl at the moon".

These problems, however, do not force us to abandon Black's *basic idea*, that the analysis of metaphors consists in the transfer of "associations" from the secondary subject to the primary subject. What we do need, however, is a more sophisticated view on the nature of this metaphorical transfer.

A first concept we have to clarify is the so-called "system of associated commonplaces". As I argued in more detail in [D'Hanis, 2000], the problems concerning the vagueness of Black's terminology can be solved by replacing the idea of a "system of associated commonplaces" by Barsalou's definition of *concepts* (as presented in [Barsalou, 1989]). Barsalou states that a concept is a collection of information. It is important to note that, on his view, the content of a concept is not fixed. The actual collection of information depends on the *context*.

Barsalou distinguishes three types of information: context independent information (CI), context dependent information (CD) and recent context dependent information (CD_{rec}). CI-information is information we immediately associate with a certain idea. CD-information is information with a lower level of accessibility. This information is obtained by going trough actual reasoning *processes*, whereas CI-information is immediately retrieved. The third type of information is of no importance in this paper. The idea I developed in [D'Hanis, 2000] is that analyzing a metaphor consists in transferring *CI-information* about the secondary subject to the primary subject.[1] What a metaphorical projection or transfer is and how this exactly proceeds will be clarified by the logic I will present further in this paper.

The other important problem is the *relevance problem*. A first question we have to ask ourselves is how we get absurd conclusions. When we turn to metaphor (2) again, we see that we derive such conclusions when we attribute properties of wolves to men, for which we know that men cannot have them. It is absurd to conclude from (2) that men howl at the moon because we *know* that humans don't do this. So, it seems that we derive problematic conclusions when we transfer information that is contradicted by information we have about the primary subject or from the context of the metaphor. This seems to indicate that the primary subject functions as a filter for the transfer. This, however, presupposes that much more information is included in the primary subject than Black seems to allow for. The natural way out is that we associate a *set*

---

[1]The transfer of this type of information is typical for the analysis of *metaphors*. In the case of an analogy, we also transfer CD-information. This distinction between CI-information and CD-information is a first way of distinguishing metaphors and analogies. I refer to my [D'Hanis, 2000] for a further discussion.

of information with the primary subject too. So, what this comes to is that we start the analysis of a metaphor with two full sets of information, a set of information on the primary subject and a set of information on the secondary subject. The information we have about the primary subject *filters* the information about the secondary subject. It selects the information that is consistent with the information the primary subject already contains. This idea seems circular, but the specific properties of the logical system I will present in the next sections can overcome this circularity.

Most researchers agree that understanding a metaphor proceeds in three stages. A first stage is what I will call the *recognition* of a metaphor. In this stage, we figure out what the expression is about. This means that we also determine what the two subjects of the metaphor are. A second stage I will call the *analysis* of the metaphor. In this stage, we consider the information we have on the two subjects the metaphor consists of. We transfer the information we have on the secondary subject to the primary subject. Those pieces of information that are contradicted by the primary subject or the context of the metaphor are excluded from transfer. A third and last stage, I will call the *interpretation* of the metaphor. The result of the transfer in the second stage is that the information we have about the primary subject is extended. Those elements that seem unproblematic – at that moment – are added to the primary subject. This can cause a reorganization of the information that was already present in the concept.

There is a fundamental difference between, on the one hand, the revision of the interactionist view I presented here and on the other hand the original interactionism and the comparison view. According to the view I defended here, the interpretation of a metaphor depends on the information an individual associates with the subjects. This view excludes that metaphors have a final and unique meaning. An interpretation is always contextual, never absolute. A concept is a contextual collection of information and the context of the metaphor is also important for the analysis of the metaphor. This means that the interpretation of a metaphor can differ according to the context.

Another important characteristic of the view I defended here is that it captures the dynamics involved in the analysis of metaphors. It is clear that adding new information on the primary subject can render certain transfers invalid. This type of dynamics is external – it is caused by changing the premises – and is called non-monotonicity. There is also an internal kind of dynamics. This occurs when a further analysis of the concept causes the withdrawal of certain conclusions.

These two types of dynamics are crucial in the analysis of metaphors I presented here. Neither Millers model, nor Black's model can account for these two types of dynamics.

The adaptive logic I will present in the next sections will solve certain problems (such as the relevance problem and the problems connected with this) that the original interactionist view faces and will also demonstrate how a metaphorical transfer exactly proceeds. It is therefore an indispensable element of a metaphor theory.

4.   The advantages of adaptive logics

The logical system I present in this paper is an *adaptive logic*.[2] Adaptive logics are extremely suited to handle the specific type of dynamics that occur in the analysis of metaphors.

There are other examples of logical approaches to metaphors such as the systems presented by Van Dijk [Van Dijk, 1975], Guenthner [Guenthner, 1975], and Bergmann [Bergmann, 1979]. Their systems, however, cannot account for the specific dynamics involved in a metaphorical transfer such as non-monotonicity. The question these authors ask themselves is how metaphors – expressions that seem false at first sight – can get a truth value. They develop a special type of semantics that is capable of dealing with this property. Furthermore, these systems are based on the idea that a metaphor is a decoded comparison. Therefore, their systems cannot account for the innovative power of metaphors. They also see the meaning of a metaphor as fixed and presuppose that a metaphor only operates on a *linguistic level*. According to these authors, a metaphor causes a temporary reorganization of the meaning of certain words, but is not capable of leaving more permanent traces such as conceptual changes. They have a very static view on metaphors and language use and reasoning in general. Therefore, it is not surprising that their systems cannot account for any dynamics that are inherent to metaphorical reasoning.

The metaphor theory I present in this paper, does not imply that metaphors are deviations from normal language use. They are important and inherent elements of our language and they are even capable of causing conceptual changes. To understand metaphors, we have to go through dynamical reasoning processes. Both types of dynamics I discussed in the previous section are important in the analysis of metaphors. Both types can force us to withdraw previously derived conclusions. This

---

[2]For further information on this type of logics, I refer the reader to [Batens, 1999] and [Batens, 2000].

implies that we can never reach a final conclusion and therefore, that the interpretation of a metaphor can never be fixed. Therefore, the logical approaches available today are not suited to grasp metaphorical analysis.

Adaptive logics are capable of grasping these types of dynamics and therefore do not presuppose that the meaning of a metaphor is fixed. They are non-monotonic and can also account for the internal type of dynamics. Adaptive logics moreover have the advantage that they are based on classical first order logic (henceforth **CL**) and hence, that they have a proper proof theory and semantics (for which soundness and completeness proofs can be formulated). The logic I will present, **ALM**, is based on **CL**, but it is *non-monotonic* and has a *dynamic* proof theory.

I add a few remarks before we proceed to the next section. There are some constraints on the system I will present. At the moment, only metaphors of the form "X is Y" (for example metaphor (2)) can be analysed because in these cases we have two subjects that are clearly distinguished.

Another constraint is that the set of premises may contain only *one* metaphor. The simultaneous analysis of a series of metaphors is possible, but in that case we need a more complex proof structure. This proof theory should for instance avoid that the information we have on the subjects of one metaphor merges with the information we have on other metaphors and thus gives rise to unwanted conclusions.

Finally, the system I present here has to be situated at the second stage of understanding a metaphor, or what I called the *analysis*. As we shall see, the logic presupposes that we already know that a certain expression is a metaphor and that we identified its primary and secondary subject. The logic can only provide us with an interpretation of the metaphor in that particular context.

5.   ALM, an adaptive logic for metaphors

5.1   The general idea

A first problem we have to tackle when developing a logic for the analysis of metaphors is to grasp the idea of non-literal language. I shall do this by constructing a formal language $\mathcal{L}^*$, in which we have for each so called "literal" predicate π, a "metaphorical" predicate π^*. This means that all predicates in this language are "doubled". The secondary subject is formalized by means of a metaphorical predicate, the primary one, as usual, by means of a literal predicate. Metaphor (2) for example, is formalized as $(\forall x)(Mx \supset W^*x)$.

A question we have to ask ourselves at this point concerns the relation between the letters for literal predicates and for non-literal ones. There are two extremes. The first extreme is that we consider them as totally unrelated, that we consider, for instance, π and π^* as different letters for different predicates. The second extreme is that we consider π and π^* as different letters for the same predicate. In the latter case, but not in the former, the letters for literal and for metaphorical predicates can be substituted for one another and hence, only in the latter case are we able to infer, for instance, $(\forall x)(Mx \supset Cx)$ ("men are cruel") from $(\forall x)(Mx \supset W^*x)$ and $(\forall x)(Wx \supset Cx)$.

To see that neither of these extremes can lead to an adequate logic, consider again metaphor (2). When analysing this metaphor, we start from the information we have on normal, everyday wolves. What we want to find is a subset of this information that determines the meaning of the metaphorical concept in this particular case. As (2) is a metaphor, we know that not all information about wolves can be included in the metaphorical concept. Otherwise, the expression would be a plain literal expression. On the other hand, we know that it should be possible to include at least some information about wolves. If it were not, the metaphorical expression would be meaningless. As we want our interpretation of the metaphors to be as rich as possible, we want the subset of information that is included to be as large as possible. What the latter comes to is that we want to interpret the literal predicate *as metaphorically as possible*. In the logic I present here, this will be realized by allowing that the letters for the literal predicates are replaced *as much as possible* by the letters for the corresponding metaphorical predicates.

The only remaining question is what I mean by *as much as possible*. What we want is that all information about the secondary subject is transferred to the primary subject *except* for those pieces of information that are contradicted by the context or by the information we have on the primary subject. Hence, replacing letters for literal predicates by letters for metaphorical predicates will be allowed for *unless and until* this leads to unwanted conclusions (conclusions that contradict information about the context or about the primary subject).

5.2   The outlines of ALM

An adaptive logic is characterized by three elements: an upper limit logic (henceforth **ULL**), a lower limit logic (henceforth **LLL**) and an adaptive strategy. The **ULL** incorporates a set of (logical) presuppositions. The **LLL**, a subset of the **ULL**, drops some of these presuppositions. The idea behind an adaptive logic is that a set of premises is interpreted *as much as possible* in accordance with the presuppositions of the **ULL**. The adaptive strategy tells us how we have to interpret the expression *as much as possible*.

Let us now see what this amounts to in the case of **ALM**. The **ULL** presupposes that *all* information on the secondary subject can be transferred to the primary subject. So, the **ULL** enables one to replace π by π^*. When this presupposition is violated, the logic leads to triviality. The **ULL** can be seen as the logic that grasps Black's version of interactionism. It assumes that everything we know about *wolves* also applies to *wolves**. This implies that all the information can be transferred to the primary subject. This comes down to Black's original formulation of interactionism. The **LLL** presupposes that *no* information can be transferred to the primary subject. Therefore it is not possible to replace π by π^*.

When a set of premisses violates the presupposition that all information on the secondary subject can be transferred to the primary subject, it is said to behave abnormally with respect to the **ULL**. It is important to note that "abnormality" refers to properties of the application context – to presuppositions that are considered desirable, but that may be overruled. The adaptive logic follows as much as possible the **ULL**, only when abnormalities are derived, it switches to the **LLL**. Thus the adaptive logic *oscillates* between the two systems. This way it overcomes difficulties other formal approaches to metaphors have to deal with. The logic **ALM** decides, in view of the premisses, which elements can be transferred. As a consequence, there is no need to modify the set of premisses to avoid triviality. In most other approaches, unwanted conclusions are avoided by not adding the problematic information to the set of premisses. **ALM** isolates itself – without interference of the user – problematic information. Therefore, this system is closer to human reasoning than the systems that the premisses are analysed beforehand.

The system presented in this paper has to be seen against the background of a recent development in the adaptive logic program. The first adaptive logics were *corrective logics*. Examples of this type of logics are **ACLuN1**, **ACLuN2** (see [Batens, 1999] and [Batens, 2000]) and **ANA** (see [Meheus, 2000]). In corrective adaptive logics, the **ULL** determines the standard of reasoning and specific deviations from this standard are minimized. They adapt themselves to specific *violations* of presuppositions of **CL** such as inconsistencies.[3] The system I present in this paper is an *ampliative logic* like the more recent systems presented in [Batens and Meheus, 2000], [Meheus, to appear], [Batens and Meheus,

---

[3] All current corrective adaptive logics have **CL** as their **ULL**.

to appear] and [Meheus et al., to appear]. In ampliative logics, the standard of reasoning is determined by the **LLL** and specific extensions of this standard are maximized. The consequence set one obtains with an ampliative logic is thus richer than the one obtained with **CL**. It will easily be observed from the next section that the latter also holds true for **ALM**.

5.3 The proof theory

Lines of a proof in an adaptive logic have five elements: (i) a line number, (ii) the formula A that is derived, (iii) the line numbers of the formulas from which A is derived, (iv) the rule by which A is derived and (v) a condition. The condition determines which formulas have to behave normally in order for A to be valid. A wff is said to be derived *unconditionally* iff it is derived on a line with an empty fifth element.

The generic proof format of adaptive logics consists of three elements: an unconditional rule, a conditional rule and a marking rule. The **LLL** determines the unconditional rule, the **ULL** determines the conditional rule and the marking rule is determined by the adaptive strategy. A line that is marked is no longer considered to be a part of the proof.

The **LLL** of **ALM** is **CL**, with an extended language $\mathcal{L}^*$. Where $\mathcal{L}$ is the standard predicative language and $\mathcal{P}^r$ is the set of predicates of rank r, $\mathcal{L}^*$ is obtained by extending $\mathcal{P}^r$ to $\mathcal{P}^{r*}$ by stipulating that $\pi^* \in \mathcal{P}^{r*}$ iff $\pi \in \mathcal{P}^r$.

The **ULL** of **ALM** is **CL***. This is an extended version of **CL**. The following definitions are needed:

Definition 1 $A(\pi) =_{df} \pi$ *occurs in* A.

Definition 2 $A(\pi^*/\pi) =_{df}$ *the formula we obtain when we replace the predicate* π *in* A, *systematically by* π^*.

We obtain **CL*** syntactically by extending **CL** with this rule:

$$\text{If} \quad A(\pi^*/\pi), B(\pi^*) \vdash_{\mathbf{CL}} C, \quad \text{then} \quad A(\pi), B(\pi^*) \vdash_{\mathbf{CL}^*} C$$

Finally, we arrive at the adaptive strategy. The adaptive strategy for **ALM** is the *simple strategy*. According to this strategy, we have to mark a line as soon as it is derived that one of the members of its fifth element behaves abnormally.

Now we can define the proof theoretical rules.

PREM   At any stage of a proof one may add a line consisting of (i) an appropriate line number, (ii) a premise, (iii) a dash, (iv) "PREM", and (v) "∅".

RU   If $A_1, \ldots, A_n \vdash_{\mathbf{CL}} B$, and $A_1, \ldots, A_n$ $(n \geq 0)$ occur in the
          proof on the conditions $\Delta_1, \ldots, \Delta_n$ respectively, then one may
          add to the proof a line with an appropriate line number, B as
          its second element and $\Delta_1 \cup \ldots \cup \Delta_n$ as its fifth element.

RC   If $A(\pi^*/\pi), B(\pi^*) \vdash_{\mathbf{CL}} C$, and $A(\pi)$ and $B(\pi^*)$ occur uncondi-
          tionally in the proof, then one may add a line with an appro-
          priate line number, C as its second element, and $\{\sim C\}$ as its
          fifth element.

MARK   A line i is marked iff where Δ is the fifth element of i,
          (1) C is unconditionally derived for some $C \in \Delta$, or
          (2) There is a D such that for some $C \in \Delta$, $C \vdash_{\mathbf{CL}} D$ and D is
          unconditionally derived.

The second part of the marking rule is related to the fact that $A \supset B \vdash_{\mathbf{CL}} A \supset (B \vee C)$ for an arbitrary C. Thus, in the case of metaphor (2), RU enables one to derive $(\forall x)(Wx \supset (Hx \vee Tx))$ ("wolves howl at the moon or have 36 teeth") from $(\forall x)(Wx \supset Hx)$. But then, in view of RC, we can derive that "men howl at the moon or have 36 teeth". Given the premises at issue, it may be impossible to derive that the latter is not the case. This is why we need the second clause of the marking rule. It allows us to mark the line that has $(\forall x)(Mx \supset (Hx \vee Tx))$ as its second element as soon as $\sim(\forall x)(Mx \supset Hx)$ is derived.

Definition 3 *A is finally derived in a proof from Γ iff A is derived at line that is not marked and will not be marked in any extension of the proof.*

Definition 4 $\Gamma \vdash_{\mathbf{ALM}} A$ *(A is finally derivable from Γ) iff A is finally derived in a proof from Γ.*

I will illustrate with an example what the proof theory amounts to. Consider for instance the metaphor

(4) John is a donkey.

Let us assume that we hear this metaphor in a conversation. We have already recognized it as a metaphorical expression and we have already decided that this expression says something about *John* rather than about donkeys. At this point starts the second stage, or the analysis of the metaphor. Let us assume we associate the following properties with donkeys:

- Donkeys have long ears. (E)

- Donkeys are stupid. (S)

- Donkeys are stubborn. (T)

- Donkeys bray. (B)

Suppose also that we know that John is human (H) and that we know that humans don't have long ears and that they don't bray.

1	D^*j	$-$; PREM	$\emptyset$
2	$(\forall x)(Dx \supset Ex)$	$-$; PREM	$\emptyset$
3	$(\forall x)(Dx \supset Sx)$	$-$; PREM	$\emptyset$
4	$(\forall x)(Dx \supset Tx)$	$-$; PREM	$\emptyset$
5	$(\forall x)(Dx \supset Bx)$	$-$; PREM	$\emptyset$
6	Hj	$-$; PREM	$\emptyset$
7	$(\forall x)(Hx \supset \sim Ex)$	$-$; PREM	$\emptyset$
8	$(\forall x)(Hx \supset \sim Bx)$	$-$; PREM	$\emptyset$
9	$(\forall x)(D^*x \supset Ex)$	1, 2; RC	$\{\sim(\forall x)(D^*x \supset Ex)\}$
10	$(\forall x)(D^*x \supset Sx)$	1, 3; RC	$\{\sim(\forall x)(D^*x \supset Sx)\}$
11	$(\forall x)(D^*x \supset Tx)$	1, 4; RC	$\{\sim(\forall x)(D^*x \supset Tx)\}$
12	$(\forall x)(D^*x \supset Bx)$	1, 5; RC	$\{\sim(\forall x)(D^*x \supset Bx)\}$
13	$D^*j \supset Ej$	9; RU	$\{\sim(\forall x)(D^*x \supset Ex)\}$
14	$D^*j \supset Sj$	10; RU	$\{\sim(\forall x)(D^*x \supset Sx)\}$
15	$D^*j \supset Tj$	11; RU	$\{\sim(\forall x)(D^*x \supset Tx)\}$
16	$D^*j \supset Bj$	12; RU	$\{\sim(\forall x)(D^*x \supset Bx)\}$
17 $\checkmark$	Ej	1,13; RU	$\{\sim(\forall x)(D^*x \supset Ex)\}$
18 $\checkmark$	Sj	1,14; RU	$\{\sim(\forall x)(D^*x \supset Sx)\}$
19	Tj	1,15; RU	$\{\sim(\forall x)(D^*x \supset Tx)\}$
20 $\checkmark$	Bj	1,16; RU	$\{\sim(\forall x)(D^*x \supset Bx)\}$
21	$Hj \supset \sim Ej$	7; RU	$\emptyset$
22	$\sim Ej$	6, 22; RU	$\emptyset$
23	$Hj \supset \sim Bj$	8; RU	$\emptyset$
24	$\sim Bj$	6, 24; RU	$\emptyset$
25	$D^*j \wedge \sim Ej$	1, 22; RU	$\emptyset$
26	$(\exists x)(D^*x \wedge \sim Ex)$	25; RU	$\emptyset$
27	$\sim(\forall x)(D^*x \supset Ex)$	26; RU	$\emptyset$
28	$D^*j \wedge \sim Bj$	1, 24; RU	$\emptyset$
29	$(\exists x)(D^*x \wedge \sim Bx)$	28; RU	$\emptyset$
30	$\sim(\forall x)(D^*x \supset Bx)$	29; RU	$\emptyset$
31	Pj	$-$; PREM	$\emptyset$
32	$(\forall x)(Px \supset \sim Sx)$	$-$; PREM	$\emptyset$
33	$Pj \supset \sim Sj$	32; RU	$\emptyset$
34	$\sim Sj$	31, 33; RU	$\emptyset$
35	$D^*j \wedge \sim Sj$	1, 34; RU	$\emptyset$
36	$(\exists x)(D^*x \wedge \sim Sx)$	35; RU	$\emptyset$

37     $\sim(\forall x)(D^*x \supset Sx)$ 36; RU $\emptyset$

On the first eight lines of this example, we find the premises. The first line is the formalization of the metaphor. The next lines consist of the information we have on the subjects of the metaphor. It is needless to say that the set of premises that is presented here is not the only possible one for this metaphor. Especially the set of premises on the primary subject will not appear that way in actual human reasoning. The inferences to derive this type of information are in reality a lot more complex than the way they are represented in the proof, but for reasons of brevity and simplicity, they are presented in this form.

On lines 9 to 12, we derive conditionally that everything we know about about literal donkeys is also information about metaphorical donkeys. This derivation is conditional because at this stage, we cannot be sure that we can replace every instance of D with D^*. As soon as the condition on which the line is derived is violated, the line will be deleted.

On lines 13 to 16 we can find instantiations of the lines nine till twelve. Since these are lines derived on a certain condition, we have to write these conditions also on the lines thirteen till sixteen.

Line 17 to 20, we can derive by means of modus ponens on the first line and the previous four lines. Also here, we have to carry over the fifth elements of the previous lines, since they are used to derive these lines.

On lines 21 to 24 a further analysis of the premisses can be found. Line 25 and 27 are simple conjunctions of information. They permit us to derive the lines 26 and 29 in an unconditional way. These lines allow us to derive lines 27 and 30 in an unconditional way and these lines are the fifth elements of lines 17 and 20. This means that every line which has this as its condition has to be marked and that these lines at this stage are no longer considered to be elements of the proof. This demonstrates the first type of dynamics, the internal one. Further analysis of the premises causes us to withdraw conclusions we derived earlier in the proof.

At this point we can decide that the person who used the metaphor wanted to say that John is stupid and stubborn. Now, suppose that later in the conversation, we hear that John is a professor in physics and that we have the idea that professors in physics are not stupid. On line 31, we introduce the new premise that John is a professor in physics and on line 32, we have the premise that professors in physics are not stupid. This allows us to derive at line 33, that if John is a professor in physics, he is not stupid. and that John is not stupid on line 35 and by extension that John is a donkey and John is not stupid on line 36. This allows us to derive in an unconditional way that there is an entity that is a metaphorical donkey and that is not stupid and this comes down to line 37. At this point we have obtained the condition on which line 10 is derived, in an unconditional way. The result is that we have to mark each line that is derived using the information on line 10.

At this point, we adjust our interpretation of the metaphor. Now we interpret it as "John is stubborn". The last part of the example shows that the system is also *non-monotonic*. When we add more premises, certain previously derived conclusions can be marked and thus become invalid.

This example illustrates how the proof theory works. Both types of dynamics, the internal and the external dynamics, are covered by **ALM**. This system has some further advantages. As soon as the necessary information is provided, the system can derive the conclusions by its own means. So it is *the logic itself* that can exclude certain elements from transfer. We don't have to do this before adding the premises to the proof.

6.   Conclusions

Although most philosophers of science recognize the importance of metaphors in scientific research, they do not provide any theoretical background on what metaphors are. Metaphor theories are available in philosophy of language, but most of the theories exclude that metaphors can have a cognitive function. Interactionism does allow for it, but there are some serious problems with the basic theory. The logical system I presented in this paper plays an important role in providing a solution for these problems and in general in clarifying the idea of interaction between the two parts of a metaphor. We can conclude that a logic for metaphors is a central element of a metaphor theory.

The system I presented here is a significant improvement of some of the existing formal approaches to metaphors and to model based reasoning in general. It is very close to actual human reasoning, since the system itself can account for dynamics that are crucial in the analysis of metaphors. The adaptive approach offers the method for doing this. It seems to be the only approach to metaphors that allows one to start from full sets of premises, without filtering them before we start the analysis. These logics can also account for the types of dynamics that are considered to be crucial for human reasoning.

There are still some important open problems. In this paper, I only presented the proof theory of **ALM**. It is obvious that also the semantics and the meta-proofs have to be provided. This seems, however, not to cause any fundamental problems for the proof theory I presented in this paper. A second element is that **ALM** is only a basic system. It is suited to analyse relatively simple metaphors. In cases of more complex metaphors (as most metaphors that are used in scientific development) we need a finer instrument than **ALM**. We need, for example, a way of distinguishing which information about the secondary subject is central in that context and which information is not. This way we can avoid that we transfer information that is not explicitly contradicted, but nevertheless irrelevant for the primary subject.

Acknowledgment

Research Assistant of the Fund for Scientific Research – Flanders (Belgium). I want to thank Joke Meheus for her helpful comments while writing this paper. The research for this paper was supported by the Fund for Scientific Research – Flanders (Belgium)(F.W.O. - Vlaanderen).

References

Barsalou, L.W., 1989, Intra-concept similarity and its implications for inter-concept similarity, in: *Similarity and Analogical Reasoning*, S. Vosniadou and A. Ortony, eds., Cambridge University Press, Cambridge, pp. 76–121.

Batens, D., 1999, Inconsistency-adaptive logics, in: *Logic at Work. Essays Dedicated to the Memory of Helena Rasiowa*, E. Orłowska, ed., Physica Verlag (Springer), Heidelberg, New York, pp. 445–472.

Batens, D., A survey of inconsistency-adaptive logics, in: [Batens et al., 2000, pp. 49–73].

Batens, D. and Meheus, J., 2000, The adaptive logic of compatibility. *Studia Logica* 66:327–348.

Batens, D. and Meheus, J., Adaptive logics of abduction, to appear.

Batens, D., Mortensen, C., Priest, G., and Van Bendegem, J.P., eds., 2000, *Frontiers of Paraconsistent Logic*, Research Studies Press, Baldock, UK.

Bergmann, M., Metaphor and formal semantic theory, *Poetics* 8:213–230.

Black, M., 1962, Metaphor, in: *Models and Metaphors*, Cornell University Press, Ithaca, pp. 25–48.

Black, M., More about metaphor, in: [Ortony, 1993, pp. 19–42], pp. 19–42.

Brüning, R. and Lohmann, G., 1999, Charles Sanders Peirce on creative metaphor: A case study on the conveyor belt metaphor in oceanography, *Foundations of Science* 4(3):237–270.

D'Hanis, I., 2000, Metaforen vanuit een taalfilosofisch, wetenschapsfilosofisch en logisch perspectief, Master's thesis, RUG.

Van Dijk, T., 1975, Formal semantics of metaphorical discourse, *Poetics*, 4:173–195.

Guenthner, F., 1975, On the semantics of metaphor, *Poetics* 4:199–220.

McReynolds, P., 1980, The clock metaphor in the history of psychology, in: *Scientific Discovery: Case Studies*, T. Nickles, ed., Reidel, Dordrecht, pp. 97–112.

Meheus, J., An extremely rich paraconsistent logic and the adaptive logic based on it, in: [Batens et al., 2000, pp. 189–201].

Meheus, J., An inconsistency-adaptive logic based on Jaśkowski's **D2**, to appear.

Meheus, J., Verhoeven, L., Vandijk, M., and Provijn, D., Ampliative adaptive logics and the foundation of logic-based approaches to abduction, to appear.

Miller, G.A., Images and models, similes and metaphors, in: [Ortony, 1993, pp. 357–401].

Ortony, A., ed., 1993, *Metaphor and Thought*, Cambridge University Press, Cambridge, second edition.

AMPLIATIVE ADAPTIVE LOGICS AND THE FOUNDATION OF LOGIC-BASED APPROACHES TO ABDUCTION

Joke Meheus, Liza Verhoeven, Maarten Van Dyck, and Dagmar Provijn
Centre for Logic and Philosophy of Science, Ghent University,
Blandijnberg 2, 9000 Ghent, Belgium
{Joke.Meheus, Liza.Verhoeven, Maarten.Vandyck, Dagmar.Provijn}@rug.ac.be

Abstract   In this paper, we propose a reconstruction of logic-based approaches to abductive reasoning in terms of ampliative adaptive logics. A main advantage of this reconstruction is that the resulting logics have a *proof theory*. As abductive reasoning is non-monotonic, the latter is necessarily dynamic (conclusions derived at some stage may at a later stage be rejected). The proof theory warrants, however, that the conclusions derived at a given stage are justified in view of the insight in the premises at that stage. Thus, it even leads to justified conclusions for undecidable fragments. Another advantage of the proposed logics is that they are much closer to natural reasoning than the existing systems. Usually, abduction is viewed as a form of "backward reasoning". The search procedure by which this is realized (for instance, some form of linear linear resolution) is very different from the search procedures of human reasoners. The proposed logics treat abduction as a form of "forward reasoning" (Modus Ponens in the "wrong direction"). As a result, abductive steps are very natural, and are moreover nicely integrated with deductive steps. We present two new adaptive logics for abduction, and illustrate both with some examples from the history of the sciences (the discovery of Uranus and of Neptune). We also present some alternative systems that are better suited for non-creative forms of abductive reasoning.

L. Magnani, N.J. Nersessian, and C. Pizzi (eds.),
Logical and Computational Aspects of Model-Based Reasoning, 39–71.
© 2002 Kluwer Academic Publishers. Printed in the Netherlands.

1.   Aim and survey

The importance of logic-based approaches to abduction can hardly be overstated. Especially in the domain of Artificial Intelligence, they recently led to an impressive number of systems for a wide variety of application contexts (such as diagnostic reasoning, text understanding, case-based reasoning, and planning).[1] But also in the domain of cognitive science and philosophy of science they proved very fruitful. Examples are Hintikka's analysis of abduction in terms of the interrogative model of inquiry (see especially [Hintikka, 1998]), Aliseda's approach to abduction in terms of semantic tableaux (see [Aliseda, 1997]), Magnani's integration of results on diagnostic reasoning and scientific reasoning (see [Magnani, 2001]), and Thagard's reconstruction of several important discoveries in the history of science and in the history of medicine (see, for instance, [Thagard, 1988] and [Thagard, 1999]).

The aim of this paper is to propose a reconstruction of logic-based approaches to abduction in terms of ampliative adaptive logics. This reconstruction has several advantages: the resulting logics have a proper proof theory, they lead to justified conclusions even for undecidable fragments, they are much closer to natural reasoning than the existing systems, and they nicely integrate deductive and abductive inferences.

We shall present two adaptive logics for abductive reasoning. **CP1** enables one to generate explanations for novel facts (facts not entailed by, but consistent with one's background assumptions). **CP2** is a generalization of **CP1** that enables one to generate explanations for novel facts as well as for anomalous facts (facts not consistent with one's background assumptions). **CP2** moreover enables one to abduce explanatory hypotheses from a possibly inconsistent theory.

The importance of the generalized system **CP2** requires little explanation. Abductive inferences are frequently triggered by facts that contradict one's background assumptions. Examples are ubiquitous: physicians explaining why a person, after exposure to a highly contagious disease, did not become ill (as he or she should have), astronomers explaining why the orbit of a certain body exhibits certain aberrations, experimental physicists explaining why their experimental results do not coincide with the theoretical predictions, etc. But even when dealing with novel facts, one may have to reason from inconsistencies. The reason is that theories often turn out to be inconsistent, and that resolving the inconsistencies in a satisfactory way may take several years. In the

---

[1] For an interesting overview of AI-approaches to abduction, see [Paul, 2000].

meantime, however, these knowledge systems and theories are used to generate explanations.[2]

The techniques that led to **CP1** and **CP2** derive from the adaptive logic programme. The first adaptive logic was designed around 1980 by Diderik Batens (see [Batens, 1989]), and was meant to handle in a sensible and realistic way inconsistent sets of premises. This logic was followed by several other inconsistency-adaptive systems (see, for instance, [Priest, 1991; Meheus, 2000], and [Meheus, 200+b]). Later the idea of an adaptive logic was generalized to other forms of logical abnormalities, such as negation-incompleteness and ambiguity (see, for instance, [Batens, 1999]), and several inconsistency-handling mechanisms that proceed in terms of maximal consistent subsets were reconstructed in terms of adaptive logics (see [Batens, 2000] and [Verhoeven, 200+]).[3]

Until recently, all adaptive logics were *corrective* logics (they can handle logically abnormal theories – for instance, inconsistent ones). An important new development concerns the design of ampliative adaptive logics (see [Meheus, 1999a] and [Meheus, 200+a] for an informal introduction). These logics are meant to handle various forms of ampliative reasoning.[4] At the moment, (formal) results are available on adaptive logics for compatibility [Batens and Meheus, 2000], pragmatic truth [Meheus, 200+d], the closed world assumption and negation as failure [Vermeir, 200+], diagnostic reasoning [Weber and Provijn, 200+] and [Batens et al., 200+b], induction [Batens, 200+], inference to the best explanation [Meheus, 200+e], question generation [Meheus and De Clercq, 200+], and the analysis of metaphors [D'Hanis, 2002]. An informal discussion of adaptive logics for analogies can be found in [Meheus, 1999a].

As we shall see, both **CP1** and **CP2** are based on Classical Logic (henceforth **CL**). However, to make their proof theory as simple as

---

[2]It is traditionally accepted that inconsistent theories are not and cannot be used to generate meaningful explanations. Case studies indicate, however, that scientists frequently deal with inconsistent theories, and that they prefer to work with these theories rather than to resolve the inconsistencies in a premature way – for examples, see [Brown, 1990; Smith, 1988; Norton, 1987; Norton, 1993; Meheus, 1999b; Nersessian, 2002; da Costa and French, 2002]. Recent results in non-standard logic enable one to judge the soundness of such inferences in pretty much the same way as inferences from consistent sets of premises – this is illustrated in [Meheus, 1993] and [Meheus, 200+c] with some examples from nineteenth century thermodynamics.

[3]Consequence relations that proceed in terms of maximal consistent subsets were first proposed by [Rescher and Manor, 1970], and are today very popular in AI-applications (see [Benferhat et al., 1997] and [Benferhat et al., 1999] for an overview). Reconstructing these consequence relations in terms of adaptive logics has the advantage that it thus becomes possible to design a proof theory for them.

[4]An inference pattern is called ampliative if it leads to conclusions that "go beyond" the information contained in the premises.

possible, both systems are defined with respect to a modal adaptive logic that is based on (a bimodal version of) **S5**. In view of the results from [Batens and Vermeir, 200+], it is possible to design a direct proof theory for **CP1** and **CP2** – this is, one that proceeds directly in terms of **CL**. However, given the peculiarities of abductive reasoning processes, the indirect proof theory presented here is much more transparent.

We shall proceed as follows. In section 2, we briefly discuss the basic ideas behind logic-based approaches to abduction and show why the proposed reconstruction is important. In section 3, we elaborate on the main characteristics of abductive reasoning processes. This enables us to introduce, in an intuitive way, the basic mechanisms behind **CP1** and **CP2** (sections 4 and 5), and to present the technical details for **CP1** in section 6. The generalization to **CP2** is intuitively discussed in section 7, and the technical details are presented in section 8. In section 9, **CP1** and **CP2** are illustrated by means of some examples from the history of the sciences (the discovery of Uranus and of Neptune), and in section 10 we discuss some alternatives for **CP1** and **CP2** that are better suited for standard forms of abductive reasoning. We end the paper with some conclusions and open problems (section 11).

In order to keep the length of this paper within reasonable bounds, the presentation of **CP1** and **CP2** will be confined to the proof theory. It is important to note, however, that both systems are "decent" logics: a proper semantics can be designed for them, and soundness and completeness can be proven.

2.   Why the reconstruction is important

Within logic-based approaches, abductive inferences are perceived as falling under the following argumentation scheme:

(†) $A \supset B, B \mathbin{/} A$

This scheme, which is generally known as Affirming the Consequent, is evidently not deductively valid. Hence, as the framework of most logic-based approaches is a deductive one, the above scheme is not implemented directly. Instead, abductive inferences are specified as a kind of "backward reasoning": given a theory T and an *explanandum* B, find an A such that

(1) $T \cup \{A\} \vdash B$.
(2) $T \nvdash B$.
(3) $T \nvdash \neg A$.
(4) $B \nvdash A$.
(5) A is "minimal".

The first of these requirements needs little explanation. Note only that, if the underlying logic is **CL** (as is the case for most logic-based approaches to abduction), then (1) comes to $T \vdash A \supset B$. Also the next two requirements are straightforward: (2) warrants that the *explanandum B* is not explained by the background theory, and (3) that the explanatory hypothesis A is compatible with T.[5] (4) is needed to rule out degenerate cases. For instance, we do not want to abduce B as an explanation for itself. Also, if $T \cup \{A\} \vdash B$, then $T \cup \{A \vee B\} \vdash B$, but we do not want $A \vee B$ as an explanation for B. Cases like this are ruled out by requiring that the truth of the explanatory hypothesis is not warranted by the truth of the *explanandum* – this is what (4) comes to. (5) is related to the fact that, when trying to explain some *explanandum*, one is interested in explanations that are as parsimonious as possible. Hence, in view of $A \supset B \vdash_{\mathbf{CL}} (A \wedge D) \supset B$, one needs to prevent that $A \wedge D$ can be abduced, whenever A can. This can be realized by requiring that the explanatory hypothesis is "minimal". We shall see below that this notion can be defined in different ways – one may, for instance, consider an explanatory hypothesis as minimal if no alternative is available that is logically weaker. For the moment, however, the only important thing to remember is that minimality is a *comparative* notion: whether some explanatory hypothesis A is minimal with respect to some *explanandum B* depends on the available alternatives.

As mentioned in the previous section, reconstructing logic-based approaches to abduction in terms of adaptive logics has several advantages. A first one is that the resulting logics (unlike the systems available today) have a *proof theory*. As we shall see below, this proof theory is dynamic (conclusions derived at some stage may be rejected at a later stage), but it warrants that the conclusions derived at a given stage are justified in view of the insight in the premises at that stage. This is especially important as, at the predicative level, abductive reasoning is not only undecidable, there even is no positive test for it[6] (see section 3).

Another advantage of the proposed logics is that they are much closer to natural reasoning than the existing systems. As we mentioned before, abduction is usually viewed as a form of backward reasoning – "find an A that satisfies the requirements (1)–(5)". The search procedure by which this is realized in the existing systems (for instance, some form of linear resolution) is very different from the search procedures of human reasoners. The logics proposed in this paper treat abduction as a form of

---

[5]A formula A is said to be compatible with a set of premises Γ iff $\Gamma \nvdash \neg A$.

[6]Even if A follows abductively from a theory T and an *explanandum B*, there need not exist any finite construction that establishes this.

"forward reasoning": they are ampliative systems that directly validate inferences of the form (†).

The third advantage is related to this: deductive and abductive steps are nicely integrated into a single system. As a consequence, the logics not only enable one to generate explanatory hypotheses, but also to infer predictions on the basis of explanatory hypotheses and the background theory. This is highly important from the point of view of applications. In medical diagnosis, for instance, explanatory hypotheses are typically used to derive predictions which, in turn, may lead to a revision of the original hypotheses.

3.   Main characteristics of abductive reasoning

In order to explain how the proposed reconstruction works, we first have to discuss in some more detail the main characteristics of abductive reasoning processes.

As mentioned in the previous section, abductive inferences lead to conclusions that are compatible with the premises. It is important to note, however, that the conclusions are not necessarily *jointly* compatible with the premises. None of the requirements (1)–(5) excludes that different explanations are incompatible with each other. As we shall see below, this raises the question how one can avoid, in a classical framework, that the generation of contradicting explanations leads to triviality.

A second characteristic is that abduction is a *non-monotonic* form of reasoning: conclusions that follow abductively from some theory T may be withdrawn when T is extended to $T \cup T'$. This characteristic is related to the fact that some of the requirements for abductive inferences are *negative*. This not only holds true for (2)–(4), but also for (5). Indeed, as minimality is a comparative notion, (5) entails:

(5') $T \cup \{C\} \nvdash_{\mathbf{CL}} B$, for every C that satisfies (2)–(4) and in view of which A is not minimal.

To see the relation between the negative requirements and the non-monotonic character of abductive inferences more clearly, consider the following simple example:

(6)   John has a fever and small red spots on his face and body.
(7)   Everybody who has rubeola (the measles) has a fever and small red spots on the face and body.
(8)   Nobody has rubeola more than once.

Suppose that (6) is the *explanandum* B, and (7)–(8) the theory T. From T, we can derive by **CL**:

(9)   If John has rubeola, then he has a fever and small red spots on his face and body.

Hence, as requirements (1)–(5) are evidently fulfilled for B, T, and the antecedent of (9), we can abduce:

(10)   John has rubeola.

as an explanation for (6). If, however, we add to T:

(11)   John had rubeola last year.

(10) has to be withdrawn as an explanation for (6). The reason is that from T together with (11) we can derive:

(12)   John does not have rubeola.

which violates requirement (2). The same mechanism applies for the other negative requirements. Suppose, for instance, that we define "A is a minimal explanation for B" as "A is the logically weakest explanation for B", and that we add to T:

(13)   Everybody who has rubella (the German measles) has a fever and small red spots on the face and body.

From (7) and (13), we can derive by **CL**:

(14)   If John has rubeola *or* rubella, he has a fever and small red spots on the face and body.

But then, as "John has rubeola or rubella" is logically weaker than (10), (10) has again to be withdrawn.

The final characteristic is that abductive reasoning processes, at the predicative level, do not have a *positive test*. This is related to the fact that first-order predicate logic is undecidable – if some conclusion A does *not* follow from a set of premises Γ, we may not be able to establish this. Hence, as abductive inferences are partly defined in terms of negative requirements, it immediately follows that, for undecidable fragments, they lack a positive test. Suppose, for instance, that for some theory T, some *explanandum* B, and some sentence A, (1) is satisfied. In that case, it seems reasonable to conclude that A follows abductively from T. However, if one is unable to establish that also (2)–(5) are satisfied, no reasoning can warrant that this conclusion is not erroneous.

There are different ways to deal with the lack of a positive test. The most common one is to consider only decidable fragments of first-order logic. The rationale behind this is clear: when dealing with decidable fragments, one may be sure that, for arbitrary theories T and *explananda* B, there is an algorithm for (2)–(5), and hence, that a decision method can be designed for "follows abductively from". From the point of view of applications, however, this is an enormous restriction: many interesting theories are undecidable.

An alternative way is to allow that inferences are made, not on the basis of absolute warrants, but on the basis of one's best insights in the premises. When this second option is followed, abductive reasoning processes not only exhibit an *external* form of dynamics (adding new information may lead to the withdrawal of previously derived conclusions), but also an *internal* one (the withdrawal may be caused by merely analysing the premises). Suppose, for instance, that for some theory T, some *explanandum* B, and some sentence A, one established that (1) is satisfied, and one did not establish that one of (2)–(5) is violated. In that case, it seems rational to conclude that A follows abductively from T. This conclusion, however, is provisional. If at a later moment in time, one is able to show that one of the negative requirements is violated (for instance, because one established that $\neg A$ follows from T), A has to be withdrawn as an explanation for B.

There are several arguments in favor of this second option. The first is that unwanted restrictions are avoided: abduction can be defined for *any* first-order theory. A second argument is that the conclusions of abductive reasoning processes are defeasible anyway. Whether the withdrawal of a conclusion is caused by an external factor or an internal one does not seem to be essential. The third, and most important argument is that, even for decidable fragments, it is often unrealistic to require absolute warrants. Even if a decision method is available, reasoners may lack the resources to perform an exhaustive search, and hence, may be forced to act on their present best insights.

The logics presented in this paper follow the second option. This has the advantage that, even for undecidable fragments, they enable one to come to justified conclusions. These conclusions are tentative and may later be rejected, but they constitute, given one's insight in the premises at that moment, the best possible estimate of the conclusions that are "finally derivable" from the premises.[7]

---

[7] Roughly speaking, an "abductive conclusion" A is finally derivable from a theory T if the requirements (1)–(5) are satisfied – see section 6 for a precise definition of this notion.

4. The general format

As explained in the previous section, abduction is an ampliative type of reasoning that leads to conclusions that are compatible with some theory T, and that satisfy some additional requirements (for instance, that they explain some other sentence). As this notion of compatibility is crucial to understand the general format of the proof theories, we briefly elaborate on it.

From a semantic point of view, compatibility is a simple notion: A is compatible with a set of premises Γ iff A is true in *some* **CL**-model of Γ. Evidently, if A is true in some **CL**-models of Γ, and $\neg A$ in some others, then both A and $\neg A$ are compatible with Γ. If A is true in *all* **CL**-models of Γ, then $\neg A$ is incompatible with Γ.

When trying to capture these notions in a consequence relation, and a corresponding proof theory, matters seem at first somewhat more complicated. As both A and $\neg A$ may be compatible with some set Γ, restrictions are needed to avoid that mutually inconsistent conclusions lead to triviality. Formulating these restrictions in the language of **CL** leads to a proof theory that is quite complex and not very transparent. The problem can easily be solved, however, by moving to a modal approach. That A is true in *some* **CL**-model of Γ can naturally be expressed as "A is *possible*" ($\Diamond A$), and that A is true in *all* of them as "$\neg A$ is *impossible*" ($\neg \Diamond \neg A$), or, in other words, "A is *necessary*" ($\Box A$).

This modal approach is used in [Batens and Meheus, 2000] to design an adaptive logic for compatibility that is called **COM**. **COM** is based on **S5**, and is defined with respect to sets of the form $\Gamma^\Box = \{\Box A \mid A \in \Gamma\}$. As is shown in [Batens and Meheus, 2000], **COM** has the interesting property that $\Gamma^\Box \vdash_{\mathbf{COM}} \Diamond A$ iff $\Gamma \nvdash_{\mathbf{CL}} \neg A$, and hence, iff A is compatible with Γ. Note that a modal approach immediately solves the problem that mutually inconsistent conclusions lead to triviality (in view of $\Diamond A, \Diamond \neg A \nvdash_{\mathbf{S5}} B$).

The logics presented here share their general format with **COM**: premises are treated as necessarily true, and conclusions arrived at by abduction as possibly true. So, the proof theory will not rely on argumentation scheme (†), but on its modal translation, namely

(15) $\Box(A \supset B), \Box B \; / \; \Diamond A$

There is, however, a small complication. This is related to the fact that $\Box B \vdash_{\mathbf{S5}} \Box(A \supset B)$, for arbitrary A, and that $\Box \neg A \vdash_{\mathbf{S5}} \Box(A \supset B)$, for arbitrary B. So, if (15) would be validated without further restrictions, one would be able to generate arbitrary explanations.

We shall solve this problem by making a distinction between two different sets of premises – a set of observational statements Γ_1 and a set of background assumptions Γ_2 – and require that (15) can only be applied if $\Box(A \supset B)$ is **S5**-derivable from $\Gamma_2^\Box$, whereas $\Box B$ and $\Box\neg A$ are not. In order to realize all this in the easiest way possible, we shall rely on a bimodal version of **S5** that we shall call **S5²**. Its language includes two necessity operators ("$\Box_1$" and "$\Box_2$"), and two possibility operators ("$\Diamond_1$" and "$\Diamond_2$"). To simplify things, we shall only consider modal formulas of first degree – a modal formula is said to be of first degree if it contains one or more modal operators, but none of these is inside the scope of any other modal operator. A formula that does not contain any modal operator will be said to be of degree zero.

The operator "$\Box_1$" will be used for observational statements, and "$\Box_2$" for background assumptions. **S5²** is defined in such a way that $\Box_1 A$ is derivable from $\Box_2 A$, but not *vice versa*. It is thus warranted that the consequences from both sets can be conjoined (to derive predictions, for instance). At the same time, however, it is possible to recognize which sentences are derivable from the background assumptions alone.

So, this is the general idea. The logics **CP1** and **CP2** are adaptive logics based on **CL**. Their proof theories, however, are defined with respect to a modal adaptive logic based on **S5²**. These modal adaptive logics will be called **MA1** and **MA2**. It will be stipulated that A is a **CP1**-consequence (respectively **CP2**-consequence) of some theory T iff $\Diamond A$ is an **MA1**-consequence (respectively **MA2**-consequence) of the modal translation of T.

Let us end this section with two small remarks. The first is a notational one. In the subsequent sections, Γ_1 and Γ_2 will be used to refer to sets of closed formulas of some first-order language. It will always be assumed that Γ_1 is a set of observational statements, and Γ_2 a set of background assumptions. Σ will refer to an ordered set $\langle \Gamma_1, \Gamma_2 \rangle$, and $\Sigma^\Box$ to $\{\Box_1 A \mid A \in \Gamma_1\} \cup \{\Box_2 A \mid A \in \Gamma_2\}$.

The second remark concerns the selection of *explananda*. In order to keep the logics as general as possible, they are designed in such a way that several *explananda* can be considered at the same time. Actually, whenever $\Box_1 B$ is **S5²**-derivable from the modal translation of Σ, B functions as an *explanandum*. Evidently, the set of *explananda* may be restricted in various ways. The aim of this paper, however, is to present a generic format that can easily be adapted to the needs of a specific application context.

5.   Introducing the dynamics

In section 3, we explained that abductive reasoning processes exhibit two different kinds of dynamics, namely an external and an internal one.

As we shall see below, handling inconsistencies in a sensible way leads to similar forms of dynamics. To keep matters simple, however, we first discuss the consistent case, and hence, concentrate for the moment on **MA1**.

The main question is how the dynamics that is typical of abductive inferences can be incorporated in a proof theory. The trick is actually extremely simple, and can be applied to any form of ampliative reasoning:[8] *if an inference rule is defined in terms of a combination of positive and negative requirements, consider the* fulfillment *of the former as a sufficient condition to* validate *applications of the rule, and the* violation *of the latter as a sufficient condition to* invalidate *such applications*.

In line with this very simple idea, abductive inferences in **MA1** will be governed by one rule (RC) and a set of criteria. The rule RC incorporates the positive requirements for abductive inferences. It enables one to add $\Diamond_2 A$ to an **MA1**-proof whenever $\Box_1 B$ and $\Box_2(A \supset B)$ are derived. At each new stage of the proof (with each new line added), the criteria are evoked. They ensure that, if one of the negative requirements is explicitly violated for some conclusion $\Diamond_2 A$ (relative to some *explanandum B*), then $\Diamond_2 A$ is withdrawn. Technically, this is realized by *marking* the line on which $\Diamond_2 A$ occurs. This indicates that the formula at issue is no longer derivable.

In addition to abductive inferences, **MA1** validates a series of deductive inferences. These are governed by a generic rule that is called RU, and that enables one to add A to an **MA1**-proof from $\Sigma^\Box$ whenever $\Sigma^\Box \vdash_{\mathbf{S5^2}} A$.

In order to illustrate these basic ideas, we give a very simple example of an **MA1**-proof (the fifth column can safely be ignored for the moment):

1	$\Box_1 p$	–	PREM	$\emptyset$
2	$\Box_2(q \supset (p \wedge r))$	–	PREM	$\emptyset$
3	$\Box_2(q \supset s)$	–	PREM	$\emptyset$
4	$\Box_2(s \supset \neg q)$	–	PREM	$\emptyset$
5	$\Box_2(q \supset p)$	2	RU	$\emptyset$

Lines 1–4 are introduced by the premise rule PREM. As is clear from the indices, the formula on line 1 is an observational statement, and those on lines 2–4 are background assumptions. At stage 5, the rule RU is applied – the formula on line 5 is $\mathbf{S5^2}$-derivable from that on line 2. Next, the rule RC can be applied to the formulas on lines 1 and 5:

---

[8] It is typical of ampliative reasoning processes that their definition necessarily contains some negative requirements.

6    $\Diamond_2 q$ 1, 5   RC $\langle q, p \rangle$

The fifth element of line 6 is the condition that has to be satisfied in order for the formula to be derivable. Intuitively, $\langle q, p \rangle$ is read as "provided that q is a good explanation for p". As may be expected, this condition is satisfied iff none of the negative requirements is explicitly violated. At the present stage of the proof, the condition of line 6 is satisfied. Suppose, however, that we continue the proof as follows:

7    $\Box_2 \neg q$ 3, 4   RU $\emptyset$

It now becomes clear that q is not compatible with one's background assumptions, and hence that it is not a good explanation for p. In view of this, line 6 is *marked* as no longer derivable.[9] The only formulas considered as derived at stage 7 are those on lines (1)–(5) and (7).

6.   The logics MA1 and CP1

Let $\mathcal{L}$ be the standard predicative language of **CL**, and let $\mathcal{L}^M$ be obtained from $\mathcal{L}$ by extending it with "$\Box_1$", "$\Box_2$", "$\Diamond_1$", and "$\Diamond_2$". Let the set of wffs of $\mathcal{L}^M$, $\mathcal{W}^M$, be restricted to wffs of degree zero and first degree. The relation between **MA1** and **CP1** is given by the following definition:

Definition 1 $\Sigma \vdash_{\textbf{CP1}} A$ *iff* $\Sigma^\Box \vdash_{\textbf{MA1}} \Diamond_2 A$.

The proof theory for **MA1** proceeds in terms of the bimodal logic $\textbf{S5}^2$. Syntactically, $\textbf{S5}^2$ is obtained by extending an axiomatization of the full predicative fragment of **CL** with every instance (for $i \in \{1, 2\}$) of the following axioms, rule, and definition:[10]

A1 $\Box_i A \supset A$
A2 $\Box_i(A \supset B) \supset (\Box_i A \supset \Box_i B)$
A3 $\Box_2 A \supset \Box_1 A$
NEC   if $\vdash A$ then $\vdash \Box_i A$
D$^\Diamond$ $\Diamond_i A =_{df} \neg\Box_i\neg A$

As is usual for adaptive logics, **MA1**-proofs consist of lines that have five elements: (i) a line number, (ii) the formula A that is derived, (iii) the line numbers of the formulas from which A is derived, (iv) the rule

---

[9]The exact format of this marking will become clear in Section 6. For the moment, just imagine that some mark is added to the utmost right of the line.
[10]As $\mathcal{W}^M$ contains only wffs of degree zero and first degree, no axiom is needed for the reduction of iterated modal operators.

by which A is derived, and (v) a condition. The condition has to be satisfied in order for A to be so derivable.

The condition will either be $\emptyset$ or a couple $\langle A, B \rangle$. As indicated already in the previous section, a line of the form

$$i \quad C \quad \ldots \quad \ldots \quad \langle A, B \rangle$$

will be read as "C provided that A is a good explanation for B". The interpretation of the ambiguous phrase "is a good explanation for" will be fixed by the marking criteria. A wff is said to be derived unconditionally iff it is derived on a line the fifth element of which is empty.

Here are the generic rules that govern **MA1**-proofs from $\Sigma^{\square}$:

PREM   If $A \in \Sigma^{\square}$, then one may add to the proof a line consisting of (i) the appropriate line number, (ii) A, (iii) "$-$", (iv) "PREM", and (v) $\emptyset$.

RU   If $A_1, \ldots, A_n \vdash_{\mathbf{S52}} B$ ($n \geq 0$), $A_1, \ldots, A_n$ occur in the proof on the conditions $\Delta_1, \ldots, \Delta_n$ respectively, and at most one of the Δ_i is non-empty, then one may add to the proof a line consisting of (i) the appropriate line number, (ii) B, (iii) the line numbers of the A_i, (iv) "RU", and (v) $\Delta_1 \cup \ldots \cup \Delta_n$.

RC   If $\square_1 B$ and $\square_2 (A \supset B)$ occur in the proof, then one may add to the proof a line consisting of (i) the appropriate line number, (ii) $\Diamond_2 A$, (iii) the line numbers of $\square_1 B$ and $\square_2 (A \supset B)$, (iv) "RC", and (v) $\langle A, B \rangle$.

Note that RU enables one to conjoin formulas that are conditionally derived to formulas that are unconditionally derived. This is important, because it warrants that explanatory hypotheses can be used to derive predictions from the background assumptions. The rule RU does not enable one, however, to conjoin explanatory hypotheses to one another (it cannot be applied if more than one of the A_i occurs on a non-empty condition). By using a somewhat more complex format for the condition, the rule RU can easily be generalized to such inferences. However, as we are not interested in **MA1** itself, and as the generalization would not lead to a richer consequence set for **CP1** (in view of $\Diamond_i A, \Diamond_i B \nvdash_{\mathbf{S52}} \Diamond_i (A \wedge B)$ and Definition 1), this would only complicate matters.

From the generic rules RU and RC, several rules can be derived that lead to proofs that are more interesting from a heuristic point of view. We list only two such rules that will be useful in the examples below:

RD1   If $\square_1 B(\beta)$ and $\square_2 (\forall \alpha)(A(\alpha) \supset B(\alpha))$ occur in the proof, then one may add to the proof a line that has $\Diamond_2 A(\beta)$ as its second element, "RD1" as its fourth, and $\langle A(\beta), B(\beta) \rangle$ as its fifth.

RD2   If $\square_1 B_1(\beta), \ldots, \square_1 B_n(\beta)$ and $\square_2(\forall\alpha)(A(\alpha) \supset B_1(\alpha)), \ldots, \square_2(\forall\alpha)$
   $(A(\alpha) \supset B_n(\alpha))$ occur in the proof, then one may add to the proof
   a line that has $\diamondsuit_2 A(\beta)$ as its second element, "RD2" as its fourth,
   and $\langle A(\beta), B_1(\beta) \wedge \ldots \wedge B_n(\beta)\rangle$ as its fifth.

Let us now turn to the marking criteria. As explained above, these criteria determine when the condition of a line is violated. Put differently,
they determine which requirements a hypothesis A and an *explanandum*
B should meet so that A is a *good* explanation for B. If some condition is
no longer satisfied, the marking criteria warrant that the lines on which
it occurs are marked.

The first two criteria are straightforward. The first, that for T-
marking, is related to ruling out (consequences of) trivial explanations:

CMT   A line that has $\langle A, B\rangle$ as its fifth element is T-marked iff some
   line in the proof has $\square_2 B$ or $\square_2 \neg A$ as its second element.

The second criterion warrants that (all consequences of) empirically falsified explanations are withdrawn:

CME   A line that has $\langle A, B\rangle$ as its fifth element is E-marked iff some
   line in the proof has $\square_1 \neg A$ as its second element.

The third criterion is related to ruling out (partial) self-explanations.
In view of (4), this presupposes that one is able to recognize, in the proof,
that A is entailed by B. We shall assume that this is done by deriving
$\square_i(B \supset A)$ on a line j such that, at line j, the path of $\square_i(B \supset A)$ does
not include any premise. Expressions of the form $\pi_j(A)$ will refer to the
path of a formula A as it occurs on line j. To keep things simple, we
first define a set $\pi_j^\circ(A)$:

Definition 2 *Where A is the second element of line j, $\pi_j^\circ(A)$ is the
smallest set Λ that satisfies:*
(i) $j \in \Lambda$, and
(ii) if $k \in \Lambda$, and $l_1, \ldots, l_n$ is the third element of line k, then $l_1, \ldots, l_n \in \Lambda$.

Definition 3 $\pi_j(A) = \{B \mid B$ *occurs as the second element of line i
and $i \in \pi_j^\circ(A)\}$.*

Here is the third marking criterion:

CMS   A line that has $\langle A, B\rangle$ as its fifth element is S-marked iff some
   line j in the proof has $\square_i(B \supset A)$ as its second element and
   $\pi_j(\square_i(B \supset A)) \cap (\Gamma_1 \cup \Gamma_2) = \emptyset$.

The final marking criterion, that for *P*-marking, requires a bit more explanation. As mentioned in section 2 the problem of selecting the minimal explanatory hypotheses (the most parsimonious ones) can be realized in different ways.

A first one is to warrant that, whenever *A* and *C* explain the same *explanandum*, and *C* is (according to one's best insights) logically weaker than *A*, *A* is rejected in favor of *C*. This has the advantage that recognizing the minimal explanatory hypotheses is extremely simple. The disadvantage is, however, that it becomes impossible to generate alternative explanations for the *explanandum* (in view of $A \supset C$, $B \supset C \vdash_{\mathbf{CL}}$ $(A \vee B) \supset C$).

A second way is to warrant that, if two explanations *A* and *C* are available for an *explanandum B*, and *C* is logically weaker than *A*, then *A* is withdrawn in favor of *C*, *unless C* is **CL**-equivalent to a formula $A \vee D$ such that *D* includes some "bits of information" not included in *A*. This second option enables one to generate alternative explanations. For instance, if both $p \wedge q$ and p are alternative explanations for the same *explanandum*, then the former is withdrawn in favor of the latter. However, if the alternative explanations are $p \wedge q$ and $(p \vee r) \wedge (q \vee r)$, then $p \wedge q$ is not withdrawn. The reason is that $(p \vee r) \wedge (q \vee r)$ is **CL**-equivalent to $(p \wedge q) \vee r$, and *r* is a new piece of information with respect to $p \wedge q$. In order to keep the proof theory of **MA1** as generic as possible, it is based on this second notion of minimality. In section 10, however, we shall show how **MA1** can easily be transformed to a system for the first notion.

The second option raises the question how to compare the "building blocks" of different formulas. Obviously, this cannot be done by referring to the atoms as they occur in the formulas at issue (because, for instance, $A \supset B$ is equivalent to $\neg A \vee B$). Instead, we first have to associate each formula with one or more formulas in some kind of normal form, and next refer to the atoms as they occur in the latter. This is why, below, we shall associate each formula with a set of formulas in prenex disjunctive normal form. Where $\mathcal{F}^a$ is the set of atoms of the standard predicative language (primitive formulas – schematic letters and primitive predicative formulas – and their negations), *A* is said to be in prenex disjunctive normal form (PDNF) iff it is of the form

(†) $\quad \mathbf{Q} \bigvee \{\bigwedge \Sigma_1, \ldots, \bigwedge \Sigma_n\}$

in which **Q** is a sequence of quantifiers and $\Sigma_1 \cup \ldots \cup \Sigma_n \subset \mathcal{F}^a$.

There is a further complication, namely that even a formula in PDNF may contain some "irrelevant" atoms – where *A* is in PDNF, we say that an atom *B* occurs irrelevantly in *A* iff *A* is equivalent to a formula *C* (in PDNF) in which B does not occur. For instance, q and $\neg q$ occur irrelevantly in $(p \wedge q) \vee (p \wedge \neg q)$ – in view of $\vdash_{\mathbf{CL}} ((p \wedge q) \vee (p \wedge \neg q)) \equiv p$. This problem can be solved by associating each formula with a unique formula in PDNF. However, to keep the proof theory as realistic as possible, we shall proceed in a somewhat different way. The idea will be to associate each formula with a set of formulas in PDNF that are, at a certain stage in the proof, *recognized* as being equivalent to A, and to refer to the atoms that occur in each of these.[11]

In order to make the criterion for P-marking as transparent as possible, we define, for each A, four sets. The first of these, $\phi_s(A)$, contains all formulas that are, at stage s, recognized as being logically weaker than A. The second, $eq_s(A)$, contains all formulas in PDNF that, at stage s, are recognized as being equivalent to A. The third, $at_s(A)$, selects those atoms that occur in every member of $eq_s(A)$. The fourth set, $\psi_s(A)$, contains all formulas that are recognized at stage s as being more parsimonious than A. Let $\mathcal{W}$ be the set of wffs of $\mathcal{L}$.

Definition 4 $\phi_s(A) = \{B \in \mathcal{W} \mid \Box_i(A \supset B)$ *occurs on a line j such that* $\pi_j(\Box_i(A \supset B)) \cap (\Gamma_1 \cup \Gamma_2) = \emptyset$; $\Box_i(B \supset A)$ *does not occur on a line j such that* $\pi_j(\Box_i(B \supset A)) \cap (\Gamma_1 \cup \Gamma_2) = \emptyset\}$.

Definition 5 $eq_s(A)$ *is the smallest set Λ that satisfies:*
(i) *if $\Diamond_i A$ or $\Box_i A$ occurs in the proof at stage s, and A is in PDNF, then $A \in \Lambda$, and*
(ii) *if $\Box_i(A \equiv B)$ occurs in the proof at stage s on a line j, B is in PDNF, and $\pi_j(\Box_i(A \equiv B)) \cap (\Gamma_1 \cup \Gamma_2) = \emptyset$, then $B \in \Lambda$.*

Definition 6 $at_s(A) = \{B \in \mathcal{F}^a \mid B$ *occurs in C, for every $C \in eq_s(A)\}$.*

Definition 7 $\psi_s(A) = \{B \in \mathcal{W} \mid at_s(B) \neq \emptyset$; $at_s(B) \subseteq at_s(A)\}$.

Here is the criterion for P-marking:

CMP   A line that has $\langle A, B \rangle$ as its fifth element is P-marked iff some line in the proof that is not $(T\text{-}, E\text{-}, S\text{-})$marked has $\langle C, B \rangle$ as its fifth element, $C \in \phi_s(A)$, and $C \in \psi_s(A)$.

Let us add three final remarks to this. First, CMP warrants that formulas recognized as being equivalent are not marked in view of each

---

[11]This is more in line with the general philosophy behind adaptive logics, namely that the application of inference rules and criteria is relative to the distinctions that, at a certain stage, are made by the reasoner. The easiest way to realize this is to refer to formulas that are actually written down in the proof.

other (in view of the definition of $\phi_s(A)$). Next, CMP prevents that explanatory hypotheses are marked simply because the logical form of some alternative explanation has not yet been recognized (in view of the first clause in the definition of $\psi_s(A)$). Finally, formulas that are P-marked at some stage may at a later stage be unmarked.

The criterion for P-marking is illustrated in Figure 1. At stage 7, line 5 is P-marked: both $q \wedge s$ and q are derived at stage 7 as explanations for p, but q is logically weaker than $q \wedge s$ ($q \in \phi_7(q \wedge s)$ in view of line 7, and the absence of an analogous line for $q \supset (q \wedge s)$), and moreover q is more parsimonious than $q \wedge s$ ($q \in \psi_7(q \wedge s)$ in view of $at_7(q) = \{q\} \subseteq \{q, s\} = at_7(q \wedge s)$). Note that line 6 is not P-marked in view of lines 9 and 10 – $at_9(q \vee r) \not\subseteq at_9(q)$.

1	$\Box_1 p$	–	PREM	$\emptyset$	
2	$\Box_2(q \supset p)$	–	PREM	$\emptyset$	
3	$\Box_2(r \supset p)$	–	PREM	$\emptyset$	
4	$\Box_2((q \wedge s) \supset p)$	2	RU	$\emptyset$	
5	$\Diamond_2(q \wedge s)$	1, 4	RC	$\langle q \wedge s, p \rangle$	$\sqrt{}^{P7}$
6	$\Diamond_2 q$	1, 2	RC	$\langle q, p \rangle$	
7	$\Box_2((q \wedge s) \supset q)$	–	RU	$\emptyset$	
8	$\Box_2((q \vee r) \supset p)$	2, 3	RU	$\emptyset$	
9	$\Diamond_2(q \vee r)$	1, 8	RC	$\langle q \vee r, p \rangle$	
10	$\Box_2(q \supset (q \vee r))$	–	RU	$\emptyset$	

Figure 1. A simple illustration of P-marking.

In view of the marking criteria, two forms of derivability can be defined. We say that a line is *marked* iff it is T-marked, E-marked, S-marked or P-marked. If the mark of a line is removed, we say that it is unmarked.

Definition 8 *A is* derived at a stage *in an **MA1**-proof from $\Sigma^\Box$ iff A is derived in the proof on a line that is not marked.*

Definition 9 *A is* finally derived *in an **MA1**-proof from $\Sigma^\Box$ iff A is derived on a line j that is not marked, and any extension of the proof in which line j is marked may be further extended in such a way that line j is unmarked.*

As is usual for adaptive logics, the consequence relation of **MA1** is defined with respect to final derivability:

Definition 10 *$\Sigma^\Box \vdash_{\mathbf{MA1}} A$ (A is finally derivable from $\Sigma^\Box$) iff A is finally derived in an **MA1**-proof from $\Sigma^\Box$.*

7.   Generalizing to the inconsistent case

As explained in the first section, abducing explanations frequently involves reasoning from inconsistencies, either because one is trying to find an explanation for an anomalous fact or because one is dealing with an inconsistent theory. One reason why such explanations are important is that they may lead to a revision of one's theory. For example, when observational data do not fit the theoretical predictions, researchers will try to explain these anomalies on the basis of the available theory. When an acceptable explanation is found, this can in turn be used to modify the theory. Also in cases where one is dealing with inconsistent theories, generating explanatory hypotheses may be useful in resolving the inconsistencies.

For reasons explained in section 4, the general format of **MA1** warrants that it can be used to generate mutually incompatible hypotheses. It is important to note, however, that **MA1** is not adequate to explain anomalous facts. Suppose, for instance, that one is dealing with the following set of premises:

$$(23)\quad \langle \Gamma_1, \Gamma_2 \rangle = \langle \{q\}, \{p,\ p \supset \neg q,\ r \supset q\} \rangle.$$

It is easily observed that the rule RU enables one to derive arbitrary conclusions from $\{\Box_1 q,\ \Box_2 p,\ \Box_2(p \supset \neg q),\ \Box_2(r \supset q)\}$, and hence, that **MA1**, in view of the anomalous fact q, does not enable one to derive sensible explanations and predictions. For similar reasons, **MA1** is inadequate to generate explanations from inconsistent theories – consider, for instance, the following example:

$$(24)\quad \langle \Gamma_1, \Gamma_2 \rangle = \langle \{q\}, \{p \supset q,\ \neg(p \supset q)\} \rangle.$$

In this section, we shall explain how **MA1** can be generalized to a system that enables one to explain anomalous facts, and moreover enables one to generate explanations on the basis of an inconsistent theory.

In view of the discussion in section 4, the generalization may seem straightforward. If the members of Γ_2 are "problematic", in the sense that they are contradicted by the members of Γ_1 or are mutually inconsistent, it seems natural to consider them as *possibly true* rather than as *necessarily true*.[12] In this way, one can express that at least some of the members of Γ_2 will have to be modified in order to resolve the inconsis-

---

[12]This idea underlies the first formal paraconsistent logic that was presented by Stanisław Jaśkowski in 1948.

tencies.[13] It is easily observed that this reinterpretation of the premises avoids that inconsistencies lead to triviality. The modal translations of (23) and (24), for instance, now are

(23') $\{\Box_1 q,\ \Diamond_2 p,\ \Diamond_2(p \supset \neg q),\ \Diamond_2(r \supset q)\}$.

(24') $\{\Box_1 q,\ \Diamond_2(p \supset q),\ \Diamond_2\neg(p \supset q)\}$.

from which no arbitrary conclusions are derivable. The reinterpretation, however, also raises two problems.

The first is that, even if it seems natural to consider the members of Γ_2 as possibly true only, we still want to use them for generating explanations. In (23), for instance, we want to abduce r as an explanation for q, notwithstanding the fact that the latter is anomalous with respect to p and $p \supset \neg q$. This problem can easily be solved by modifying the conditional rule in such a way that $\Diamond_2 A$ can be added to the proof whenever $\Box_1 B$ and $\Diamond_2(A \supset B)$ occur in it.

The second problem seems more serious. By reinterpreting the members of Γ_2, all (genuine) multi-premise rules are invalidated. As a consequence, the inferential strength as well as the explanatory power of one's theory is seriously reduced. Consider, for instance, the following example:

(25) $\langle \Gamma_1, \Gamma_2 \rangle = \langle \{q\}, \{p,\ p \supset \neg q,\ s \vee (r \supset q),\ \neg s\} \rangle$.

As $\{\Diamond_2(s \vee (r \supset q)),\ \Diamond_2\neg s\} \nvdash_{\text{S5}^2} \Diamond_2(r \supset q)$, the reinterpretation would prevent one from generating r as an explanation for q. We shall now show that this problem can be solved by relying on an idea that was first presented in [Meheus, 200+b].

Consider again (25). Assuming that one does not question the observational statement q, both p and $p \supset \neg q$ are problematic – one has to reject or modify at least one of them in order to restore consistency. Hence, as long as it is unclear what changes are justified, it seems rational to consider both premises as possibly true only. The same does not hold true, however, for $\neg s$ – the latter is totally unrelated to the inconsistency that follows from q, p and $p \supset \neg q$.

This brings us to the general idea: formulas that behave consistently should be considered as necessarily true. Obviously, this can be realized by deciding beforehand which formulas should be interpreted as possibly

---

[13]If it is clear how the set of premises should be modified in order to restore consistency, there is obviously no need for this reinterpretation. In interesting cases, however, the problem is precisely to find out which premises should be modified, and in what way this should be done.

true only and which as necessarily true. This strategy has the disadvantage, however, that the proof theory of **MA2** would be restricted to decidable fragments, and moreover would be unrealistic even for those.

This is why we shall proceed in a different way. The consequence relation of **MA2** will be defined with respect to sets of the form $\Sigma^\Diamond = \{\Box_1 A \mid A \in \Gamma_1\} \cup \{\Diamond_2 A \mid A \in \Gamma_2\}$. It will be allowed, however, that $\Box_2 A$ is derived from $\Diamond_2 A$ *on the condition* that $\Diamond_2 A$ behaves "normally" – a formula A will be said to behave normally with respect to $\Sigma^\Diamond$ iff it behaves consistently with respect to Σ. If this condition is no longer satisfied, then $\Box_2 A$ will be withdrawn.

As a first approximation, one may say that a formula A behaves abnormally iff $\Diamond_2 A \wedge \Diamond_2 \neg A$ is **S5²**-derivable from $\Sigma^\Diamond$. There are, however, three small complications. The first is that, in some cases, no abnormality is **S5²**-derivable from $\Sigma^\Diamond$, but some *disjunction* of abnormalities is. If, for instance, $\Sigma^\Diamond = \{\Diamond_2(p \vee q), \Diamond_2 \neg p, \Diamond_2 \neg q\}$, then neither $\Diamond_2 p \wedge \Diamond_2 \neg p$ nor $\Diamond_2 q \wedge \Diamond_2 \neg q$ is **S5²**-derivable from $\Sigma^\Diamond$, but $(\Diamond_2 p \wedge \Diamond_2 \neg p) \vee (\Diamond_2 q \wedge \Diamond_2 \neg q)$ is. This suggests that a formula A behaves abnormally iff $\Diamond_2 A \wedge \Diamond_2 \neg A$ is a disjunct of a "minimal" disjunction of abnormalities. A disjunction of abnormalities that is **S5²**-derivable from $\Sigma^\Diamond$ will be called a "*Dab*-consequence"; a *Dab*-consequence will be called minimal iff no result of dropping some disjunct from it is an **S5²**-consequence of $\Sigma^\Diamond$. The two other complications are that, for reasons explained in [Meheus, 200+b], the abnormalities have to be restricted to *primitive* formulas, and that, at the predicative level, abnormalities are of the form $\exists(\Diamond A \wedge \Diamond \neg A)$, where $\exists A$ abbreviates A preceded by a sequence of existential quantifiers (in some preferred order) over the variables that occur free in A.

What all this comes to is that **MA2**-proofs are dynamic, not only because of the abductive steps, but also because the premises are interpreted "as normally as possible": from $\Diamond_2 A$ one may derive $\Box_2 A$, but only on the condition that all primitive formulas in A behave normally. In line with this, we shall introduce an additional conditional rule (RC1) and an additional marking criterion (CMI). A line added on the condition that some formulas $A_1, \ldots, A_n$ behave consistently will be I-marked iff this condition is explicitly violated. The conditional rule for abductive inferences (RC2), the marking criteria for abductive inferences, and the unconditional rule will be as for **MA1**, except for some small changes that result from the new format and the interplay between RC1 and RC2.

The example in Figure 2 illustrates the basic mechanisms. In **MA2**, the fifth element is an ordered set of two conditions. The first of these is related to RC1, the second to RC2. The fifth element of line 16,

1	$\square_1 q$	–	PREM	$\langle \emptyset, \emptyset \rangle$
2	$\lozenge_2 p$	–	PREM	$\langle \emptyset, \emptyset \rangle$
3	$\lozenge_2(p \supset \neg q)$	–	PREM	$\langle \emptyset, \emptyset \rangle$
4	$\lozenge_2(s \vee (r \supset q))$	–	PREM	$\langle \emptyset, \emptyset \rangle$
5	$\lozenge_2 \neg s$	–	PREM	$\langle \emptyset, \emptyset \rangle$
6	$\lozenge_2(r \supset t)$	–	PREM	$\langle \emptyset, \emptyset \rangle$
7	$\square_2 p$	2	RC1	$\langle \{p\}, \emptyset \rangle \quad \sqrt{I^{11}}$
8	$\square_2(p \supset \neg q)$	3	RC1	$\langle \{p, q\}, \emptyset \rangle \sqrt{I^{11}}$
9	$\square_1 \neg q$	7, 8	RU	$\langle \{p, q\}, \emptyset \rangle \sqrt{I^{11}}$
10	$\neg \square_1 \neg q$	1	RU	$\langle \emptyset, \emptyset \rangle$
11	$(\lozenge_2 p \wedge \lozenge_2 \neg p) \vee (\lozenge_2 q \wedge \lozenge_2 \neg q)$	1, 2, 3	RU	$\langle \emptyset, \emptyset \rangle$
12	$\square_2 \neg s$	5	RU	$\langle \{s\}, \emptyset \rangle$
13	$\lozenge_2(r \supset q)$	4, 12	RU	$\langle \{s\}, \emptyset \rangle$
14	$\lozenge_2 r$	1, 13	RC2	$\langle \{s\}, \langle r, q \rangle \rangle$
15	$\square_2(r \supset t)$	6	RC1	$\langle \{r, t\}, \emptyset \rangle$
16	$\lozenge_2 t$	14, 15	RU	$\langle \{s, r, t\}, \langle r, q \rangle \rangle$

Figure 2.   A simple example of an **MA2**-proof.

for instance, is intuitively read as "provided that s, r and t behave consistently, and that r is a good explanation for q".

The observational statement is on line 1, the background assumptions on lines 2–6. On line 7, the formula of line 2 is "upgraded" from possibly true to necessarily true, on the condition that p behaves consistently. Analogously, the formula on line 8 is obtained from that on line 3 on the condition that both p and q behave consistently. This enables one to derive, on line 9, a formula that contradicts the observational statement of line 1 (compare also lines 9 and 10). However, as it can be derived that p or q behaves inconsistently (line 11), lines 7–9 are I-marked, and triviality is avoided. Still, the theory is interpreted in a way that is as rich as possible: all formulas that do not rely on inconsistently behaving formulas can be upgraded and will not be marked in any extension of the proof (see, for instance, lines 12 and 15). As a consequence, the explanatory power as well as the predictive power of the theory is as great as possible. For example, as $\square_2 \neg s$ is derivable (line 12), so is the formula on line 13, and hence, r can be generated as an explanation for the anomalous fact q (line 14). As $\square_2(r \supset t)$ is derivable, the prediction t can be derived (line 16).

8.　　The logics MA2 and CP2

The language of **CP2** is $\mathcal{L}$, and that of **MA2** is $\mathcal{L}^M$. The relation between the two systems is as that between **CP1** and **MA1**:

Definition 11 $\Sigma \vdash_{\mathbf{CP2}} A$ *iff* $\Sigma^\diamond \vdash_{\mathbf{MA2}} \diamond_2 A$.

The proof format of **MA2** is as that for **MA1**. The fifth element of a line in an **MA2**-proof is of the form $\langle \Theta, \emptyset \rangle$ or $\langle \Theta, \langle A, B \rangle \rangle$. A wff is said to be derived unconditionally iff it is derived on a line the fifth element of which is $\langle \emptyset, \emptyset \rangle$.

The premise rule is as for **MA1**, except that all references to $\Sigma^\square$, respectively $\emptyset$, are replaced by references to $\Sigma^\diamond$, respectively $\langle \emptyset, \emptyset \rangle$. As in **MA1**, different explanations cannot be conjoined to one another. So, the application of the unconditional rule is restricted to consequences of the premises and at most one explanation. For the same reason, the conditional rule RC1 (that governs the "upgrading" of formulas) is restricted to formulas that occur on a condition of the form $\langle \Theta, \emptyset \rangle$. Let $\mathcal{F}^p$ be the set of primitive formulas of $\mathcal{L}$, and let $pr(A)$ refer to the set of members of $\mathcal{F}^p$ that occur in A.

RU†　If $A_1, \ldots, A_n \vdash_{\mathbf{S52}} B$ ($n \geq 0$) and $A_1, \ldots, A_n$ occur in the proof on the conditions $\langle \Theta_1, \Phi_1 \rangle, \ldots, \langle \Theta_n, \Phi_n \rangle$ respectively, and at most one Φ_i is non-empty, then one may add to the proof a line consisting of (i) the appropriate line number, (ii) B, (iii) the line numbers of the A_i, (iv) "RU†", and (v) $\langle \Theta_1 \cup \ldots \cup \Theta_n, \Phi_1 \cup \ldots \cup \Phi_n \rangle$.

RC1　If $\diamond_2 A$ occurs in the proof on the condition $\langle \Theta, \emptyset \rangle$, then one may add to the proof a line consisting of (i) the appropriate line number, (ii) $\square_2 A$, (iii) the line number of $\diamond_2 A$, (iv) "RC1", and (v) $\langle \Theta \cup pr(A), \emptyset \rangle$.

The second conditional rule is as the one for **MA1**, except for some evident changes related to the new format:

RC2　If $\square_1 B$ and $\diamond_2 (A \supset B)$ occur in the proof, on the conditions $\langle \Theta_1, \emptyset \rangle$ and $\langle \Theta_2, \emptyset \rangle$ respectively, then one may add to the proof a line consisting of (i) the appropriate line number, (ii) $\diamond_2 A$, (iii) the line numbers of $\square_1 B$ and $\diamond_2 (A \supset B)$, (iv) "RC2", and (v) $\langle \Theta_1 \cup \Theta_2, \langle A, B \rangle \rangle$.

In order to present the additional marking criterion, we need some additional definitions. Let $Dab(\Delta)$ refer to $\bigvee \{ \exists (\diamond A \wedge \diamond \neg A) \mid A \in \Delta \}$ provided $\Delta \subset \mathcal{F}^p$. $Dab(\Delta)$ is a minimal Dab-consequence of $\Sigma^\diamond$ at stage

s iff it is unconditionally derived at that stage in the proof and $Dab(\Delta')$ is not derived unconditionally at that stage for any $\Delta' \subset \Delta$.

According to the additional marking criterion, a line is I-marked iff one of the formulas that should behave normally is a member of a minimal Dab-consequence:

CMI   A line that has $\langle \Theta, \Phi \rangle$ as its fifth element is I-marked at stage s iff for some minimal Dab-consequence $Dab(\Delta)$ at that stage, $\Theta \cap \Delta \neq \emptyset$.

As the deductive steps in **MA2** are not only governed by RU, but also by RC1 and CMI, the marking criteria for abductive steps have to refer to CMI. (If a formula A, at a certain stage in the proof, only occurs on lines that are I-marked, then A is, at that stage, not considered as derived, and hence, should not play a role in deciding whether the requirements for abductive steps are fulfilled.) The only other changes to the marking criteria are that they have to be adjusted to the new format of the condition, and that, in order to avoid (consequences of) trivial explanations, CMT has to be adapted to the "weaker" RC2:

CMT†   A line that has $\langle \Theta, \langle A, B \rangle \rangle$ as its fifth element is T-marked iff some line in the proof that is not I-marked and that has $\langle \Theta', \emptyset \rangle$ as its fifth element has $\Diamond_2 B$ or $\Diamond_2 \neg A$ as its second element.

CME†   A line that has $\langle \Theta, \langle A, B \rangle \rangle$ as its fifth element is E-marked iff some line in the proof that is not I-marked and that has $\langle \Theta', \emptyset \rangle$ as its fifth element has $\Box_1 \neg A$ as its second element.

CMS†   A line that has $\langle \Theta, \langle A, B \rangle \rangle$ as its fifth element is S-marked iff some line j in the proof has $\Box_i (B \supset A)$ as its second element and $\pi_j (\Box_i (B \supset A)) \cap (\Gamma_1 \cup \Gamma_2) = \emptyset$.

CMP†   A line that has $\langle \Theta, \langle A, B \rangle \rangle$ as its fifth element is P-marked iff some line in the proof that is neither I-marked nor (T-, E-, S-)marked has $\langle \Theta', \langle C, B \rangle \rangle$ as its fifth element, $C \in \phi_s(A)$, and $C \in \psi_s(A)$.

Note that a line that is I-marked may at a later stage be unmarked (because a "shorter" Dab-consequence was derived). As the other marking criteria refer to CMI,[14] the easiest way to perform the marking is as follows: whenever a line is added to the proof, remove all marks; check next which lines have to be marked according to the marking criteria *in the order that they are listed here.*

---

[14] As CMS† only refers to the derivation of theorems, the reference to CMI can be omitted.

Definition 12 *A is* derived at a stage *in an* **MA2**-*proof from* $\Sigma^\diamond$ *iff A is derived in the proof on a line that is not marked.*

Definition 13 *A is* finally derived *in an* **MA2**-*proof from* $\Sigma^\diamond$ *iff A is derived on a line j that is not marked, and any extension of the proof in which line j is marked may be further extended in such a way that line j is unmarked.*

As for **MA1**, the consequence relation of **MA2** is defined with respect to final derivability:

Definition 14 $\Sigma^\diamond \vdash_{\mathbf{MA2}} A$ *(A is finally derivable from* $\Sigma^\diamond$*) iff A is finally derived in an* **MA2**-*proof from* $\Sigma^\diamond$.

9.   Two examples from the history of astronomy

In this section, we illustrate the logics **CP1** and **CP2** with two examples from the history of the sciences. The first concerns the explanation of a novel fact (that led to the discovery of Uranus), the second that of an anomalous fact (that eventually led to the discovery of Neptune). For reasons of space, we shall restrict ourselves to the main inference steps as they occur in standard accounts of these discoveries.[15]

9.1   The discovery of Uranus

In [Kuhn, 1977], Kuhn presents the following account of the discovery of Uranus:

> On the night of 13 March 1781, the astronomer William Herschel made the following entry in his journal: "In the quartile near Zeta Tauri ... is a curious either nebulous star or perhaps a comet." [...] Between 1690 and Herschel's observation in 1781 the same object had been seen and recorded at least seventeen times by men who took it to be a star. Herschel differed from them only in supposing that, because in his telescope it appeared especially large, it might actually be a *comet!* Two additional observations on 17 and 19 March confirmed that suspicion by showing that the object he had observed moved among the stars. As a result, astronomers throughout Europe were informed of the discovery, and the mathematicians among them began to compute the new comet's orbit. Only several months later, after all those attempts had repeatedly failed to square with observation, did the astronomer Lexell suggest that the object observed by Herschel might be a planet. And only when additional computations, using a planet's rather than a

---

[15]A further constraint is that the proof format presented here is a generic one. Designing a proof system that is heuristically more interesting is straightforward in view of the results from [Batens et al., 200+a], and would lead to a more realistic reconstruction of the examples.

comet's orbit, proved reconcilable with observation was that suggestion
generally accepted [Kuhn, 1977, pp. 171–172].

As we shall now show, the dynamics of this discovery process can
adequately be captured by **CP1**. Let us agree that objects can have
the following properties: occurring at a certain time and place (O),
appearing large (L), moving (M), moving along a certain trajectory (T_i),
being a nebulous star (S), being a comet (C), and being a planet (P).

Following Kuhn's account, Herschel started with four premises: (1)
the observed object (call it a) appeared large, (2) a occurred at a certain
time (say, t_0) at a certain place (say, r_0), (3) nebulous stars appear large,
(4) comets appear large. From these, he abduced that a "is a curious
either nebulous star or perhaps a comet". In **MA1**, this can be modelled
as follows:[16]

1	$\Box_1 La$	–	PREM	$\emptyset$
2	$\Box_2 Oar_0 t_0$	–	PREM	$\emptyset$
3	$\Box_2 (\forall x)(Sx \supset Lx)$	–	PREM	$\emptyset$
4	$\Box_2 (\forall x)(Cx \supset Lx)$	–	PREM	$\emptyset$
5	$\Diamond_2 Sa$	1, 3	RD1	$\langle Sa, La \rangle$
6	$\Diamond_2 Ca$	1, 4	RD1	$\langle Ca, La \rangle$

At this stage of the proof both hypotheses are justified. However,
after adding the observational statement that a moves among the stars,
and the background assumption that stars do not move, the explanation
that a is a star has to be withdrawn:[17]

...

5	$\Diamond_2 Sa$	1, 3	RD1	$\langle Sa, La \rangle \; \sqrt{}_{E^9}$
6	$\Diamond_2 Ca$	1, 4	RD1	$\langle Ca, La \rangle$
7	$\Box_1 Ma$	–	PREM	$\emptyset$
8	$\Box_2 (\forall x)(Sx \supset \neg Mx)$	–	PREM	$\emptyset$
9	$\Box_1 \neg Sa$	7, 8	RU	$\emptyset$

At line 10, it becomes clear that the hypothesis on line 5 is falsified.
Hence, line 5 has to be E-marked in view of CME. Using the background
assumption that comets move, the hypothesis that a is a comet can again
be abduced. This time, however, on the condition that it is a good
explanation for the fact that a is large *and* that a moves:

---

[16]As the second premise does not function as a possible *explanandum*, we shall include it
among the background assumptions.

[17]To illustrate the dynamical character of the proofs, we shall each time give the complete
proof, but, for reasons of space, omit lines that are not central to see the dynamics.

. . .

10	$\Box_2(\forall x)(Cx \supset Mx)$	–	PREM	$\emptyset$
11	$\Box_2(\forall x)(Cx \supset (Lx \wedge Mx))$	4, 10	RU	$\emptyset$
12	$\Box_1(La \wedge Ma)$	1, 7	RU	$\emptyset$
13	$\Diamond_2 Ca$	11, 12	RD1	$\langle Ca, La \wedge Ma \rangle$

What this indicates is that, at stage 13 of the proof, the hypothesis that a is a comet is better confirmed than it was at stage 6.

Still, this is not the end of the story. Using the background knowledge on the behavior of comets, "mathematicians [...] began to compute the new comet's orbit". Let us assume that the possible trajectories are $T_1, \ldots, T_n$[18] (some of the r_i, respectively t_i, may be identical to each other):

. . .

14	$\Box_2(\forall x)((Cx \wedge Oxr_0t_0) \supset (T_1x \vee \ldots \vee T_nx))$	–	PREM	$\emptyset$
15_1	$\Box_2(T_1a \supset Oar_{1'}t_1)$	–	PREM	$\emptyset$
16_1	$\Box_1 Oar_1t_1$	–	PREM	$\emptyset$
17_1	$\Box_2(Oar_1t_1 \supset \neg Oar_{1'}t_1)$	–	PREM	$\emptyset$
18_1	$\Box_1 \neg Oar_{1'}t_1$	16_1, 17_1	RU	$\emptyset$
19_1	$\Box_1 \neg T_1a$	15_1, 18_1	RU	$\emptyset$
. . .				
15_n	$\Box_2(T_na \supset Oar_{n'}t_n)$	–	PREM	$\emptyset$
16_n	$\Box_1 Oar_nt_n$	–	PREM	$\emptyset$
17_n	$\Box_2(Oar_nt_n \supset \neg Oar_{n'}t_n)$	–	PREM	$\emptyset$
18_n	$\Box_1 \neg Oar_{n'}t_n$	16_n, 17_n	RU	$\emptyset$
19_n	$\Box_1 \neg T_na$	15_n, 18_n	RU	$\emptyset$
20	$\Box_1(\neg T_1a \wedge \ldots \wedge \neg T_na)$	19_1–19_n	RU	$\emptyset$
21	$\Box_1 \neg(Ca \wedge Oar_0t_0)$	14, 20	RU	$\emptyset$
22	$\Box_1 \neg Ca$	2, 21	RU	$\emptyset$

This new confrontation with experimental results leads to the falsification of the hypothesis that a is a comet (see line 22). Hence, lines 6 and 13 are E-marked:

. . .

6	$\Diamond_2 Ca$	1, 4	RD1	$\langle Ca, La \rangle$ $\sqrt{}_{C^{22}}$

. . .

---

[18]The orbit has to be an ellipse or a parabola that passes through r_0 at t_0 – and through further positions, observed later, which we leave out of the story to simplify it – and which has our sun at (one of) its focal point(s). But, most importantly, if it is a comet (and the trajectory an ellipse), the eccentricity will have to be close to one. This means that only a limited class of trajectories is possible.

13   $\diamond_2 Ca$ 11, 12   RD1 $\langle Ca, La \wedge Ma \rangle \; \sqrt{}_{C^{22}}$

. . .

22   $\square_1 \neg Ca$ 2, 21   RU $\emptyset$

At this point, Lexell's hypothesis can be introduced. Its confrontation with the observed positions leads to a well confirmed hypothesis (on line 24 $\Xi_1 = \langle Pa, La \wedge Ma \rangle$ and on line 27, $\Xi_2 = \langle Pa, Oar_1 t_1 \wedge \ldots \wedge Oar_n t_n \rangle$; 16_1^n and 26_1^n stand for 16_1–16_n and 26_1–26_n):

. . .

23    $\square_2(\forall x)(Px \supset (Lx \wedge Mx))$ –   PREM $\emptyset$
24    $\diamond_2 Pa$ 12, 23   RD1 Xi_1
25_1 $\square_2(\forall x)((Px \wedge Oxr_0 t_0) \supset Oxr_1 t_1)$ –   PREM $\emptyset$
26_1 $\square_2(\forall x)(Px \supset Oxr_1 t_1)$ 2, 25_1 RU $\emptyset$

. . .

25_n $\square_2(\forall x)((Px \wedge Oxr_0 t_0) \supset Oxr_n t_n)$ –   PREM $\emptyset$
26_n $\square_2(\forall x)(Px \supset Oxr_n t_n)$ 2, 25_n RU $\emptyset$
27    $\diamond_2 Pa$ $16_1^n, 26_1^n$ RD2 Ξ_2

It is interesting to notice that the logic **CP1** captures different aspects of scientific methodology. As was observed by Peirce, every abduction step is followed by deduction and (eliminative) induction. These steps are clearly reflected in the example. After abducing the hypotheses that the object is a star or a comet, testable consequences are deduced. The inductive part consists in the marking or not marking of the hypotheses.[19] That the logic **CP1** captures this multi-layered view of methodology is only possible because, as already mentioned, it nicely integrates abductive and deductive steps. Note also that approaches in which abduction is characterized as a kind of "backward reasoning" are not suited to model the actual reasoning leading up to the discovery of Uranus.

9.2   The discovery of Neptune

Soon after Uranus was *accepted* to be a planet – meaning that this was now considered part of the background knowledge – further complications arose. No matter how great the efforts, it turned out that Uranus' orbit could not be accurately predicted from Newtonian theory. We will simplify this by saying that it deviated in a well-known way from an ellipsoidal trajectory (reading $T_e' u$, with u denoting Uranus, and T_e standing for a well-behaved ellipsoidal trajectory). Of course,

---

[19]Of course this is only a very crude first approximation. To capture the inextricabilities of confirmation, the logic should be modified in several ways.

this anomaly did not stop astronomers to use the theory and trust the observation reports. However, once a contradiction is derived, in this case between predictions and observations, **CP1** turns the theory into the trivial one. Hence, in order to describe how astronomers could go on making meaningful abductions in this situation, we have to turn to the logic **CP2**.

Let us first reinterpret the main elements of the theory presented in the previous section, and add the new premises to it. G will be standing for the proposition that Newton's inverse square law for gravitation holds.

1	$\Box_1(La \wedge Ma)$	–	PREM	$\langle \emptyset, \emptyset \rangle$
2	$\Diamond_2 Oar_0 t_0$	–	PREM	$\langle \emptyset, \emptyset \rangle$
3	$\Diamond_2(\forall x)(Sx \supset (Lx \wedge \neg Mx))$	–	PREM	$\langle \emptyset, \emptyset \rangle$
4	$\Diamond_2(\forall x)((Cx \vee Px) \supset (Lx \wedge Mx))$	–	PREM	$\langle \emptyset, \emptyset \rangle$
5	$\Diamond_2 Pa$	–	PREM	$\langle \emptyset, \emptyset \rangle$
6	$\Diamond_2 a = u$	–	PREM	$\langle \emptyset, \emptyset \rangle$
7	$\Box_1 T_e' u$	–	PREM	$\langle \emptyset, \emptyset \rangle$
8	$\Box_1(T_e' u \supset \neg T_e u)$	–	PREM	$\langle \emptyset, \emptyset \rangle$
9	$\Diamond_2 G$	–	PREM	$\langle \emptyset, \emptyset \rangle$
10	$\Diamond_2(G \supset T_e u)$	–	PREM	$\langle \emptyset, \emptyset \rangle$

The new premises occur on lines 5–10. As indicated above, the former hypothesis that a is a planet now forms part of the background knowledge. The formula on line 6 expresses that a is identical to (the object now called) Uranus.

In view of the rule RU and RC1, this proof can be extended as follows:

...

11	$\Box_1 \neg T_e u$	7, 8	RU	$\langle \emptyset, \emptyset \rangle$
12	$\Box_2 G$	9	RC2	$\langle \{G\}, \emptyset \rangle$
13	$\Diamond_2 T_e u$	10, 12	RU	$\langle \{G\}, \emptyset \rangle$
14	$\neg \Box_1 \neg T_e u$	13	RU	$\langle \{G\}, \emptyset \rangle$

which clearly indicates that the set of premises is inconsistent (compare lines 11 and 14). However, from 7–10, one can derive the following minimal Dab-consequence ("!A" abbreviates "$(\Diamond_2 A \wedge \Diamond_2 \neg A)$"):

15	$!G \vee !T_e' u \vee !T_e u$	7–10	RU	$\langle \emptyset, \emptyset \rangle$

As soon as this line is added to the proof, lines 12–14 have to be I-marked, and triviality is avoided:

...

11	$\Box_1 \neg T_e u$	7, 8	RU	$\langle \emptyset, \emptyset \rangle$

12	$\Box_2 G$	9	RC2	$\langle\{G\},\emptyset\rangle$ $\surd_{I^{15}}$
13	$\Diamond_2 T_e u$	10, 12	RU	$\langle\{G\},\emptyset\rangle$ $\surd_{I^{15}}$
14	$\neg\Box_1\neg T_e u$	13	RU	$\langle\{G\},\emptyset\rangle$ $\surd_{I^{15}}$
15	$!G\vee!T_e'u\vee!T_e u$	7–10	RU	$\langle\emptyset,\emptyset\rangle$

Intuitively, it is very nice to see that at this point $\Diamond_2 G$ cannot be strengthened. For a time, astronomers were indeed seriously considering that there might be something wrong with this law at great distances from the sun.

Now, let us turn back to the anomalous observation at line 7, which was a fact that needed explaining. It was conjectured during the beginning of the nineteenth century that there could be an as far unnoticed planet in Uranus' vicinity that was responsible for the deviations of its orbit from the predicted one. In 1845 and 1846 Adams and Leverrier showed independently that the existence of a planet at a given position could indeed explain Uranus' observed orbit. So, the new theoretical statement would look something like this:

...

16	$\Diamond_2((G\wedge(\exists x)(Px\wedge Oxr_a t_a))\supset T_e'u)$	–	PREM	$\langle\emptyset,\emptyset\rangle$

This formula cannot be strengthened because of the minimal *Dab*-consequence at line 15.[20] However, rule RC2 still allows us to perform the abductive step. So, at this point we can introduce the hypothesis that there is a new planet with a given position ($\Xi = \langle\emptyset, \langle G\wedge(\exists x)(Px\wedge Or_a t_a), T_e'u\rangle\rangle$):

...

17	$\Diamond_2(G\wedge(\exists x)(Px\wedge Oxr_a t_a))$	7, 16	RC2	Ξ
18	$\Diamond_2(\exists x)(Px\wedge Oxr_a t_a)$	17	RU	Ξ

Evidently, it has to be checked whether the conditions of lines 17 and 18 hold. This means, most importantly, that extra observations are needed to find out whether $(\exists x)Oxr_a t_a$ holds true for some large object. As is well known, Galle observed in 1846 an object at the predicted position, and the hypothesis could be retained. Neptune was discovered.

At this point some readers may be wondering whether the possibility to strengthen members of Γ_2 is ever useful. Two remarks are important here. The first is that resolving the inconsistencies was far less evident

---

[20]To put it more accurately: we could strengthen the formulas, using RC1, but the lines would have to be *I*-marked immediately.

than we can imagine with hindsight. In the meantime, it was important that "unproblematic" parts of the theory could be localized, and that these could be used in their full strength. For example, one is only able to infer that a is large and moves among the stars, if the formula on line 4 or that on line 5 is strengthened. The second remark is that we presented a clearly oversimplified version of the theory. For instance, it is quite easily observed that the formulas on lines 3–6 are unproblematic, and hence, that it should be possible to strengthen these. However, the historical record gives us every reason to believe that the original theory was formulated in such a way that localizing the unproblematic parts was far less obvious.

10.   Some alternatives

For some application contexts, the logics presented here are unnecessary complex. For this reason, we spell out two alternatives that are easily obtained by slightly modifying **MA1**. Similar alternatives may be obtained from **MA2**.

The first alternative concerns reasoning contexts in which one is interested in the logically weakest explanation. This holds true, for instance, for medical diagnosis: whenever two or more explanations can be abduced for the same set of symptoms, one will only accept their disjunction.[21] This alternative can be obtained by simplifying CMP:

CMP*   A line that has $\langle A, B \rangle$ as its fifth element is P-marked iff some line in the proof that is not $(T$-, E-, S-)marked has $\langle C, B \rangle$ as its fifth element, and $C \in \phi_s(A)$.

The second alternative is suited for standard forms of abductive reasoning in which a set of so-called abducibles (abducible hypotheses) is given beforehand – see [Paul, 2000] for a discussion of this notion. Medical diagnosis may again serve as an example: whenever abduction is used to explain some set of symptoms, the set of possible diseases (this is, the set of possible explanations) is given in advance. In situations like this, abductive inferences can only lead to conjunctions of abducible hypotheses. Implementing this in the format of **MA1**, requires that the consequence relation is defined with respect to a set $\Sigma^\square$ and a set of abducibles Π, and that the application of RC is restricted in the following way:

---

[21] Evidently, one will try to strengthen this explanation by asking further questions, or doing further tests. The fact remains, however, that at any moment in time one only accepts the weakest explanation.

RC*   If $\Box_1 B$ and $\Box_2((A_1 \wedge \ldots \wedge A_n) \supset B)$ occur in the proof, and if
       $A_1, \ldots, A_n \in \Pi$, then one may add to the proof a line consisting
       of (i) the appropriate line number, (ii) $\Diamond_2(A_1 \wedge \ldots \wedge A_n)$, (iii) the
       line numbers of $\Box_1 B$ and $\Box_2((A_1 \wedge \ldots \wedge A_n) \supset B)$, (iv) "RC*",
       and (v) $\langle A_1 \wedge \ldots \wedge A_n, B \rangle$.

As the abducible hypotheses are given beforehand, it is avoided that
arbitrary atoms enter the explanatory hypotheses. Hence, in this alternative too, CMP can be replaced by the simpler CMP*. If one is
interested in alternative explanations for the same set of *explananda*,
this can be realized by restricting the set of abducibles to individual
explanations, while not including their disjunctions. If also their disjunctions are included, a unique explanation is obtained for each set of
explananda (namely, the logically weakest one).

11.   Conclusion

In this paper, we showed how logic-based approaches to abduction can
be reconstructed in terms of ampliative adaptive logics, and discussed
the main advantages of this reconstruction. We presented two new adaptive logics for abduction, and showed their adequacy in modeling some
historical examples of scientific discoveries.

Important open problems concern the further elaboration of the two
systems (for instance, the design of their semantics), the design and
elaboration of alternatives (for instance, systems that can handle not
only inconsistent background assumptions, but also inconsistent empirical findings), and the further application of these systems to examples
from the history of the science.[22]

Acknowledgment

The research for this paper was supported by the Fund for Scientific
Research–Flanders, by the Research Fund of Ghent University, and in-
directly by the Flemish Minister responsible for Science and Technology
(contract BIL98/73). The first author is a Postdoctoral Fellow of the
Fund for Scientific Research–Flanders (Belgium), the second author is a
Research Assistant of the same Fund.

---

[22]Unpublished papers in the reference section are available at the internet address
http://logica.rug.ac.be/centrum/writings.

References

Aliseda, A., 1997, Seeking Explanations: Abduction in Logic, Philosophy of Science and Artifical Intelligence, PhD thesis, Institute for Logic, Language and Computation (ILLC), University of Amsterdam, The Netherlands.

Batens, D., 1989, Dynamic dialectical logics, in: *Paraconsistent Logic. Essays on the Inconsistent*, G. Priest, R. Routley, and J. Norman, eds., Philosophia Verlag, München, pp. 187–217.

Batens, D., 1999, Zero logic adding up to classical logic, *Logical Studies*, 2:15, (Electronic Journal: http://www.logic.ru/LogStud/02/LS2.html).

Batens, D., 2000, Towards the unification of inconsistency handling mechanisms, *Logic and Logical Philosophy* 8:5–31, appeared in 2002.

Batens, D., 200+, On a logic for induction, to appear.

Batens, D., De Clercq, K., and Vanackere, G., 200+a, Simplified dynamic proof formats for adaptive logics, to appear.

Batens, D., and Meheus, J., 2000, The adaptive logic of compatibility, *Studia Logica* 66:327–348.

Batens, D., Meheus, J., Provijn, D., and Verhoeven, L., 200+b, Some adaptive logics for diagnosis, *Logique et Analyse*, in print.

Batens, D., and Vermeir, T., 200+, Direct dynamic proofs for the Rescher–Manor consequence relations: The flat case, to appear.

Benferhat, S., Dubois, D., and Prade, H., 1997, Some syntactic approaches to the handling of inconsistent knowledge bases: A comparative study. Part 1: The flat case, *Studia Logica* 58:17–45.

Benferhat, S., Dubois, D., and Prade, H., 1999, Some syntactic approaches to the handling of inconsistent knowledge bases: A comparative study. Part 2: The prioritized case, in: *Logic at Work. Essays Dedicated to the Memory of Helena Rasiowa*, E. Orłowska, ed., Physica Verlag (Springer), Heidelberg, New York, pp. 473–511.

Brown, B., 1990, How to be realistic about inconsistency in science, *Studies in History and Philosophy of Science* 21:281–294.

da Costa, N.C.A. and French, S., 2002, Inconsistency in science. A partial perspective, in: J. Meheus [Meheus, ed., 2002], pp. 105–118.

D'Hanis, I., 2002, A logical approach to the analysis of metaphors, this volume.

Hintikka, J., 1998, What is abduction? The fundamental problem of contemporary epistemology, *Transactions of the Charles S. Peirce Society* 34:503–533.

Kuhn, T.S., 1977, *The Essential Tension*, The University of Chicago Press, Chicago.

Magnani, L., 2001, *Abduction, Reason, and Science. Processes of Discovery and Explanation*, Kluwer Academic/Plenum Publishers, New York.

Meheus, J., 1993, Adaptive logic in scientific discovery: The case of Clausius, *Logique et Analyse* 143–144:359–389, appeared in 1996.

Meheus, J., 1999a, Deductive and ampliative adaptive logics as tools in the study of creativity, *Foundations of Science* 4.3:325–336.

Meheus, J., 1999b, Clausius' discovery of the first two laws of thermodynamics. A paradigm of reasoning from inconsistencies, *Philosophica* 63:89–117, appeared in 2001.

Meheus, J., 2000, An extremely rich paraconsistent logic and the adaptive logic based on it, in: *Frontiers of Paraconsistent Logic*, D. Batens, C. Mortensen, G. Priest, and J.P. Van Bendegem, eds., Research Studies Press, Baldock, pp. 189–201.

Meheus, J., ed., 2002, *Inconsistency in Science*, Kluwer, Dordrecht.

Meheus, J., 200+a, Inconsistencies and the dynamics of science, *Logic and Logical Philosophy*, in print.

Meheus, J., 200+b, An adaptive logic based on Jaśkowski's **D2**, to appear.

Meheus, J., 200+c, Inconsistencies in scientific discovery. Clausius's remarkable derivation of Carnot's theorem, in: *History of Modern Physics*, G. Van Paemel et al., eds., Brepols, in print.

Meheus, J., 200+d, An adaptive logic for pragmatic truth, in: *Paraconsistency. The Logical Way to the Inconsistent*, W.A. Carnielli, M.E. Coniglio, and I.M.L. D'Ottaviano, eds., Marcel Dekker, in print.

Meheus, J., 200+e, Empirical progress and ampliative adaptive logics, to appear.

Meheus, J. and De Clercq, K., 200+, Dynamic proof theories for erotetic inferences, *Logique et Analyse*, to appear.

Nersessian, N.J., 2002, Inconsistency, generic modeling, and conceptual change in science, in: J. Meheus [Meheus, ed., 2002], pp. 197–211.

Norton, J., 1987, The logical inconsistency of the old quantum theory of black body radiation, *Philosophy of Science* 54:327–350.

Norton, J., 1993, A paradox in Newtonian gravitation theory, *PSA 1992* 2, pp. 421–430.

Paul, G., 2000, AI approaches to abduction, in: *Handbook of Defeasible Reasoning and Uncertainty Management Systems. Volume 4: Abductive Reasoning and learning*, D.M. Gabbay and P. Smets, eds., Kluwer Academic Publishers, Dordrecht, pp. 35–98.

Priest, G., 1991, Minimally inconsistent **LP**, *Studia Logica* 50:321–331.

Rescher, N. and Manor, R., 1970, On inference from inconsistent premises, *Theory and Decision* 1:179–217.

Smith, J., 1988, Inconsistency and scientific reasoning, *Studies in History and Philosophy of Science* 19:429–445.

Thagard, P., 1988, *Computational Philosophy of Science*, MIT Press/Bradford Books, Cambridge, MA.

Thagard, P., 1999, *How Scientists Explain Disease*, Princeton University Press, Princeton.

Verhoeven, L., 200+, Proof theories for some prioritized consequence relations, to appear.

Vermeir, T., 200+, Two ampliative adaptive logics for the closed world assumption, to appear.

Weber, E. and Provijn, D., 200+, A formal analysis of diagnosis and diagnostic reasoning, *Logique et Analyse*, to appear.

DIAGRAMMATIC INFERENCE
AND GRAPHICAL PROOF

Luis A. Pineda

Instituto de Investigaciones en Matemáticas Aplicadas y Sistemas
(IIMAS), UNAM, Circuito Interior S/N, Ciudad Universitaria, México
luis@leibniz.iimas.unam.mx

Abstract   In this paper we present a diagrammatic inference scheme that can
be used in the proof and discovery of diagrammatic theorems. First,
we present the theory of abstraction markers and notational keys for
explaining how abstraction can be incorporated in the interpretation of
graphics through both syntactic and semantic means. We also explore
how the process of reinterpretation of graphics is essential for learning
and proving graphical theorems. Then, we present the diagrammatic
inference scheme; it is illustrated with the proof of the theorem of the
sum of the odd numbers. The paper concludes with a discussion on
the relation between abstraction, visualization, interpretation change
and learning, applied to understand a purely diagrammatic proof of the
Theorem of Pythagoras.

1.   Introduction

Diagrammatic proofs are usually easier to learn and understand than
the corresponding proofs expressed in mathematical or logical notation.
In diagrammatic proofs, proof procedures involve a limited number of
operations that transform diagrams representing the premises of a theorem into a diagram representing its conclusion; the proof of the Theorem
of Pythagoras in Figure 1 is an example of this kind of proofs.

Diagrammatic proofs have also been used as logical reasoning systems; for instance, Euler circles and Venn diagrams have been used to
reason about syllogisms. In these kinds of systems it can be more easily
appreciated that there is a set of valid operations that can be applied
to produce the diagram representing the conclusion out of the diagrams
representing the premises of a logical argument. Consider the Euler cir-

L. Magnani, N.J. Nersessian, and C. Pizzi (eds.),
Logical and Computational Aspects of Model-Based Reasoning, 73–91.
© 2002 Kluwer Academic Publishers. Printed in the Netherlands.

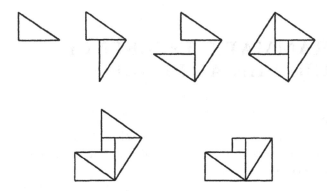

Figure 1.   Proof of the Theorem of Pythagoras.

cle representation of the syllogism *All A are B, All B are C, All A are C* which is shown in Figure 2.

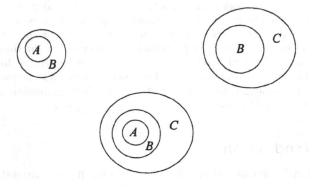

Figure 2.   Syllogism representation with Euler circles.

Diagrams have also been used to illustrate proofs of arithmetic theorems that are normally proven through mathematical induction, such as the theorem of the sum of odd numbers, $1 + 3 + 5 + \ldots + (2n - 1) = n^2$, illustrated in Figure 3. A sample of theorems of this kind can be found in [Nelsen, 1993].

The figure represents the theorem because the leftmost L's of all the squares whose upper-right corner is at the upper-right corner of the grid can be interpreted as odd numbers, such that an L with a side of size n can be interpreted as the odd number $2n - 1$. Any sequence of n consecutive L's (from right to left, starting from the upper-right) can be interpreted as the sum of the corresponding odd numbers and, as the

Figure 3. Theorem of the sum of the odds.

area (number of dots) covered by this sequence is equal to the area of a square of size n^2, the diagram represents the theorem.

As this sample of proofs suggests, diagrams can be used effectively to present and assess the validity of arguments that would be much harder to understand through logical or natural language representations. The study of these proofs is important, as they provide a paradigmatic case of study for the use of graphical representations in more open forms of graphical reasoning and problem-solving tasks. The study of diagrammatic reasoning is not only of theoretical concern; it also has important applications in computational technology, both for enhancing the effectiveness of visual displays and for providing a scientific base for the construction of representations that can be stored and manipulated by computers. This is relevant to several fields of research, such as human-computer interaction, multimodal communication, visual programming, artificial intelligence and, in general, to any discipline which deals which the effective presentation and use of graphical information.

2.   Abstraction markers

Diagrams used in diagrammatic proofs are *analogical* [Sloman, 1985] or *direct* [Hayes, 1985] representations of the mathematical objects they denote. Unlike propositional representations in which the relation between symbols and their denotations has a conventional character, in analogical representations, the medium in which symbols and expressions are embedded has the same kind of structure of the represented world (e.g. 2-D space). In this sense, symbols in analogical representations denote objects of the world directly, and properties and relations between symbols denote properties and relations between the corresponding entities in the world. In the case of diagrammatic proofs, diagrams can be thought of as reflecting directly the abstract structure of the corresponding mathematical objects. Intuitively, people take advantage of this morphism between representation and thing for interpreting effectively diagrammatic proofs and, more generally, for interpreting analogical representation.

The notion of representation medium leads to the notion of levels of representation, as has been pointed out by Hayes [Hayes, 1985], as the medium of a representation can itself be represented by a representation in a different medium; for instance, when a drawing in computer graphics is represented through a logical representation, as it is commonly done in AI programs using diagrams (e.g. [Pineda, 1992]). This observation has set the fate for diagrammatic representations used in problem-solving and theorem-proving in AI, as the real computational objects in these programs are the underlying propositional representations of the diagrams (e.g. [Chandrasekaran, 1997]).

Diagrams are also concrete representations. Unlike propositional representations in which there are explicit symbols marking or signaling that some expressions must be interpreted as general statements, like variables and quantifiers in First Order Logic (FOL), no such overt markers appear in diagrammatic proofs. In this regard, it is often noticed that this concrete character of graphics supports effective interpretation; for instance, in the theory of the Specificity of Graphics (TSG) of Stenning and Oberlander [Stenning and Oberlander, 1995], it is argued that graphics limit abstraction and thereby aid processibility.

However, diagrammatic proofs seem to challange this assumption: if the diagrams used in diagrammatic proofs are formal objects representing universal mathematical statements, the abstraction required for the generalization needs to be unlimited, and it should be marked or introduced somehow in the interpretation process. To investigate further this seeming conflict, next we explore how abstraction can be expressed in diagrams. In the present discussion we take TSG as a reference framework.

In logical representations, the paradigmatic case of propositional representations, expressions can be ordered in a hierarchy according to the kind of abstraction that they can express. For the present discussion, we place FOL at the top of this hierarchy: it has the means to express general statements through quantifiers and variables. Here, we will refer to these symbols as *unlimited abstraction markers*, as the expressive power permitted by these kinds of markers is similar to the expressive power of the so-called *Unlimited Abstraction Representational Systems* (UARS) in TSG.

Expressions used for syllogistic reasoning (i.e. *All A are B, some A are B, etc.*) have markers for universal and existential quantification, but variables are not used in this formalism; this restriction places a limit on the kind of generalizations that can be expressed in relation to full FOL, and expressions used in syllogistic reasoning are one level down from FOL in the hierarchy of abstraction. For this reason, we will refer to quantifiers used in this kind of expressions as *limited abstraction markers*, relating the abstraction that can be expressed through these markers with *Limited Abstraction Representational Systems* (LARS) in TSG.

Propositional logic and quantifier-free sentences of predicate logic can also express abstractions but ones of a more limited character. The use of disjunction allows us to express that one of two statements is true without telling which one, and negation can be used to express that something is not the case without telling what it is. Material implication can be reduced to negation and disjunction, and can express some limited abstractions too. For these reasons, we will also call the negation, disjunction and implication symbols *limited abstraction markers*.

Expressions formed using conjunction and negation have a minimal abstract import as their interpretation or truth value can be known by checking the truth value of the propositions involved, and applying the functions denoted by the conjunction and negation symbols. For this reason, we will call the conjunction and negation symbols as *minimal abstraction markers*, after the *Minimal Abstractions Representational Systems* (MARS) in TSG. Expressions of the propositional logic in the conjunctive normal form, for instance, are concrete representations expressing minimal abstraction.

Negation, as can be seen, can be used as a limited or as a minimal abstraction marker according to whether the expression of incomplete information is permitted in the representational language, as will be explained below.

Now, with this hierarchy of abstraction on hand, we can investigate the kinds of abstraction that can be expressed through graphics in terms of the kinds of abstraction markers involved in graphical representations. In particular, if the arithmetic or logical expression of a theorem requires an unlimited abstraction marker, but at same time there is a graphical representation of such a theorem, then it is important to investigate how the unlimited abstraction can be expressed through graphical means. Next, we explore the question of how abstraction markers can be realized in graphical representations.

We take as a basic intuition that diagrams express conjunctive information in a natural way. The superposition or addition of graphical information on the representational medium plays the role of the conjunction symbol in logical representations and signals minimal abstraction for the interpretation process. For instance, the intersection of areas representing types in Euler circles represents the conjunction of such types.

The introduction of negation in diagrammatic representations is subtler. We consider two possibilities: marking negation by absence and using an explicit marker for this purpose. The absence of a symbol, a property of a symbol or a relation between symbols can be interpreted as stating that the individual denoted by the missing symbol is not there, and that an individual does not have a property that is not represented graphically, and does not stand in a relation that does not hold in the diagram. To interpret omission or absence as negation is also to adopt the close-world assumption in the interpretation of diagrams, or to stay that diagrams convey complete information. Marking negation by absence is equivalent to the use of negation as a minimal abstraction marker in logical representations. This kind of negation can only be used in comprehensive configurations stating a fully determined state of affairs of the world. For instance, consider a world characterized by two predicate symbols P and Q and the names a, b and c. A comprehensive configuration states whether the individuals named by the three names have or not have the properties named by the two predicates (assuming the unique names assumption); an instance is shown in (1):

$$P(a) \wedge \neg P(b) \wedge P(c) \wedge Q(a) \wedge Q(b) \wedge \neg Q(c) \tag{1}$$

We turn now to the second option for expressing negation. Here a marker for this purpose is introduced in the representational system explicitly (e.g. a mark for prohibiting cars to park in a given place). This additional expressive device permits the expression of incomplete information. Suppose that we have the same world described by the symbols used in (1), but we have no information about whether b has or not has P nor whether c has or not Q, as shown in (2):

$$P(a) \wedge P(c) \wedge Q(a) \wedge Q(b) \tag{2}$$

If we adopt the closed world assumption, (2) would be interpreted as stating that b has not P and c has not Q; but if we don't, (2) is not comprehensive as there are some facts about the world that are not expressed. This can be better appreciated if we consider the equivalent expression (3) where the lack of information about the world is expressed through disjunction and negation, as follows:

$$P(a) \wedge (P(b) \vee \neg P(b)) \wedge P(c) \wedge Q(a) \wedge Q(b) \wedge (Q(c) \vee \neg Q(c)) \tag{3}$$

Disjunction can be realized graphically as the union of areas, as it is normally illustrated with Euler circles. However, while graphical conjunction is a simple superposition operation with a concrete interpretation, the interpretation of graphical disjunction permits limited abstraction. While a configuration formed by a mark included in a circle is normally interpreted as stating that the individual represented by the mark has that property represented by the circle, a configuration formed by a mark included in the union of two circles can also be interpreted as stating that the individual denoted by the mark has either of the properties represented by the circles, without specifying which one. Here again, configurations can be taken as specifying complete information, forcing the interpretation that if a mark is in a circle the individual denoted by the mark has the property denoted by the circle, or incomplete information, leaving that information uncertain.

Quantification can also be expressed graphically as illustrated by Euler circles. In the graphical representation of syllogisms, existential and universal quantification are marked through the intersection and inclusion relations between the areas representing types. However, as was mentioned above, in syllogistic representations no variables are associated to quantifiers, limiting the expressive power of this kind of representations; quantifiers used in syllogistic reasoning are also limited abstraction markers, because, although they permit the expression of quantification, this kind of quantification ranges over a domain composed of a finite number of types, and hence, is limited. This is consistent with the analysis in TSG where syllogistic reasoning is considered a reasoning task with limited abstraction. In similar lines, Johnson-Laird's mental models theory [Johnson-Laird, 1983] allows the expression of limited abstraction only and, in this regard, TSG and mental models are equivalent theories. This can be verified by studying the abstraction markers used in this latter formalism.

The expression of the Theorem of Pythagoras and the theorem of the sum of the odds, on the other hand, requires additional means to express the general abstraction. In the first case, the general statement requires the use of universal quantifiers with bound variables, and in the second, the ellipsis symbol [...] is normally used to express the generality of sum. However, there are not unlimited or even limited abstraction markers in the diagrams in Figures 1 and 3, and there seems to be no way to mark this abstraction in a natural way. The question is then, if no unlimited abstraction markers are employed in diagrammatic proofs, on the one hand, but people do read the theorems out of such diagrams, on the other, how unlimited abstraction is introduced in the interpretation process. To address this question we move to the semantic ground.

3.   Notational keys

In order to read a pattern of marks on the paper as a representation one needs to know how the symbols or configurations must be interpreted. Here, we refer to this kind of knowledge as *representational keys*. Following loosely the notion of terminological knowledge (stored in the so-called TBOX) of KL-ONE knowledge representation formalisms [Brachman and Schmolze, 1985], the theory of the Specificity of Graphics introduces the notions or *Key terminology*. Expressions in the so-called TBOX of KL-ONE represent concepts with attributes and values that do not have an assertive import. These expressions are similar to dictionary entry definitions, in that when one needs to know the meaning of a word, one can look up its definition, but to know the concept expressed by a word does not imply that there is an object in the world that is denoted by that word. The fact that one knows the meaning of the word *chair* does not imply that any particular chair is being referred to, or even that there is a chair in the interpretation domain. Terminological keys state simply the meaning of the constant symbols, logical and non-logical, of the representational language. According to the theory of the Specificity of graphics, the meaning of expressions in MARS and LARS is defined only in terms of terminological keys.

In similar vein, following the notion of assertive knowledge (stored in the so-called ABOX) of KL-ONE, TSG introduces the notion of *key assertions*. Unlike key terminology, which only states meanings, key assertions do assert facts about the world. To understand key assertions one has to know the meaning of the terminology employed, but names and predicates used in assertive statements do refer to specific individuals and properties about the interpretation domain. With the notion of key assertions in hand, TSG introduces the so-called Unlimited Abstraction Representational Systems; TSG also claims that the distinguishing feature of UARS, in opposition to LARS which only use key terminology, is that UARS use key assertions in addition to key terminology.

This contrasts with the theory of the abstraction markers presented here: we claim that unlimited abstraction can be expressed in external representations if there are unlimited abstraction markers in the representational system, like quantifiers and variables of FOL. From our point of view, to know that an expression makes an universal claim, either about logic, the mathematical world, or the physical world, is to know something about the meaning of the expression, regardless what is been referred to or asserted about the world through the expression. If abstraction cannot be expressed externally it has to be stated in the semantics of the representational system if it is there at all. Next, we turn to this possibility.

In propositional representations, terminological keys are primitive associations between symbols and their meanings or denotations stated by convention. This is a general property of propositional representations. In particular, names in a representational system keep no relation to each other. However, notational systems seem to be an exception to this rule. Monadic, binary and decimal notation, for instance, define rules through which strings made out of symbols of the alphabet are assigned to the numbers they denote in a systematic fashion. Notational systems resemble analogical representations in that the structure of representational objects reflects directly the structure of the objects that are being represented. Also when a notational system is employed, the symbols are embedded in the medium of the representation, and it is the structure of the medium what resembles the structure of the world. Monadic notation is a paradigmatic extreme case in which a string of n 1s refers to the number n. However, all notational systems share this property because the rules for forming and interpreting strings as numbers are defined as a function not only of the interpretation of the symbols of the alphabet, but also of the locations of the symbols in the representation medium.

The rules defining the form of names in a notational system also resemble key terminology in that they constitute the knowledge that permits to interpret a particular string as its corresponding number. The string "111" can be interpreted as three, seven or one hundred an eleven, depending on whether the representation key for monadic, binary or decimal notation is employed in the interpretation process. To know how to interpret a numeral is a question of meaning. However, unlike terminological keys, which are specific associations between a symbol and its meaning, notational rules are general statements about notation. These rules can be thought of as a vehicle to assign a meaning to an infinite number of symbols through a single statement. Here, we will refer to the rules for forming or interpreting names of a notational system as *notational keys*.

Notational keys can also be defined for the interpretation of other kinds of representations; for instance, in the definition of graphical languages in relation to the Graflog system [Pineda, 1989], general statements about the interpretation of graphical configurations are used in the dynamic definition of graphical languages. In this system, every graphical expression makes an assertion about the interpretation domain, but the notational keys required to interpret the drawings, that are also stated interactively, have themselves no assertional import.

The definition of notational keys for interpreting expressions in a representational formalism allows the interpretation of configurations as unlimited abstraction representational systems, but unlike the use of abstraction markers, which are syntactic devices, the abstraction introduced through notational keys has a semantic character. Notational keys are already general statements that have to be kept in mind to carry on with the interpretation process. Furthermore, there are *truths of meaning* that follow necessarily from the interaction between notational keys and the representation medium, as will be explained below in this paper. These statements will not be evident when a notation system is defined, but they can be discovered when the properties of the representation system are analyzed.

4.   The syntactic effect and reinterpretation

The choice of medium has an effect on how easily a configuration can be interpreted by people. The information expressed through the logical representation in (1) in section 2, for instance, can also be expressed through a tabular representation as shown in Table 1.

	a	b	C
P	1	0	1
Q	1	1	0

Table 1.

The meaning of both the logical and the tabular configuration is the same, but the interpretation of the table has a more holistic, synoptic, character. This property of the interpretation process that depends on the medium of the representation is called in the theory of the Specificity of Graphics the *Syntactic effect*.

Tabular representations can also express incomplete information. While Table 1 is a comprehensive configuration expressing complete knowledge, the information expressed through Table 2 is not. If the close-world assumption is not considered in the interpretation of (2) in section 2, the equivalent representation in Table 2 expresses the same uncertainty, as it is not known whether b has P and whether c has Q.

	a	b	C
P	1		1
Q	1	1	

Table 2.

Figure 4.   Interpretation shift.

Although it is extremely difficult to characterize what properties of the perceptual system are employed in the interpretation process, we can use the architecture of standard Turing Machines (TMs) as a reference for studying some possible computational mechanisms involved in this process. TMs compute functions. The arguments and values of these functions are placed on the TM's tape at the initial and end state of the computation, and need to be interpreted in relation to the standard input and output interpretation conventions [Boolos and Jeffrey, 1990]. The function itself is defined indirectly through an algorithm, which is specified, in turn, through the set of states and transitions of the TM. TM computations have also a discrete character as they deal with atomic symbols, stored in discrete cells, all or nothing, that are read all or nothing in each scanning operation, by a local scanning device, which inspects one cell at a time; the content of the rest of the tape is irrelevant for the local computational step which depends only on the state of the TM and the content of the cell that is inspected by the scanning device. Expressions (1) to (3) in section 2, for instance, are composed by atomic symbols that are placed in a lineal tape and are scanned locally; however, the corresponding configurations in Tables 1 and 2 need not be scanned locally, as the synoptic view of the configurations can be grasped by a single holistic, global operation by people.

Evidence that people do scan pictures in a global fashion comes also from the interpretation of ambiguous pictures like the well-known duck-rabbit picture popularized by Wittgenstein [Wittgenstein, 1953, part II, 10], in Figure 4. This picture can be interpreted as a duck or as a rabbit. In both interpretations the external representation on the paper is the same, and what varies is the output of the interpretation process. These interpretation processes can hardly be thought of in terms of a local scanning device and a compositional process in which the meaning of the whole can be worked out as a function of the meaning of the parts and their mode of composition. The process of interpretation shift or *reinterpretation* is more common than it seems at first sight, as it occurs often in design and creativity processes. It is also very important for verifying, and also for discovering, the graphical proofs in Figure 1 and 3, as will be shown below in this paper.

Global scanning devices for inspecting graphical symbols and configurations can be implemented computationally, as shown by the Whisper system [Funt, 1980]. In Whisper, a computational retina is used to scan and interpret diagrams through a small number of specific perceptual processes, which are composed out of a finite set of perceptual primitives. Perceptual primitives are the basic operations that can be performed by the retina. Reinterpretation can be achieved by inspecting the same picture with different perceptual processes.

5.   A diagrammatic inference scheme

In this section we present the structure of diagrammatic inference scheme through which theorems can be verified and learnt. For the definition of this scheme we use the theoretical notions presented in sections (2) to (4). In this section, the theorem of the sum of the odds is used as a case study; in section 6, we address the proof of the Theorem of Pythagoras.

The theorem of the sum of the odd numbers is normally proved through mathematical induction. In the arithmetic, informal, statement of the theorem (i.e., $1 + 3 + 5 + \ldots + (2n - 1) = n^2$), the generality of the sum is marked through the ellipsis symbol $(\ldots)$. However, this representation has some ambiguity, as it is not stated explicitly that the elements of the sum are all and only the odd numbers. In informal mathematics, this knowledge is just in the mind of the person who interprets the expression and carries on with the proof. In formal mathematics, on the other hand, the ellipsis is removed and the theorem is expressed through a different abstraction marker: the sign for summatories $\sum$, but even using this sign some ambiguity will remain[Bundy and Richardson, 1999]. For this reasons, in automatic theorem proving in AI, the expression is first translated into a recursive formulation in which the ambiguity is resolved, and the theorem is expressed as a function which returns the value of the sum for any given n [Jamnik, 1999]. An alternative semantics for the ellipsis in diagrammatic proofs is given in [Foo et al., 1999]. The abstraction represented by the ellipsis in the informal external representation is replaced by another abstraction marker in formal mathematics, and by a notational key, which is coded internally in the conventions for interpreting the theorem, in AI programs.

Now we explore how the numbers, the arithmetic symbols and the abstraction markers of the propositional representation of the theorem of the sum of the odds are expressed graphically in Figure 3. It can easily be seen that L-shape forms express odd numbers, square-shape forms squares numbers, and the graphical alignment of Ls express the sum. More precisely, the area covered by two L-shapes aligned represents the sum of the numbers represented by each of the L-shapes. This seems to be an instance of a more general principle, with wide application in the interpretation of graphical representations and diagrammatic proof, as follows:

- Principle of interpretation of composite area: The interpretation of a composite area is a function of the interpretation of its (non overlapping) parts.

In the case at hand, the function is the sum: the area formed by the union of the areas of two adjacent L-shapes represents the value of the sum.

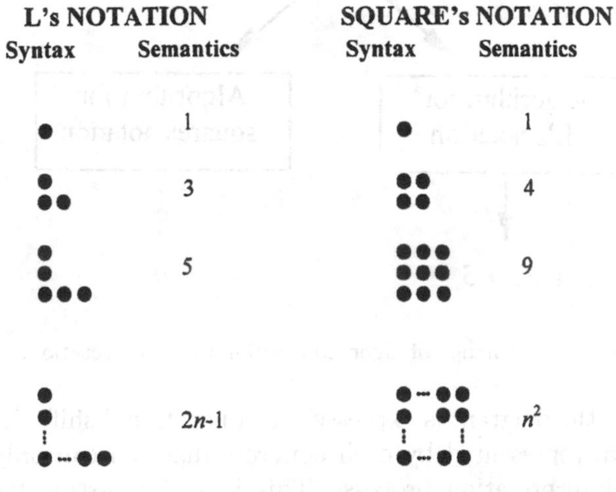

L's NOTATION

Syntax	Semantics
•	1
(L-shape)	3
(L-shape)	5
(L-shape)	$2n-1$

SQUARE's NOTATION

Syntax	Semantics
•	1
(square)	4
(square)	9
(square)	n^2

Figure 5. Bidimensional monadic notations.

It is not that easy to see, however, how the equality sign and the ellipsis are expressed. To investigate the properties of this representation we first ask what is the notation. As can be see in Figure 5, there are really two notational systems involved. These resemble standard monadic notation of TMs, but unlike the linear case, which extends over one dimension, the *L-shape monadic notation* and the *square-shape monadic notation* extend over the 2-D plane. As can also be seen, the notational keys are general statements defining the interpretation of an L of side n or a square of size n, for any n; these notational keys introduce the abstraction in the interpretation process.

In standard computational processes and TMs, algorithms are defined in relation to the notation and shape of the representational medium, and once a notation is defined, it is used along the computational process. However, if more than one notation is involved, changing the focus from one to the other is a major computational shift. It amounts to changing the algorithm under which the representation is being interpreted, and the whole system of conventions associated with the corresponding notations, as illustrated in Figure 6. Interpreting the diagram under the algorithm defined for L-shape notation returns an expression, but interpreting the same diagram with the algorithm for square notion returns a number.

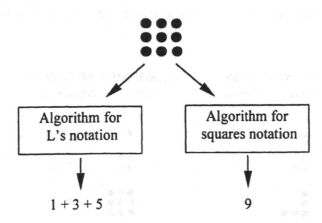

Figure 6.   Change of algorithm during the interpretation.

Equality in the diagram is expressed as a notational shift. The equality of the sum represented by each concrete diagram can only be seen through a reinterpretation process. This is an interesting fact signaling that the two notational systems are related; here, a relation of this kind will be referred to as a *meaning relation*. We can also see that the meaning relation depends on the realization of a graphical relation in the representational media, which we call *medium relation*. The process of reinterpretation requires realizing the corresponding medium relation first.

Next, we move into the pragmatics of theorem-proving. Unlike propositional representations, where a notation and medium is normally adopted beforehand, once and for all when the process is defined, in analogical representations we can reason about the notations and the properties of the representational medium. The propositional account takes a view entirely *within* a system of representation but to reason about graphics requires us to take a view *from* or *above* the representational system [Gurr et al., 1998]. This second view permits also to realize truths that follow from the choice of medium and notation, and the abstraction embodied in the interpretation process.

Meaning relations are interesting, but it is even more interesting if the representations involved stand not just for themselves, but also for a number of cases, possibly infinite. Similarly, specific medium relations can be instances of general patterns that can be realized by reasoning about medium relations. In the interpretation, general patterns will refer to a set, possibly infinite, of meaning relations. We will refer to these latter patterns as *universal medium relations*, and to the generalization of meaning relation as *universal meaning relations*. As we will see, graphical theorems are universal meaning relations. To induce such universal meaning relations we suggest the following inference process that we call *diagrammatic inference scheme*. The scheme is illustrated with the theorem of the sum of the odds.

Diagrammatic inference scheme:

1 Given the notational keys:

   Any L-shape of size n stands for the number $2n - 1$

   Any square of size n stands for the number n^2

2 Given the principle of interpretation of composite area

3 Given a medium relation (realized through reinterpretation):

   A specific sequence of L-shapes aligned in relation to a reference point (the right-top corner) makes a square (e.g., a specific square of size 3)

4 Induce a meaning relation (from (1), (2) and (3):

   A specific sum of odd numbers (e.g., $1 + 2 + 3$) is equal to a specific square number (i.e., 9).

5 Induce an universal medium relation out of the given medium relation (from (3) and general geometric knowledge):

   Any sequence of n L-shapes aligned in relation to a reference point (the right-top corner) makes a square of size n

6 Induce a universal meaning relation (from (1), (2), (5):

The sum of the first n odd numbers is n^2

Step (4) consists of the instantiation of the principle of interpretation of composite area in (2) for the concrete case of the medium relation inspected in (3). This instantiation requires the notational keys in (1), and results in an instance of the theorem. Step (6) uses also (1) and (2), but instead of using the concrete relation in (3), it uses the universal medium relation in (5), producing a universal meaning relation: the theorem.

Crucial to the argument is the path from concrete thinking to abstract thought. Here, the concrete diagram is used as a pivot to launch the generalization, first through the generalization of the graphical relation in (5) and then, to the full abstraction expressed by the theorem. The generalization in (5) involves also a complex reasoning process, but one in which interpretations are not involved. The resulting statement about the medium is a fact about geometry; here, we assume that it can be proved by contradiction, with the help of visualization. Also, see [Foo et al., 1999] for two formal approaches to this kind of problems.

To stress that a diagram represents a proof only if it is interpreted with a particular notation in mind, notice that the same diagram can be read as representing different proofs. Suppose that we define a notation that we call *monadic diagonal notation* such that a diagonal of n dots represents the number n. Then, a square of size n represents the theorem $1 + 2 + ... + n + (n-1) + ... + 2 + 1 = n^2$, which also appears in Nelsen's book [Nelsen, 1993]. If such notational key is used in the diagrammatic reasoning scheme in (1) instead of the L-shape notation key, this latter theorem is produced, but not the theorem of the sum of the odds.

The theorem of the sum of the odds has recently been studied in the context of the Diamond system [Jamnik, 1999]. There, the theorem is induced out of a number of samples of the theorem, which are provided manually by human users. The general statement is produced by a kind of learning induction inference, the so-called constructive ω-rule, resulting in an "empirical" mathematical truth. In our approach, on the other hand, universal meaning relations are kinds of synthetic truths, in the Kantian sense, that necessarily follow from the notational keys, the properties of the representational media and general principles about the interpretation of the space. Universal meaning relations are not facts about the world, but truths of meaning.

We conclude this section by observing that the diagrammatic inference scheme can be applied even if the interpreter does not know the theorem. In this latter case, the inference is a discovery process. Proving and learning are two sides of the same coin: for people, learning the abstraction that the figure represents the theorem and proving the theorem are really one and the same thing.

6.   Global reinterpretation

We turn now to the proof of the Theorem of Pythagoras presented in Figure 1, which is repeated for clarity in Figure 7 (from [Bronowski, 1981]). As can be seen, the construction is performed out of a right triangle which is rotated and translated four times to form a square on the hypotenuse. The length of the side of the square inside the construction is the difference between the lengths of the two right sides of the triangle. In the last state, the triangles are rotated, and from the resulting configuration, two squares emerge. Each of these squares is on each of the right sides.

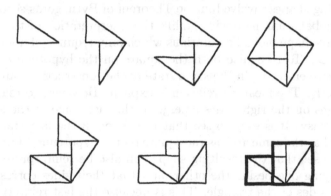

Figure 7.   The Theorem of Pythagoras.

The intuition of the validity of the proof is overwhelming, and nevertheless, some people would argue that it is not formal, and it does not constitute a valid argument. In a recent discussion on the Internet about another diagrammatic proof of the theorem, for instance, it was argued that the diagrammatic sequence would constitute an individual instance of the proof. This issue can be addressed on the light of the present discussion. First, we consider whether the triangle in the initial state of the construction stands for itself, an individual concrete object, or stands rather for any triangle, for the class of all triangles; similarly for the squares involved in the proof. The answer depends on the representational key. If people are looking at the diagram without a notation in mind, the diagram is not a representation at all. If the notation is that the triangle stands for a concrete individual object, a kind of terminological key, then the whole construction stands for a concrete instance of the proof, as only minimal abstraction is involved. In particular, the assertion that the area of the square emerging as the square on the hypotenuse is the same as the area of the two squares on the right sides of the triangle is a particular assertion holding only for the particular construction.

However, if the interpretation is such that the initial triangle stands for *any* (right-angled) triangle, a notational key, then it represents the class of all such triangles, and the particular geometric properties of the triangle are also part of the abstraction. In this reading, realizing that the area of *any* square on the hypotenuse of *any* right triangle is the same as the area of *any* two squares on the right sides of the same triangle, the statement of the theorem, is to learn a universal meaning relation.

We conclude with a note on global reinterpretation. We note that interpreting shapes involved in the Theorem of Pythagoras seems to require a global scanning device, unlike the interpretation of the L-shape and square-shape monadic notations which only requires of a local scanning device. To recognize both the square on the hypotenuse and the inner square emerging in the same state of the construction can be realized directly. The meaning relation is explicit. However, to realize that the squares on the right sides emerge in the last stage of the sequence is not so easy. It is easy to see that the shape in the last stage of the sequence has the same area as the square on the hypotenuse, but it is not so easy to see that the resulting shape can also be reinterpreted as two squares lying one beside the other, and that their sides correspond to the right sides of the triangle. This is because the last reinterpretation requires the agent looking at the construction to restructure five geometrical figures – the four triangles and the little square – into the two new squares. As this reinterpretation does not appear in the diagram, it has to be visualized by a global scanning device. We are fortunate that our retina has such a preference for square notations.

Acknowledgments

The author gratefully acknowledges the contributions to earlier versions of the this work from Gabriela Garza and John Lee, and the very stimulating conversations with Keith Stenning, B. Chandrasekaran, Ivan Bratko, Rafael Morales, and to the anonymous reviewers of this and previous versions of this work; also to Francisco Hernández, Erik Schwarz and Miguel Salas for comments and technical support. The author also acknowledges the support of Conacyt's grant 400316-5-27948A.

References

Boolos, G.S. and Jeffrey, R.C., 1990, *Computability and Logic*, Cambridge University Press.

Brachman, R.J. and Schmolze, J.G. 1985, An Overview of the KL-ONE Knowledge Representation System, *Cognitive Science* 9:171–216.

Bronowski, J., 1981, *The Ascent of Man*, BBC Corporation, London, pp. 155–188.

Bundy, A. and Richardson, J., 1999, Proofs about lists using ellipsis, in: *Proceedings of 6th International Conference, Logic for Programming and Automated Reasoning*, H. Ganzinger, D. McAllester, and A. Voronkov, eds., LNAI 1705, Springer-Verlag, pp. 1–12.

Chandrasekaran, B., 1997, Diagrammatic representation and reasoning: some distinctions, in: *Working notes on the AAAI-97 Fall Symposium Reasoning with Diagrammatic Representations II. MIT*, November 1997.

Foo, N.Y., Pagnucco, M., and Nayak, A.C., 1999, Diagrammatic proofs, in: *Proceedings of the 16th International Joint Conference on Artificial Intelligence*, IJCAI-99, Stockholm, Aug 1999, Morgan Kaufmann, pp. 378–383.

Funt, B.V., 1980, Problem-solving with diagrammatic representations, *Artificial Intelligence* 13:210–230.

Gurr, C., Lee, J., and Stenning, K., 1998, Theories of diagrammatic reasoning: Distinguishing component problems, *Minds and Machines* 8:533–557.

Hayes, J.P., 1985, Some problems and non-problems in representation theory, in: *Readings in Knowledge Representation*, R. Brachman and H. Levesque, eds., Morgan and Kaufmann, Los Altos, California, pp. 3–22.

Jamnik, M., 1999, Automating Diagrammatic Proofs of Arithmetic Arguments, Ph. D. Thesis, University of Edinburgh.

Johnson-Laird, P.N., 1983, *Mental Models: Towards a Cognitive Science of Language, Inference, and Consciousness*, Cambridge University Press, Cambridge.

Nelsen, R.B., 1993, *Proofs without Words: Exercises in Visual Thinking*, The Mathematical Association of America.

Pineda, L., 1989, Graflog: a Theory of Semantics for Graphics with Applications to Human-Computer Interaction and CAD Systems, PhD thesis, University of Edinburgh, UK.

Pineda, L.A., 1992, Reference, synthesis and constraint satisfaction, *Computer Graphics Forum*, 11(3):C-333–C-344.

Sloman, A., 1985, Afterthoughts on Analogical Representations, in: *Readings in Knowledge Representation*, R. Brachman and H. Levesque, eds., Morgan and Kaufmann, Los Altos, California, pp. 431–440.

Stenning, K. and Oberlander, J., 1995, A cognitive theory of graphical and linguistic reasoning: logic and implementation, *Cognitive Science* 19:97–140.

Wittgenstein, L., 1953, *Philosophical Investigations*, Basil Blackwell, Oxford.

A LOGICAL ANALYSIS OF
GRAPHICAL CONSISTENCY PROOFS

Atsushi Shimojima
School of Knowledge Science, Japan Advanced Institute of
Science and Technology, 923-1292, Japan
ashimoji@jaist.ac.jp

Abstract   In this paper, we investigate the semantic mechanism of graphical consistency proofs, where one expresses a certain condition in a chart, a diagram, or some other graphical representation, and uses the existence of such an graphic as a proof of the consistency of the expressed conditions. We first show that such a proof is guaranteed to be sound by a special matching of constraints between representations and represented situations. We then extend our analysis to another types of graphics-based inferences called "free rides", and show that they also rely on a matching of constraints between representations and represented situations. Comparisons of graphical consistency proofs and free rides let us see three commonalities between these two inferential procedures, and let us define the general notion of physical on-site inference, where perceptually present objects are used as inferential surrogates through physical operations. Our analysis is therefore a clarification of the exact semantic requirements for one representative way in which a visual representation participates in distributed cognition [Giere, 2001] or manipulative inferences [Magnani, 2001].

1.   Introduction

Suppose Mr. and Mrs. Murata have a small living room with a large sectional, two side tables, a center table, and a large TV cabinet. Mrs. Murata wants to rearrange the furniture to create a large pathway across the living room to the kitchen. Mr. Murata is rather reluctant about rearrangement, fearing that it may result in a less convenient setting of the TV cabinet and the sectional. Thus, Mrs. Murata needs to show that it is possible to create a desired pathway without sacrificing a good viewing angle of the TV from the sectional. For this purpose, she draws

L. Magnani, N.J. Nersessian, and C. Pizzi (eds.),
Logical and Computational Aspects of Model-Based Reasoning, 93–115.
© 2002 Kluwer Academic Publishers. Printed in the Netherlands.

a diagram depicting their living room after a would-be rearrangement (Figure 1). "Look," she says. "It's possible. Let's do it."

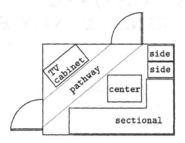

Figure 1.   Room Map Drawn by Mrs. Murata.

The type of "proof" that she has just conducted is the main subject of this paper. She has shown the possibility, or consistency, of an arrangement of furniture that satisfies specific requirements. Interestingly, she has done so *by constructing a graphical expression of the arrangement in question.* For some reason, the constructibility of such an expression is taken to guarantee the consistency of the represented conditions. Let us call this type of proof *graphical consistency proof.*

Although this procedure is quite natural and ubiquitous, its validity should not be taken for granted. A diagram is, after all, just a *representation* of something, not the thing itself. How can a construction of a representation be a proof of the consistency of what is expressed? Let Γ be the set of specifications of the desired furniture arrangement, concerning the desired viewing angle of the TV display from the sectional, the desired width and route of the pathway, and the horizontal dimensions of Mr. and Mrs. Murata's living room and furniture, the desired viewing angle of the TV display from the sectional. Let s be Mrs. Murata's map, which expresses Γ. The question is how the construction of s can be a proof of the consistency of Γ.

Note that *not every* representation that expresses Γ is taken to be a proof of Γ's consistency. Suppose you specify all the requirements in the form of a list of English sentences. This would be a representation expressing all the requirements Γ, just as the room map s is. Yet, nobody would count it as a proof of the consistency of Γ. So, there must be something in representations such as s that is missing from lists of sentences. What is it?

A quick answer to this question is "an auto-consistency property of a representation system." Roughly, an auto-consistency property of a representation system is its incapability of expressing a certain range of inconsistent conditions. For example, the system of room maps of the kind Mrs. Murata has produced cannot express spatially impossible arrangements of furniture, and in this respect, the system is auto-consistent. This means that if some arrangement of furniture can be expressed in a room map, it guarantees that the expressed arrangement is spatially possible. The system of English sentences, on the other hand, is not auto-consistent in this respect, and for this reason, expressibility of a furniture arrangement in English does not guarantee the spatial possibility of the arrangement. Auto-consistency, defined as the incapability of expressing a certain range of inconsistent conditions, is clearly responsible for a system's capacity of graphical consistency proofs.

Researchers in philosophy, logic, AI, and cognitive psychology have paid some attention to auto-consistency properties of representation systems. Although the formal notion of auto-consistency was not introduced until later, Gelernter's Geometry Machine [Gelernter, 1959] exploited the auto-consistency of geometry diagrams to short-cut the search for provable theorems. Sloman [Sloman, 1971] explicitly suggested non-expressibility of inconsistent information as a characteristic of "analogical" representations. Lindsay [Lindsay, 1988] proposed a general framework of knowledge representation that exploits the auto-consistency property. Barwise and Etchemendy [1994] formally introduced the notion of auto-consistency and showed how their system of Hyperproof diagrams exploits this property to enable graphical consistency proofs. They also proved that a sub-system of Hyperproof diagrams is in fact auto-consistent [Barwise and Etchemendy, 1995]. Stenning and Inder [Stenning and Inder, 1995] discussed the trade-off of the expressive power of a representation system and its auto-consistency property.

Despite these exceptions, studies of the phenomenon of auto-consistency have been rather scattered and cursory so far. In particular, the question still remains on the semantic mechanism *behind* an auto-consistency property of a representation system. Exactly what makes a representation system auto-consistent? What prevents a member of a representation system from expressing a certain range of inconsistencies? Is there any common semantic property shared by various systems that allow graphical consistency proofs?

This paper has two main goals. The first goal is to develop a semantic analysis adequate to answer the questions just posed. We will start with examining more examples of auto-consistent representation systems and their potentials for graphical consistency proofs (section 2). We will then propose our model of auto-consistency properties of representation systems in a simplified semantic framework of channel theory [Barwise and Seligman, 1997], and characterize, in its terms, the conditions for a graphical consistency proof to be valid in a representation system (section 3). As it turns out, a special type of matching of constraints in the source and the target of the system is responsible for auto-consistency. This indicates an important connection of the phenomenon of "free ride" discussed in the literature of diagrammatic reasoning and the phenomenon of graphical consistency proof, and our analysis motivates the general concept of *physical on-site inference* that covers both types of inferences.

The second goal of this paper is then to formulate this concept as clearly as possible. After giving an analysis of the exact procedures and requirements for free rides, we will highlight three characters shared by free rides and graphical consistency proofs: both procedures utilize *perceptually accessible* objects, such as graphics on a sheet of paper, as *inferential surrogates* by applying *physical operations* on them (section 4). Thus, seen in connection with the on-going research on model-based reasoning, this paper defines one exact sense in which perceptual objects are used as *models* for inference in manipulative processes [Magnani, 2001]. In connection with the framework of distributed cognition, this paper is a case study of an important class of inferences that use visual representations as parts of distributed cognitive systems [Giere, 2001].

2.   Examples

To get a surer grasp of our target, let us examine various examples of representation systems with auto-consistency properties and the kinds of graphical consistency proofs allowed in the systems.

Example 1 Mrs. Murata's room map is a member of an auto-consistent system of representation, since no representation of this type can express a spatially inconsistent arrangement of furniture. Imagine that the sectional in the living room were twice as large. Then it would be impossible even to lay out, without stacking, all the furniture in their small living room. *Correspondingly*, it is impossible to express this condition in a room map, for it is impossible to lay out, without overlapping, furniture icons of appropriate shapes and sizes in a small rectangular that stands for the living room.

This auto-consistency property of the system of room maps in turn allows us to conduct graphical consistency proofs, such as the one conducted by Mrs. Murata. You arrange furniture icons in a bounded area on a room map, and the expressed arrangement is thereby guaranteed to be spatially consistent.

It is important to note that the auto-consistency of a representation system is always relative to some particular range of inconsistencies. A representation system that is auto-consistent for some range of inconsistencies can fail to be some other range of inconsistencies. For example, although the furniture arrangement expressed by a room map is always spatially possible, it may well be psychologically impossible for Mr. Murata, or culturally impossible in a Japanese community. Accordingly, a graphical consistency proof with a room map can establish only spatial consistency, and not psychological or cultural consistency.

Example 2 Consider ordinary drawings of people such as Figure 2. Clearly, one cannot be taller than oneself – one's being taller than oneself is an inconsistent condition. And in fact, we cannot draw a line drawing of a person with such a height. That would amount to drawing a personal figure that is longer than itself. Similarly, it is impossible for a person A to be taller than a person B who is taller than A, nor for A to be taller than B who is taller than C who is taller than A. Correspondingly, we cannot draw line drawings of people of such relative heights. In terms of Figure 2, expressing the first condition amounts to drawing a figure that is both longer and shorter than the "B"-figure, and expressing the second condition amounts to drawing a personal figure that is longer than the "B"-figure but shorter than the "C"-figure. Clearly, both are impossible endeavors.

Figure 2. A drawing of people. Try to add a personal figure that is both longer and shorter than the left figure, or try to add a personal figure that is longer than the left figure but shorter than the right figure.

Generally, a representation system that expresses relative magnitude of one sort (relative height or percentage) by means of relative magnitude of another sort (relative length, size, or height) is auto-consistent against violations of the quasi-linearity of the expressed relation. Thus, the systems of bar charts, line graphs, pie charts, and other graphics used for quantitative analysis have analogous auto-consistency properties.

Such an auto-consistency property of course gives the system potential for simple graphical consistency proofs. For example, consider if it is possible for the following conditions to hold together:

(1) A is not taller than B.

(2) B is not taller than C.

(3) C is not taller than A.

Using the auto-consistency of the system of line drawings, one can draw a line drawing expressing all these conditions and make it a proof of their consistency. Figure 3 shows one of the many line drawings that would serve this purpose.

Figure 3. A line drawing that serves as a consistency proof of the conditions (1), (2), and (3).

In contrast, we can easily express the inconsistent conditions of people's height with different types of representations. You can write them up in English sentences such as: "A is taller than himself", "A is taller than B but shorter than B", and "A is taller than B, B is taller than C, and C is taller than A". You could also use first-order sentences to express these conditions. Or you can use a directed graph, such as Figure 4, where the edge denotes the *taller-than* relation. Thus, the systems of English sentences, first-order sentences, and directed graphs are not generally auto-consistent for people's relative height.

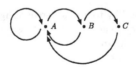

Figure 4. A directed graph expressing inconsistent conditions of people's height.

Example 3 Returning to an example of auto-consistency, consider a route map of a subway system of the kind shown in Figure 5. In common subway maps, a line with a particular pattern stands for a particular subway line (Jubilee line, Midosuji line, etc.). A connection of two stations via a particular subway line is then expressed by an corresponding type of line segment between two station icons, while a non-connection is expressed by the absence of such line segments. For example, the black line directly connecting the circles labeled "*J*" and "*K*" in Figure 5 indicates that Line *U* connects the stations *J* and *K* directly, with no station in between. On the contrary, the absence of a line segment directly connecting the circles labeled "*J*" and "*E*" indicates the absence of a direct connection between the stations *J* and *E*. Yet, the "*J*" circle and the "*E*" circle is indirectly connected by two black line segments, and this indicates that Line *U* connects the stations *J* and *E* indirectly, with one station in between. There is not even such an indirect connection between the circles labeled "*J*" and "*D*", and this indicates that no single subway line connects the stations *J* and *D*.

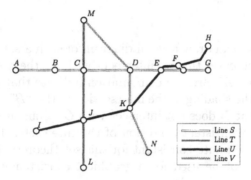

Figure 5. A route map of a subway system.

Now, it is not possible for two stations to be both connected and disconnected by the same subway line, nor for a station not to be connected to another station when the first station is connected to a station that is connected to the second station, nor for five stations to be connected to each other with less than four direct connections. *Correspondingly*, it is not possible to draw a route map that expresses any of these conditions: line segments cannot both connect and disconnect two circles, line segments of a particular color or pattern must connect two circles if line segments of that kind connect both circles to a third circle, and it is not possible to connect five circles to each other with less than four line segments. The system of route maps is auto-consistent against a certain type of graph-theoretic inconsistencies.

Example 4 The system of Euler diagrams is auto-consistent for certain set-theoretic inconsistencies. For instance, it is not possible for a set *B* to intersect with a proper subset *A* of a set *C* without intersecting with the superset *C*. Correspondingly, we cannot draw an Euler diagram that depicts sets in such relationship: try to draw a circle that intersects with the circle labeled "*A*" in Figure 6, and your circle will necessarily intersect with the circle labeled "*B*". Thus, however hard you may try, you cannot express a set disjoint from *B* that intersects with its proper subset *A*.

Figure 6.   An Euler diagram. Try to add a circle that overlaps with the "*A*"-circle but not with the "*B*"-circle.

Interestingly, we can draw a Venn diagram of such a set. In Figure 7, the *x*-sequence indicates that *B* intersects with *A*, the *y*-sequence and the shading in the "*A*"-circle in combination indicate that *A* is a proper subset of *C*, and the shading in the intersection of the "*B*"- and the "*C*"-circle indicates that *B* does not intersect with *C*. Thus, it expresses the set-theoretically inconsistent condition of the three sets. The system of Venn diagrams is *not* auto-consistent for this set-theoretic inconsistency and therefore cannot be used for a graphical consistency proof in the standard set-theoretic domain.

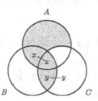

Figure 7.   Venn diagram expressing an inconsistent condition.

3.   Analysis

These examples should have given an adequate evidence that a wide variety of representations systems have auto-consistency properties, allowing us graphical consistency proofs. What is then the semantic mechanism behind auto-consistency properties?

We will explore this issue with mathematical tools of channel theory [Barwise and Seligman, 1997], although we will keep a part of our discussions informal. Mathematically oriented readers should consult chapter 20 and related chapters of [Barwise and Seligman, 1997] for formal details of the concepts used in this paper, such as "constraint", "indication", and "representation system".

Intuitively, auto-consistency involves some correspondence between inconsistent conditions in the domain of representations and inconsistent conditions in the domain of things represented. Take the example of the system of Euler diagrams (Example 4). Due to some spatial constraints holding in the domain of Euler diagrams, a certain arrangement of Euler circles is just impossible, and this impossibility corresponds to the impossibility of a certain inclusion or jointness relation in the domain of sets.

Let us make this intuition more precise. What exactly is impossible in the domain of representations in the case of Euler diagrams? Due to spatial constraints, it is impossible that the following three conditions hold together in a single Euler diagram:[1]

(4*) A circle labeled "A" is inside a circle labeled "B".

(5*) A circle labeled "C" is outside a circle labeled "B".

(6*) A circle labeled "C" overlaps with a circle labeled "A".

Due to the semantic conventions associated with the system of Euler diagrams, the conditions (4*), (5*), and (6*) respectively indicate the following conditions on the represented sets:

(4) Set A is a proper subset of set B.

(5) Set C is disjoint from the B.

(6) Set C has an intersection with the A.

Corresponding to the mutual inconsistency of (4*), (5*), and (6*), these indicated conditions cannot hold together in a single situation. That is, conditions (4), (5), and (6) are also mutually inconsistent. Figure 8 depicts this situation schematically.

Now, (4*), (5*), and (6*) must hold in any old Euler diagram expressing (4), (5), and (6), but (4*), (5*), and (6*) cannot hold together in a single Euler diagram. It follows that no Euler diagram can express the conditions (4), (5), and (6), which are mutually inconsistent. Here we see the correspondence of inconsistent sets of conditions through the

---

[1]We are assuming that no distinct circles can have the same label in a well-formed Euler diagram, making distinct sets denoted by distinct circles. "One-one denotation" rules of this sort apply to objects in many different kinds of graphical representations, including circles in Venn diagrams and station icons in subway route maps.

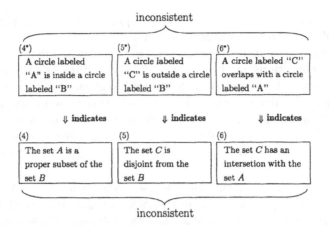

Figure 8.   Correspondence of inconsistencies in the system of Euler diagrams.

indication relation, and it accounts for the inability of Euler diagrams to express a particular inconsistent set of conditions. The system of Euler diagrams obviously involves many other cases of semantic correspondence of inconsistent sets, and they combine to define a significant range of inconsistent sets of conditions that cannot be expressed in Euler diagrams.

Figure 9 shows another instance of corresponding inconsistent sets in the system of route maps (Example 3). The conditions listed in the upper part of the figure must hold in any route map if it is to express the conditions in the lower part. Yet the set of conditions in the upper part is inconsistent. Hence the incapability of the system to express the inconsistent set of conditions in the lower part.

How can we characterize these correspondences of inconsistent sets in more general terms? Let us introduce certain terminology to ease our analysis. By *source types*, we mean conditions that (potentially) hold in a representation, such as the three conditions of Euler diagrams listed in the upper part of Figure 8. In contrast, *target types* are conditions that (potentially) hold in a represented situation, such as the three conditions of sets A, B, and C listed in the lower part. If Ω is a set of source or target types, we call Ω *inconsistent* if there is no possible situation in which all members of Ω hold. Let us introduce the notion of "projections of sets of source types" in the following sense:

Definition 1 (Projection of sets of source types)   A set Γ of source types is projected to a set Δ of target types in a representation system $\mathcal{R}$ if:

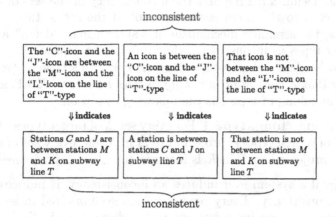

inconsistent

The "C"-icon and the "J"-icon are between the "M"-icon and the "L"-icon on the line of "T"'-type	An icon is between the "C"-icon and the "J"-icon on the line of "T"'-type	That icon is not between the "M"-icon and the "L"-icon on the line of "T"'-type
⇓ indicates	⇓ indicates	⇓ indicates
Stations C and J are between stations M and K on subway line T	A station is between stations C and J on subway line T	That station is not between stations M and K on subway line T

inconsistent

Figure 9. Correspondence of inconsistencies in the system of route maps.

- Each member of Γ indicates at least one member of Δ in $\mathcal{R}$,

- Each member of Δ is indicated by at least one member of Γ in $\mathcal{R}$.

For example, the set $\{(4^*), (5^*), (6^*)\}$ of source types is projected to the set $\{(4), (5), (6)\}$ of target types in the system of Euler diagrams. The projection is a one-one correspondence between the sets in this particular case, although the above definition does not require a one-one correspondence. Also, the set of source types listed in the upper part of Figure 9 is projected to the set of target types listed in the lower part in the system of line drawings.

With this preparation, we can now give a general characterization of "inconsistency inducement":

Definition 2 (Inducement of inconsistency)   A set Δ of target types is an *induced inconsistency in* a representation system $\mathcal{R}$ if:

- There is a set Γ of source types projected to Δ in $\mathcal{R}$,

- Every set Γ of source types projected to Δ is inconsistent.

For example, the set $\{(4), (5), (6)\}$ is an induced inconsistency in the system of Euler diagrams. For there is a set, $\{(4^*), (5^*), (6^*)\}$, projected to it, and while every set projected to it entails this set, it is inconsistent. Hence every set projected to $\{(4), (5), (6)\}$ is inconsistent. Likewise, the set of target types listed in the lower part of Figure 9 is an induced inconsistency in the system of line drawings. Each of these sets of target types cannot be expressed in the respective representation system that induces it.

Note that an induced inconsistency in a given system is not necessarily inconsistent. Definition 2 requires the inconsistency of the sets of source types projected to the target type, but not of the set of target types itself. Thus, the semantic mechanism of a system may "deem" a given set of target types as inconsistent, while it is in fact consistent.

On the other hand, if a system $\mathcal{R}$ never makes this type of "errors", we will say that *a constraint matching of type 1* holds in $\mathcal{R}$. That is, a constraint matching of type 1 is the following condition:

Constraint matching, type 1 For every set Δ of target types, if some set Γ of source types is projected to Δ in $\mathcal{R}$ and every set of source types projected to Δ in $\mathcal{R}$ is inconsistent, then Δ is inconsistent.

Typically, if a system ever induces an inconsistency, it induces more than one inconsistency. Every representation system cited in section 2 induces more than one inconsistency, as the discussions in that section show. So, it makes sense to talk about the *set* of inconsistencies induced in a given system $\mathcal{R}$. If, in addition, a constraint matching of type 1 holds in $\mathcal{R}$, we can think of that set as a special range of inconsistencies that are correctly "tracked" by the semantic mechanism of $\mathcal{R}$. We will call this set "$K_{\mathcal{R}}$", and call the members of this set "$K_{\mathcal{R}}$-*inconsistent*" to distinguish them from inconsistent sets of target types not tracked by $\mathcal{R}$.

We will also call any set of target types outside this set "$K_{\mathcal{R}}$-*consistent*". Thus, even when a set of target types is $K_{\mathcal{R}}$-consistent, it may be inconsistent. Being $K_{\mathcal{R}}$-consistent only guarantees that the set is consistent so far as the range $K_{\mathcal{R}}$ of inconsistencies is concerned. The set may be inconsistent with respect to some other range of inconsistencies not tracked by the system $\mathcal{R}$.

Earlier, we roughly characterized the auto-consistency property of a representation system as the inability of the system to express a certain range of inconsistent conditions. We can now refine this characterization. That is, the auto-consistency of a representation system $\mathcal{R}$ is nothing but a constraint matching of type 1, where the set $K_{\mathcal{R}}$ of induced inconsistencies corresponds to the "range of inconsistencies" that $\mathcal{R}$ cannot express. For, by the definition of $K_{\mathcal{R}}$, every member of $K_{\mathcal{R}}$ is a set of target types inexpressible in the system $\mathcal{R}$, and by the definition of type 1 matching, every member of $K_{\mathcal{R}}$ is in fact inconsistent.

More importantly, a constraint matching of type 1 holding in a system $\mathcal{R}$ guarantees the following form of inferences to be valid:

Graphical Consistency Proof Express the set Δ of information in the representation system $\mathcal{R}$. Conclude, from the success of that operation, that Δ is $K_{\mathcal{R}}$-consistent.

This procedure is valid since, if some consistent set Γ is projected to Δ in $\mathcal{R}$, then Δ is not an induced inconsistency in $\mathcal{R}$ and not a member of $K_{\mathcal{R}}$. Given the constraint matching of type 1, this just means Δ is $K_{\mathcal{R}}$-consistent. Now, one of the most efficient ways of verifying that some consistent set Γ of source types is projected to Δ in $\mathcal{R}$ is to actually construct a representation in $\mathcal{R}$ that expresses Δ, and this is exactly what is done through actual drawing of diagrams, charts, and maps. Thus, a graphical consistency proof is valid under a representation system with a constraint matching of type 1.

Note, however, the consistency result obtained through this procedure is just the $K_{\mathcal{R}}$-consistency of the set Δ, not the unconditional consistency of Δ. As we noted above, the $K_{\mathcal{R}}$-consistency of a set Δ of target sets only means that Δ is consistent with respect to the particular range of inconsistencies tracked by the system $\mathcal{R}$, and Δ may be actually inconsistent with respect to some other range of inconsistencies. And this is just natural as a model of graphical consistency proofs performed under a particular representation system. For example, the consistency of a particular furniture arrangement established in Example 1 is just the *spatial* consistency of the arrangement, and not necessarily the social or psychological or cultural consistency of the arrangement. Likewise, the consistency of a subway route established in Example 3 is just the *graph-theoretical* consistency of the expressed route, and not necessarily the commercial or economical or urbanologistic consistency. Generally, consistency established through a graphical consistency proof in a particular representation system is *limited* in nature. Our model reflects this limitation in terms of the limitation implied by the notion of $K_{\mathcal{R}}$-consistency.

4.   Physical on-site inferences

Shimojima [Shimojima, 1995a; Shimojima, 1995b] has introduced the notion of *free ride* to capture another common form of inferences typically done with graphical representations. Free rides are similar to graphical consistency proofs in that they exploit a certain type of constraint matching between representations and represented situations and that they essentially involve physical operations on representations for this exploitation. In this section, we will first review a few examples of free rides and extend the analysis given in [Shimojima, 1995a]. We will then specify the common elements of graphical consistency proofs and free rides in order to define the general notion of "physical on-site inference."

4.1   Free rides

Let us start with looking at two simple examples of free rides.

Example 5 Harry is asked to recall and describe the geographical features of the village in which he grew up, as accurately as possible. After some unsuccessful attempts of remembering the geographical features of his home village with no tools, he decides to draw an approximate map of his home town. On the basis of fragments of his memory, he draws lines and curves on a sheet of paper to represent the streets, pathways, rivers, and such. He then uses wood blocks to represent the buildings that he remembers to have existed, and places them on his map, to represent the approximate locations of those buildings. (He keeps revising and supplementing the map, and eventually obtains a map that represents his home town to the best of his memory.)

At the beginning of this procedure, Harry remembers the locations of a river, two roads, and several houses, and constructs a tentative map (Figure 10, left).

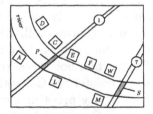 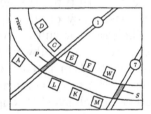

Figure 10.   A manipulation of memory maps producing free rides.

Then he recollects one more piece of information about his home village, that is:

(7) The house K was halfway between the houses L and M.

To present this new fragment of memory in his map, Harry puts a wood block standing for the house K between the wood blocks standing for the houses L and M (Figure 10, right).

As the result of this simple operation, Harry's map now presents many pieces of new information, *other than* (7), that were absent from the initial map. Among them are:

(8) The house K was across the house F over the river.

(9) The house K was closer to road 1 than the house M was.

(10) The house M was closer to the bridge S than the house K was.

(11) The house K and the house A had the road 1 in between.

To get the sense of utility of the system of Harry's memory map, imagine how many deduction steps would be needed if he tried to obtain the same results with pure thought on the basis of the principles of geometry. By operating on his map in the way described above, Harry has skipped all these complications of computation, and obtain the information (8), (9), (10), and (11) almost "for free". The operation is extremely efficient, for the purpose of updating the information content of his map toward the solution of the problem.

Example 6 We use Venn diagrams to check the validity of the following syllogism:

(12) All Cs are Bs.

(13) No Bs are As.

(14) (Therefore) no Cs are As.

We start by drawing three circles, labeled "As", "Bs", and "Cs" respectively. On the basis of the premises (12) and (13) of the syllogism, we shade the complement of the "C"-circle with respect to the "B"-circle (Figure 11, left) and shade the intersection of the "B"-circle and the "A"-circle (Figure 11, right). Observing that the intersection of the "C"-circle and the "A"-circle is shaded as a result, we read off the conclusion (14), and decide that the syllogism is valid.

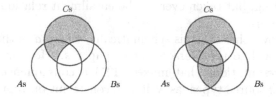

Figure 11.   A manipulation of Venn diagrams producing a free ride.

Again, we obtain a piece of information "for free" just by operating on diagrams: updating a diagram on the basis of the information (12) and (13) lets the diagram generate the information (14), which in turn lets us decide that the syllogism is valid.

In a nut shell, a free ride is an inferential procedure where we express the set Δ of information in a representation system $\mathcal{R}$, observe that this operation results in the condition σ that indicates the information θ, and conclude that Δ entails θ.

Let us analyze the free ride in example 6 in more detail. Let us assume that we start with a blank sheet of paper, say s. We apply a certain operation on s to create a Venn diagram that expresses the target types (12) and (13). Due to the semantic rules associated with Venn diagrams, this requires the resulting sheet of paper, say s', to support the following source types:

(12*) The complement of a "C"-circle to a "B"-circle is shaded.

(13*) The intersection of a "B"-circle and an "A"-circle is shaded.

These types respectively indicate (12) and (13), and make s' express these pieces of information. Interestingly, (12*) and (13*) not only indicate (12) and (13), but also *entail* another source type, namely:

(14*) The intersection of a "C"-circle and an "A"-circle is shaded.

That is, if you shade the complement of a "C"-circle to a "B"-circle and shade the intersection of the "B"-circle and an "A"-circle, you end up shading the intersection of the "C"-circle and the "A"-circle! Now, according to the semantic rules associated with Venn diagrams again, this source type indicates the target type (14). Thus, the Venn diagram s' ends up with expressing the information (14) too. Thus, one can simply read off the information (14) from s', and since (14) is entailed by the original information (12) and (13), one is making a *valid* inference in this way. The major part of the inference, however, is not done by the user's thinking, but taken over by the entailment relation from (12*) and (13*) to (14*).

Figure 12 shows this analysis schematically, where a is the operation of drawing applied to the blank sheet of paper s. Here we see a constraint on representations that makes (14*) a consequence of the set $\{(12^*),(13^*)\}$ of source types, as well as a constraint on represented situations that makes (14) a consequence of the set $\{(12),(13)\}$ of target types. Moreover, the antecedent $\{(12^*),(13^*)\}$ is projected to the antecedent $\{(12),(13)\}$ via the indication relation $\Rightarrow$ associated with the representation system, while the consequent (14*) indicates the consequent (14). Thus, our analysis implies that a free ride is also a form of inference that utilizes the semantic matching of a constraint governing representations with a constraint governing the represented situations. This much is the analysis of Example 6 given in [Shimojima, 1995a], which can be easily extended to other cases of free rides such as Example 5.

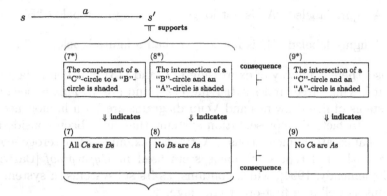

Figure 12.   Analysis of a free ride in the system of Venn diagrams.

What exactly is the type of constraint matching required for free rides then? It is slightly different from type 1 of constraint matching, required for physical consistency proofs. We therefore call it "type 2":

Constraint matching, type 2 For every set Δ of target types and every target type θ, if some set Γ is projected to Δ in the system $\mathcal{R}$ and every set of source types projected to Δ in $\mathcal{R}$ entails a source type σ that indicates θ in $\mathcal{R}$, then Δ entails θ.

Note that this condition by itself does not allow the simple inferential procedure we called "free ride". For, if we are to simply use a constraint matching of this type to conclude that Δ entails θ, we must verify that every set Γ of source types projected to Δ entails a source type σ that indicates θ in $\mathcal{R}$. This amounts to exploring *every* way of expressing Δ in the system and finding what will happen, while intuitively, a free ride consists in just expressing Δ in a *particular* way and observing what happens.

In actual cases of free rides, this gap is filled by the *homogeneity* of the relevant representation systems. Intuitively, a system is homogeneous if, for each set Δ of target types, there is at most one "line" of expressing Δ in the system. For example, when you want to express the information (15) and (16) in a line drawing, you must have three figures labeled "A", "B", and "C" satisfying the conditions (15*) and (16*), and there is no way of expressing (15) and (16) without having such figures.

(15) A is not taller than B.

(16) B is not taller than C.

(15*) A figure labeled "A" is not longer than a figure labeled "B".

(16*) A figure labeled "B" is not longer than a figure labeled "C".

Likewise, there is no way of expressing (12) and (13) in a Venn diagram without having three circles satisfying (12*) and (13*). This is because the systems of line drawings and Venn diagrams are both homogeneous systems. In fact, all representation systems that have been considered in this paper are homogeneous. A typical example of heterogeneous system is the full representation system used in *Hyperproof* [Barwise and Etchemendy, 1994], which contains, as its subsystems, a system of diagrams as well as a first-order language.

Let us define homogeneity and heterogeneity of a system more explicitly:

Definition 3 (Primitive indicator and homogeneous system)

- A set Δ of target types is *expressible* in a representation system $\mathcal{R}$ if there is a set Γ of source types projected to Δ in $\mathcal{R}$.

- A set Γ of source types is the *primitive indicator* of a set Δ of target types in a representation system $\mathcal{R}$ if Γ is projected to Δ in $\mathcal{R}$ and every set of source types projected to Δ in $\mathcal{R}$ entails Γ.

- A representation system $\mathcal{R}$ is *homogeneous* if every expressible set of target types has its primitive indicator; it is *heterogeneous* otherwise.

For example, the set $\{(15^*), (16^*)\}$ of source types is the primitive indicator of the set $\{(15), (16)\}$ of target types in the system of line drawings, and $\{(12^*), (13^*)\}$ is the primitive indicator of $\{(12), (13)\}$ in the system of Venn diagrams.

When a system is homogeneous, we can express any expressible set Δ of target types with its primitive indicator, and observing the result of expressing Δ in this way amounts to finding the *common* result of expressing Δ in every other way. Thus, we can exploit a constraint matching of type 2 without directly exploring every possible way of expressing Δ. If you express Δ with a primitive indicator Γ, and if you find new information θ expressed as the result, it means that Γ entails some source type σ indicating θ; but since Γ is entailed by every set of source types projected to Δ, it follows that σ is entailed by every set of source types projected to Δ; hence by a constraint matching of type 2, θ is guaranteed to be an entailment of Δ.

Thus, in its exact form, a free ride is the following inferential procedure:

Free ride Express the set Δ of information with its primitive indicator Γ. By observing that the operation results in the condition σ that indicates the information θ, conclude that θ is a consequence of Δ.

4.2 Three characters of physical on-site inference

So far, we have seen two different forms of inferences exploiting constraint matching between representations and their targets. These forms of inferences share several interesting properties: both use (1) perceptually accessible external representations (2) as inferential surrogates (3) through applications of physical operations on them. In view of these common properties, we call inferences in either of these forms "*physical on-site inferences*".

To make this notion more precise, let us clarify what each of these common properties are. We start with the property (2).

Inferential surrogate.   Suppose we are thinking about a particular object t. Let us say we are using another object s as an *inferential surrogate* when we exploit the matching of constraints on s and constraints on t to make an inference about t.

In this sense, external representations used in free rides, such as a Venn diagram (Example 6), a memory map (Example 5), and other graphical representations, are inferential surrogates. The target object t is a particular situation represented by the representation at hand, such as a particular situation with several sets in a certain inclusion relation (Examples 6) or a particular region with a certain geographical configuration (Example 5). The relevant constraint matching is specified as type 2, namely, the projection of constraints of the form $\Omega \vdash \beta$ to constraints of the same form. According to our analysis, a free ride is an inference that exploits this particular type of constraint matching between a representation s and the represented situation t, and hence it is an instance of inference using an inferential surrogate.

Likewise, a graphical consistency proof uses an inferential surrogate. The surrogate s is again a representation at hand, such as a room map (Example 1) or a route map (Example 3), and the target object t is a particular room with several pieces of furniture (Example 1) or a subway route with several subway lines (Example 3). The constraint matching exploited is specified as type 1, where a consistent set on s is projected to another consistent set on t through the indication relation.

In this regard, our analysis is a clarification of the semantic mechanism behind an important species of distributed cognition, namely, the case emphasized by Giere [Giere, 2001] when he says, "The visual representation is not merely an aid to human cognition; it is part of the system engaged in cognition" (p. 8). In physical on-site inferences, a cognitive burden of inference is partly transferred from human brains to physical constraints on external representations: we put the information to be processed into diagrams, charts, and others; the physical constraints on these representations "calculate" a consequence of the expressed information (free rides) or its consistency in a limited sense (graphical consistency proofs); we can then observe the results of calculation by attending to the new information expressed in representations or simply by checking whether the information to be expressed has been actually expressed. Later, we will see two more forms of physical on-site inferences that calculate non-consequence or inconsistency. Conceptually, inferential surrogates in our sense are special cases of "mediating structures" for distributed cognition [Hutchins, 1995].

Perceptual presence.   Inferential surrogates used in physical on-site inferences are objects such as Venn diagrams, Euler diagrams, route maps, memory maps, and room maps. They are all representations on a piece of paper, a computer display, or some other physical media, and they are accessible to our vision, and in certain case, to tactile perception too.

Compare these cases with analogical reasoning in general. As far as they use a particular object as a source for reasoning about another object, physical on-site inferences are a species of analogical reasoning. Yet the source object used in analogical reasoning does not have to be perceptually present, and indeed, cases typically cited as analogical reasoning involve source objects that are not perceptually present at the time of inference (a fictional event of a troop attacking a castle, planetary revolutions around the sun, and so on). In contrast, the notion of physical on-site inference emphasizes the fact that perceptually present representations, such as graphics on a piece of paper, often serve as source objects for inference. Although it is not in the scope of this book to study the exact internal processes involved in analogical inferences, it is clear that the internal process of using a perceptually present object as the source is significantly different from that of using a perceptually inaccessible object.

Physicality.   It should be clear by now that free rides and graphical consistency proofs have essential physical components. Expressing a certain information set Δ is a physical operation on a representation, and each form of inference draws a different type of conclusion from the result of that operation: a consequence conclusion when a piece of information is automatically expressed (free ride) and a consistency conclusion when Δ is successfully expressed (physical consistency proof). In each case, the physical operation and the accompanying result plays the role of verifying the existence or non-existence of a constraint on the representation, and under certain types of constraint matching, the existence or non-existence of a constraint on the representation guarantees the existence of the constraint or non-constraint to which it is projected. Thus, a physical operation plays a significant, or even dominant role in this inferential process, and it saves a significant amount of inferential task on the part of the user.

In this regard, our analysis should capture at least *some* instances of what Magnani [Magnani, 2001] calls "manipulative abductions". Magnani's focus is on production of explanatory hypotheses in scientific practices through physical manipulations of experimental devises. Yet the notion also covers manipulations of concrete models and diagrams (pp. 62–63), and our account could be considered specifications of the exact inferential procedures and semantic requirements involved in (some of) these cases. Although we have not considered any particular cases of manipulative abductions in scientific practices, how much of them are physical on-site inferences in our sense is an intriguing question.

4.3   Other forms of physical on-site inferences

Once the notion of physical on-site inference is thus clarified, it is obvious that there are at least two other forms of inferences that fit the definition. They are:

Physical non-consequence proof Express the set Δ of information in the representation system $\mathcal{R}$. By observing that the operation does not result in some condition σ that indicates the information θ, conclude that θ is not a $K_{\mathcal{R}}$-consequence of Δ.

Physical inconsistency proof Try to express the set Δ of information with its primitive indicator Γ. Conclude, from the impossibility of that operation, that Δ is inconsistent.

Remember that a free ride is the form of inference that utilizes a constraint matching of type 2 to conclude the existence of a constraint on represented situations from the existence of a constraint on representations. The first inferential procedure listed above is a flip side of this procedure, and it utilizes a constraint matching of type 2 to conclude the non-existence of a constraint on represented situations from the non-existence of a constraint on representations.

Here, the notion of $K_\mathcal{R}$-consequence is defined analogously as the notion of $K_\mathcal{R}$-inconsistency. A target type θ is an *induced consequence* of a set Δ of target types in a representation system $\mathcal{R}$ if some set Γ is projected to Δ in $\mathcal{R}$ and every set of source types projected to Δ in $\mathcal{R}$ entails at least one source type σ that indicates θ in $\mathcal{R}$; if every induced consequence is in fact a consequence of the relevant set of target types, an induced consequence in $\mathcal{R}$ is called $K_\mathcal{R}$-*consequence*. A constraint matching of type 2 is exactly this condition, and it is straightforward to prove that every physical non-consequence proof is valid in a system satisfying this condition.

As the name suggests, the second inferential procedure listed above is a flip side of graphical consistency proofs: while physical consistency proofs utilize a constraint matching of type 1 to conclude the non-existence of a constraint on represented situations from the non-existence of a constraint on representations, physical inconsistency proofs utilize a constraint matching of the same type to conclude the existence of a constraint on represented situations from the existence of a constraint on representations.

Although we do not have space to give actual examples of physical non-consequence proofs and physical inconsistency proofs, these forms of inferences are as common as free rides and graphical consistency proofs, and they are conductible on the basis of ordinary graphical representations such as Euler diagrams, route maps, line drawings, and geometry diagrams.

5.   Summary

In this paper, we investigated the semantic mechanism of graphical consistency proofs, where one constructs a chart, a diagram, or some other external representation that expresses certain conditions, and uses its existence as a proof of the consistency of the expressed conditions. We found that an auto-consistency property responsible for a system's capacity of such a proof can be characterized as a matching of constraints (specified as type 1) between representations and represented situations. We then extended our analysis to another type of graphics-based inferences called "free rides", and showed that they also rely on a matching of constraints (specified as type 2) between representations and represented situations. Comparisons of these procedures let us see three commonalities between them, and define the general notion of physical on-site inference as procedures using perceptually present objects as inferential surrogates through physical operations. Our analysis is therefore clarifications of the exact processes and semantic requirements under which a visual representation participate in distributed cognition [Giere, 2001] or manipulative inferences [Magnani, 2001].

References

Barwise, J. and Etchemendy, J., 1994, *Hyperproof*, CSLI Publications, Stanford.

Barwise, J. and Etchemendy, J., 1995, Heterogeneous logic, in: *Diagrammatic Reasoning: Cognitive and Computational Perspectives*, J.I. Glasgow, N.H. Narayanan and B. Chandrasekaran eds., MIT Press and AAAI Press, Cambridge and Menlo Park, pp. 211–234.

Barwise, J. and Seligman, J., 1997, *Information Flow: The Logic of Distributed Systems*. Cambridge University Press, Cambridge.

Gelernter, H., 1959, Realization of a geometry-theorem proving machine, in: *Computers and Thought*, E.A. Feigenbaum and J. Feldman eds., McGraw Hill, New York.

Giere, R.N., 2001, Scientific cognition as distributed cognition, manuscript to be published in: *Cognitive Bases of Science*, P. Carruthers, S. Stitch and M. Siegel, eds., Cambridge University Press, Cambridge.

Hutchins, E., 1995, *Cognition in the Wild*, MIT Press, Cambridge.

Lindsay, R.K., 1988, Images and inference, in: *Diagrammatic Reasoning: Cognitive and Computational Perspectives*, J.I. Glasgow, N.H. Narayanan and B. Chandrasekaran eds., MIT Press and AAAI Press, Cambridge and Menlo Park, pp. 111-135.

Magnani, L., 2001, *Abduction, Reason, and Science: Processes of Discovery and Explanation*, Kluwer Academic/Plenum Publishers, New York.

Shimojima, A., 1995a, Reasoning with diagrams and geometrical constraints, in: *Language, Logic and Computation*, D. Westeståhl and J. Seligman eds., CSLI Publications, Stanford.

Shimojima, A., 1995b, Operational constraints in diagrammatic reasoning, in: *Logical Reasoning with Diagrams*, J. Barwise and G. Allwein, eds., Oxford University Press, Oxford.

Sloman, A., 1971, Interactions between philosophy and AI: The role of intuition and non-logical reasoning in intelligence. *Artificial Intelligence* 2: 209–225.

Stenning, K. and Inder, R., 1995, Applying semantic concepts to analyzing media and modalities, in: *Diagrammatic Reasoning: Cognitive and Computational Perspectives*, J.I. Glasgow, N.H. Narayanan and B. Chandrasekaran eds., MIT Press and AAAI Press, Cambridge and Menlo Park, pp. 303-338.

ADAPTIVE LOGICS FOR NON-EXPLANATORY AND EXPLANATORY DIAGNOSTIC REASONING

Dagmar Provijn and Erik Weber

Centre for Logic and Philosophy of Science, Ghent University,
Blandijnberg 2, 9000 Ghent, Belgium

dagmar.provijn@rug.ac.be, erik.weber@rug.ac.be

Abstract   In this paper we discuss diagnosis of faults in systems. The latter are understood as structured wholes of components. Three types of diagnosis can be distinguished and are defined: non-explanatory, weak explanatory and strong explanatory. After the analysis of the reasoning process that leads to non-explanatory diagnosis, we argue that the predicative adaptive logic $\mathbf{D}^{nexp}$ is an adequate tool for modeling this kind of diagnostic reasoning. Subsequently, we follow the same pattern for weak and strong diagnosis and describe the logic $\mathbf{D}^{exp}$ which adequately formalizes weak diagnostic reasoning, even when *underlying theoretical knowledge* is taken into account. Finally it is argued that the same logic can be applied in the case of strong diagnostic reasoning whenever a number of conditions are fulfilled.

1.   Introduction

Diagnostic reasoning can relate to an established fault in a system or an established fault in an individual. A system is to be understood as a structured whole of components, while an individual is an object that is not characterized by an analysis into components. In this paper we confine ourselves to diagnosis of faults in systems. For diagnosis of faults in individuals (e.g. medical diagnosis), see [Weber and Provijn, in print].

Three types of diagnosis can be distinguished: non-explanatory, weak explanatory and strong explanatory. In section 2 we define these types, provide illustrations, and describe their respective functions. In section

L. Magnani, N.J. Nersessian, and C. Pizzi (eds.),
Logical and Computational Aspects of Model-Based Reasoning, 117–142.
© 2002 Kluwer Academic Publishers. Printed in the Netherlands.

3 we will describe some common characteristics of adaptive logics. In sections 4–6 we analyze the reasoning process by which non-explanatory diagnoses are constructed, and argue that the predicative adaptive logic $\mathbf{D}^{nexp}$ is an adequate tool for modeling this kind of diagnostic reasoning. In section 7 and 8 we discuss the formal analysis of weak explanatory diagnostic reasoning, respectively strong explanatory diagnostic reasoning. Sections 9–11 are dedicated to the description of the logic $\mathbf{D}^{exp}$ which adequately formalizes weak explanatory diagnostic reasoning, even if we take into account the use of *underlying theoretical knowledge*. We will also argue that the same logic leads to a strong explanatory diagnosis when certain conditions are fulfilled.

2.   Non-explanatory and explanatory diagnosis for faults in systems

As a system is understood as a structured whole of components, we think it is well defined by Reiter in [Reiter, 1987]:

Definition 1 *A system is a pair (*SD, COMP*) where:*

(a) SD, *the system description, is a set of first order sentences,*

(b) COMP, *the system components, is a finite set of constants.*

For the description of a system (this clearly departs from Reiter's formulation in terms of abnormality and its negation), we will require that it contains (i) a description of the input processing behavior of every component (i.e. of how inputs are transformed into outputs), and (ii) a description of the relations between the components (what clearly departs from Reiter's formulation in terms of abnormality and its negation).

As illustration, we introduce the following electric circuit, which contains three components (the gates P, Q, R):

A possible system description is:

> *Description of input processing behavior*
> P is an AND-gate, i.e. $output(P) = 1$ iff $input_1(P) = input_2(P) = 1$
> Q is an XOR-gate, i.e. $output(Q) = 1$ iff $input_1(Q) \neq input_2(Q)$
> R is an XOR-gate, i.e. $output(R) = 1$ iff $input_1(R) \neq input_2(R)$
>
>
> *Relation between components*
> $outputP = input_1(Q)$
> $outputP = input_2(R)$
> $input_1(R) = input_2(Q)$

which can be illustrated as follows:

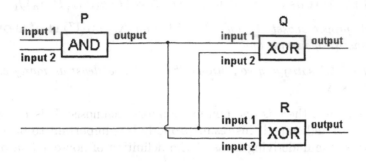

An established fault in a system is defined as follows:

Definition 2 SD ∪ OBS *is an established fault in a system (*SD, COMP*) if and only if*

(a) OBS *is a set of first-order sentences describing the observed states of the operating system*

(b) SD ∪ COMP *is inconsistent.*

As illustration, assume that in our example we observe that

$$output(Q) = 1, output(R) = 1$$

while the inputs are

$$input_1(P) = 1, input_2(P) = 1, input_1(R) = 1.$$

The conjunction of the three input values and two output values with the SD is inconsistent, because from SD and the observed inputs we can derive that

$$output(Q) = 0, output(R) = 0$$

so we have an established fault in our system.

A non-explanatory diagnosis for an established fault in a system can be defined as follows:

Definition 3 Ω *is a non-explanatory diagnosis for a fault F if and only if*

(a) *it has form* $\wedge \neg P_i \alpha_i$ *(where* α_i *is a system component and* P_i *a predicate describing the operations of a single component of the system),*

(b) SD *contains the set* $\Gamma = \{P_1 \alpha_1, \ldots, P_n \alpha_n\}$, *(where* $P_1 \alpha_1, \ldots, P_n \alpha_n$ *are the formulas whose negations are the conjuncts of* $\wedge \neg P_i \alpha_i$*),*

(c) $(\text{SD}\backslash\Gamma)\cup\Delta\cup\text{OBS}$ *is consistent (where* $\Delta = \{P_1\alpha_1,\ldots,P_n\alpha_n\}$*),*

(d) *for all proper subsets* Γ' *and* Δ' *of* Γ *and* Δ*,* $(\text{SD}\backslash\Gamma')\cup\Delta'\cup\text{OBS}$ *remains inconsistent, and*

(e) *every set* Δ'' *satisfying conditions (b)-(d) has at least as many elements as* Δ*.*

In our example there is one non-explanatory diagnosis: P is not an AND-gate. To understand this result properly it is important to have a clear view on the underlying ideas of the definition of non-explanatory diagnosis.

First of all, we have to point at the different epistemological status of the two parts of the system description (input processing behavior and relations between components): the claims about the relations between components are not doubted, even if a fault is established. The claims about input processing behavior can be given up if a fault is established; they ought to be interpreted as hypotheses which are believed to be true, but nonetheless are falsifiable through experimental observation. The observations OBS have the same epistemological status as the claims about the relations between components.

Secondly the aim of diagnosis of this type is to restore consistency: the empirical data OBS are not explained by it. Therefore we call this type of diagnosis non-explanatory.

Finally the aim of non-explanatory diagnosis is to restore consistency in a parsimonious way; which is expressed in condition (d) and (e).

Weak and strong explanatory diagnoses are defined as follows:

Definition 4 *If* $\wedge\neg P_i\alpha_i$ *is a non-explanatory diagnosis for a fault in a system, then* $\wedge Q_i\alpha_i$ *(where* Q_i *is also a predicate describing the input processing behavior) is a weak explanatory diagnosis for the same fault if and only if*

(a) $\wedge Q_i\alpha_i$ *entails* $\wedge\neg P_i\alpha_i$

(b) $(\text{SD}\backslash\Gamma)\cup E$ *entails* OBS *(where* $\Gamma = \{P_i\alpha_i\}$ *and where* $E = \{Q_1\alpha_1,\ldots, Q_i\alpha_i\}$*).*

This type of diagnosis is called explanatory because it explains the empirical data OBS. The definition entails that multiple weak explanatory diagnosis can coexist. Whenever new observations are made, it may be possible to point out a unique diagnosis. In this case we speak of a strong explanatory diagnosis. The latter is defined as follows:

Definition 5 *If* $\wedge\neg P_i\alpha_i$ *is a non-explanatory diagnosis for a fault in a system, then* $\wedge Q_i\alpha_i$ *(where* Q_i *is also a predicate describing the input* *processing behavior) is a strong explanatory diagnosis for the same fault if and only if*

(a) $\wedge Q_i \alpha_i$ *entails* $\wedge \neg P_i \alpha_i$

(b) $(\text{SD}\backslash\Gamma)\cup E$ *entails* OBS *(where* $\Gamma = \{P_i \alpha_i\}$ *and where* $E = \{Q_1 \alpha_1, \ldots, Q_i \alpha_i\}$*).*

(c) *for all sets* $F = \{R_1 \alpha_1, \ldots, R_i \alpha_i\}$ *(with* R_i *a predicate describing the input processing behavior) different from* E, $(\text{SD}\backslash\Gamma) \cup F$ *is incompatible with* OBS.

We now clarify definitions 4 and 5 by means of the circuit. As the gates have two inputs and one output, sixteen types of gates can be distinguished according to their particular input processing behavior. They are set out in the following table:

T_1: TAUT	T_2: OR	T_3	T_4: IMPL
$1 - 1 \rightarrow 1$	$1 - 1 \rightarrow 1$	$1 - 1 \rightarrow 1$	$1 - 1 \rightarrow 1$
$1 - 0 \rightarrow 1$	$1 - 0 \rightarrow 1$	$1 - 0 \rightarrow 1$	$1 - 0 \rightarrow 0$
$0 - 1 \rightarrow 1$	$0 - 1 \rightarrow 1$	$0 - 1 \rightarrow 0$	$0 - 1 \rightarrow 1$
$0 - 0 \rightarrow 1$	$0 - 0 \rightarrow 0$	$0 - 0 \rightarrow 1$	$0 - 0 \rightarrow 1$

T_5: NOT-AND	T_6: LEFT	T_7: RIGHT	T_8: EQ
$1 - 1 \rightarrow 0$	$1 - 1 \rightarrow 1$	$1 - 1 \rightarrow 1$	$1 - 1 \rightarrow 1$
$1 - 0 \rightarrow 1$	$1 - 0 \rightarrow 1$	$1 - 0 \rightarrow 0$	$1 - 0 \rightarrow 0$
$0 - 1 \rightarrow 1$	$0 - 1 \rightarrow 0$	$0 - 1 \rightarrow 1$	$0 - 1 \rightarrow 0$
$0 - 0 \rightarrow 1$	$0 - 0 \rightarrow 0$	$0 - 0 \rightarrow 0$	$0 - 0 \rightarrow 1$

T_9: XOR	T_{10}:NOT-RIGHT	T_{11}: NOT-LEFT	T_{12}: AND
$1 - 1 \rightarrow 0$	$1 - 1 \rightarrow 0$	$1 - 1 \rightarrow 0$	$1 - 1 \rightarrow 1$
$1 - 0 \rightarrow 1$	$1 - 0 \rightarrow 1$	$1 - 0 \rightarrow 0$	$1 - 0 \rightarrow 0$
$0 - 1 \rightarrow 1$	$0 - 1 \rightarrow 0$	$0 - 1 \rightarrow 1$	$0 - 1 \rightarrow 0$
$0 - 0 \rightarrow 0$	$0 - 0 \rightarrow 1$	$0 - 0 \rightarrow 1$	$0 - 0 \rightarrow 0$

T_{13}: NOT-IMPL	T_{14}	T_{15}: NEITHER	T_{16}: CONTR
$1 - 1 \rightarrow 0$	$1 - 1 \rightarrow 0$	$1 - 1 \rightarrow 0$	$1 - 1 \rightarrow 0$
$1 - 0 \rightarrow 1$	$1 - 0 \rightarrow 0$	$1 - 0 \rightarrow 0$	$1 - 0 \rightarrow 0$
$0 - 1 \rightarrow 0$	$0 - 1 \rightarrow 1$	$0 - 1 \rightarrow 0$	$0 - 1 \rightarrow 0$
$0 - 0 \rightarrow 0$	$0 - 0 \rightarrow 0$	$0 - 0 \rightarrow 1$	$0 - 0 \rightarrow 0$

Given the system description and the observations, the illustrating circuit gives rise to eight weak explanatory diagnoses: P is a NOT-AND-gate, P is an XOR-gate, P is a NOT-RIGHT-gate, P is a NOT-LEFT-gate, P is a NOT-IMPL-gate, P is a T_{14}-gate, P is a NEITHER-gate, P is a CONTR-gate. When combined with $(\text{SD}\backslash\Gamma)$ each of these will generate a new system description that is consistent with OBS.

To obtain a strong explanatory diagnosis questions are needed whose answers eliminate some of the alternative weak explanatory diagnoses.

Given our example, we can ask what happens if we run the experiment in which the inputs are changed into:

$$input_1(P) = 0, input_2(P) = 0, input_1(R) = 0.$$

If we assume that the outputs are the same as those we observed with the original inputs, four of the eight weak explanatory diagnosis are falsified: P is an XOR-gate, P is a NOT-IMPL-gate, P is a T_{14}-gate and P is a CONTR-gate. The remaining weak explanatory diagnoses are the hypotheses that need to be tested further. By means of new measurements and calculations a further elimination is possible.

It is important to pin-point the characteristic functions of each type of diagnosis. The original system description is meant to elucidate the way the normal system is. Each type of diagnosis will in its own way deal with violations of this system description.

Non-explanatory diagnosis can be applied to repair a fault in a system: the component that is diagnosed to deviate from its expected behavior, can be replaced. The aim of the replacement is to make the system work as set out by the original system description. This will be considered the primary function of non-explanatory diagnosis.

Weak and strong explanatory diagnosis give rise to different options for repairing the system. The fact that these types of diagnosis stand for a well-supported hypothesis about what is exactly wrong with the deviating component, enables one to compensate this by changing some or all of the other components or by changing their inter-relations. The results of these changes can reliably be predicted in the case of strong explanatory diagnoses. In the case of weak explanatory diagnosis this is often impossible because the different hypotheses will lead to different outcomes when an identical change is made outside the component.

Non-explanatory diagnosis and weak explanatory diagnosis have a secondary function as the former play a role in the formulation of weak explanatory diagnoses and as the latter are a useful step in finding a strong explanatory diagnosis.

3.   Adaptive logics

An adaptive logic **AL** is characterized by the oscillation between a *lower limit logic* **LLL** and an *upper limit logic* **ULL**, a process by which it adapts itself to a given set of premises. The *upper limit logic* **ULL** captures what we believe to be normal by means of a number of presuppositions, which are decided on by the application context or the kind of reasoning one wants to capture. Consequently, abnormalities are to be considered as logical falsehoods within the *upper limit logic*, which means that they will lead to triviality whenever **ULL** is applied. The *lower limit logic* **LLL** drops the presuppositions through which the abnormalities are defined. Semantically this means that the adaptive models of some Γ are a subset of the **LLL**-models of Γ, containing the models that are 'as normal as possible' in view of the **ULL**. Thus, whenever Γ contains no abnormalities, its adaptive models will be equal to its **ULL**-models. The ambiguity that is linked to the requirement 'as normal as possible', is removed from the moment on that an adaptive strategy is chosen. The latter will be discussed in due course.

One of the main characteristics of adaptive logics is their dynamic proof theory. The basic idea behind the latter is that the rules of the **LLL** apply unconditionally, whereas those of the **ULL** only apply conditionally. In a real proof this is realized through the addition of a condition at the end of each line that is derived. This condition usually refers to the formulas on whose normal behavior we rely to derive the formula on this line. When this condition is violated at a certain stage in the continuation of the proof, the line will be marked -i.e. the formula on this line is no longer derivable. This causes the dynamics of the proofs.

The limitation of this dynamics is incorporated through the notion of final derivability. A formula A is *finally* **AL** derived at line i at a stage s of a proof from Γ iff line i is not marked at stage s, and any extension of the proof in which line i is marked may be extended in such a way that line i will be unmarked. A is finally **AL** derivable from Γ by the adaptive logic iff A is finally **AL** derived at a stage of a proof from Γ.

Before we present our adaptive logic for non-explanatory diagnostic reasoning, some preliminary remarks may be useful:

First of all the epistemological difference between the two parts of SD is incorporated through the distinction between *premises that are not doubted* and *expected premises (expectancies)*.

Secondly it is assumed that the former (i.e. observations and claims about relations between components) are consistent. Inconsistencies only arise when they are used in combination with *expectancies*.

Finally the adaptive logic that is presented here is ampliative: the adaptive consequences comprise all of the **CL**-consequences of the *premises that are not doubted* and moreover comprise the **CL**-consequences of the *expectancies* in as far as the latter are compatible with the *premises that are not doubted*. From now on the set of *premises that are not* *doubted* will be denoted by Γ_0, whereas the set of *expectancies* will be denoted by Γ_1.

4. An adaptive logic for non-explanatory diagnostic reasoning

As mentioned before, we aim to add as many **CL**-consequences of a set Γ_1 to the **CL**-consequences of the consistent set Γ_0 under the condition that it will not lead to triviality. The latter occurs only when some member of Γ_1 is incompatible with some member of Γ_0.

The epistemological difference between the two sets will be incorporated through the application of the modal logic **S5** as the *lower limit logic* **LLL**, whereas **Triv** will be taken as the *upper limit logic* **ULL** since normality is defined as $\Gamma_0 \cup \Gamma_1$ being consistent. The adaptive logic $\mathbf{D}^{nexp}$ will add all **CL**-consequences of Γ_1 to those of Γ_0 as far as $\Gamma_0 \cup \Gamma_1$ determines no abnormalities.

We will apply a rather unfamiliar version of the **S5**-semantics which was introduced in [Batens and Meheus, 2000], and which was applied in [Meheus, to appear] similar to the way we will work it out here. A brief description may be useful.

Let $\mathcal{L}$ be the standard language of **CL**, and $\mathcal{F}$, $\mathcal{F}^p$, and $\mathcal{W}$ the sets of formulas, primitive formulas and wffs (closed formulas) of $\mathcal{L}$. A standard **CL**-model is represented by a domain D and an assignment function v and is symbolized as $M = \langle D, v \rangle$. To simplify the semantic metalanguage (- the clauses for the quantifiers), a non-denumerable set of pseudo-constants $\mathcal{O}$ is introduced, requiring that any element of the domain D is named by at least one member of $\mathcal{C} \cup \mathcal{O}$:

$$v : \mathcal{C} \cup \mathcal{O} \longrightarrow D \text{ (where } D = \{v(\alpha) | \alpha \in \mathcal{C} \cup \mathcal{O}\})$$

Through this operation we obtain the pseudo-language $\mathcal{L}^+$. The standard modal language $\mathcal{L}^M$ is extended to $\mathcal{L}^{M+}$ in the same way as we did for the standard language of **CL**. $\mathcal{W}^M$ refers to the set of wffs of $\mathcal{L}^M$. A **S5**-model is a couple $M^{S5} = \langle \Sigma_\Delta, M_0 \rangle$, where Δ is a set of wffs of $\mathcal{L}$, Σ_Δ is the set of **CL**-models of Δ, and $M_0 \in \Sigma_\Delta$. The valuation function determined by an **S5**-model M^{S5} is defined by the following clauses:

C1   where $A \in \mathcal{F}^p$, $v_{M^{S5}}(A, M_i) = v_{M_i}(A)$

C2 $v_{M^{S5}}(\neg A, M_i) = 1$ iff $v_{M^{S5}}(A, M_i) = 0$

C3 $v_{M^{S5}}(A \vee B, M_i) = 1$ iff $v_{M^{S5}}(A, M_i) = 1$ or $v_{M^{S5}}(B, M_i) = 1$

C4 $v_{M^{S5}}((\exists \alpha)A(\alpha), M_i) = 1$ iff $v_{M^{S5}}(A(\beta), M_i) = 1$ for at least one $\beta \in \mathcal{C} \cup \mathcal{O}$

C5 $v_{M^{S5}}(\Diamond A, M_i) = 1$ iff $v_{M^{S5}}(A, M_j) = 1$ for at least one $M_j \in \Sigma_\Delta$.

The other logical constants are defined as usual. A model M^{S5} verifies $A \in \mathcal{W}^M$ iff $v_{M^{S5}}(A, M_0) = 1$. A is valid iff it is verified by all models. A model M^{S5} is an **S5**-model of $\Sigma = \langle \Gamma_0, \Gamma_1 \rangle$ iff, for all $A \in \mathcal{L}$, M^{S5} verifies A if $A \in \Gamma_0$, and M^{S5} verifies $\Diamond A$ if $A \in \Gamma_1$. $\Sigma \models_{\mathbf{S5}} A$ denotes that all **S5**-models of Σ verify A.

The *upper limit logic* **Triv** can be defined easily by an adjustment of the **S5**-models. **Triv**-models are **S5**-models such that, for some maximally consistent subset $\Theta \subset \mathcal{W}$, $\Sigma_\Delta = \Sigma_\Theta$. In order to define our adaptive logic $\mathbf{D}^{nexp}$, we need to characterize what the possible abnormalities of $\Gamma_0 \cup \Gamma_1$ are. Although this seems obvious, as it is abnormal that an expectancy turns out false, three small complications need to be taken into account to define the form of the abnormalities. First of all it is possible that no abnormalities are derivable from the **LLL**, whereas disjunctions of abnormalities are. Secondly, **S5** spreads abnormalities. If $\Diamond p \wedge \neg p$ is verified by a model M^{S5}, $\Diamond(p \wedge q) \wedge \neg(p \wedge q)$, etc. are also true in this model. Finally we need to allow models that verify $\Diamond Po \wedge \neg Po$ for some $o \in M^{S5}$, and hence verify $(\exists x)(\Diamond Px \wedge \neg Px)$, to formulate the predicative logic in a decent way, notwithstanding the fact that diagnostic reasoning will most probably concern finite systems.

Now we are able to pin-point the form of abnormalities for $\mathbf{D}^{nexp}$. Let $\mathcal{F}^a$ be the set of atoms (primitive open and closed formulas and their negations) and let $\exists A$ abbreviate A preceded by a sequence of existential quantifiers (in some specified order) over all variables free in A. An abnormality is a formula of the form $\exists(\Diamond A \wedge \neg A)$, where $A \in \mathcal{F}^a$. $Dab(\Delta)$ refers to the disjunction $\bigvee\{\exists(\Diamond A \wedge \neg A)|A \in \Delta\}$ provided that $\Delta \in \mathcal{F}^a$. $Dab(\Delta)$ is a Dab-consequence of Σ iff all **S5**-models of Σ verify $Dab(\Delta)$. The latter is a minimal Dab-consequence iff there is no $\Delta' \subset \Delta$ such that $Dab(\Delta')$ is a Dab-consequence of Σ.

The abnormal part of a model will be defined as follows:

Definition 6 $Ab(M) =_{df} \{A \in \mathcal{F}^a | M \text{ verifies } \exists(\Diamond A \wedge \neg A)\}$.

In this part we will present the semantics of $\mathbf{D}^{nexp}$. The addition of "as many as possible" **CL**-consequences of Γ_1 can be governed by the use of different strategies, which are characterized by a specific degree of safety while adding **CL**-consequences to Γ_0.

The most cautious strategy, "Reliability" will be described in section 9. The set of unreliable formulas is defined as the set of factors of the minimal Dab-consequences of Σ:

Definition 7 $U(\Sigma) =_{df} \bigcup\{\Delta | Dab(\Delta) \text{ is a minimal } Dab\text{-consequence of } \Sigma\}$.

The "Minimal Abnormality" strategy has a stronger consequence relation as it is characterized by a stronger selection of models. It set theoretically minimizes the number of abnormalities. The selection of "Minimal abnormal" models is also definable in terms of the minimal Dab-consequences of Σ. It is useful to spell this out, as it is incorporated in the selection process of our $\mathbf{D}^{nexp}$-models. Let $f(A)$ be the result obtained by relettering the free variables in A in such a way that they occur in some standard order (the first occurring free variable is always x, the second always y, etc.), let $A \prec B$ denote that $\exists B$ follows by (non-zero applications of) existential generalization from $\exists A$, and let $g(\Delta) = \{f(A) \mid A \in \Delta;$ for no $B \in \Delta$, $f(B) \prec f(A)\}$. Let Φ_Σ^o be the set of all sets $g(\Delta)$ such that Δ contains, for each minimal Dab-consequence $Dab(\Theta)$ of Σ, at least some $A \in \Theta$, and let Φ_Σ be obtained by eliminating from Φ_Σ^o those members that are proper supersets of other members. It can easily be shown that $M^{S5} \in \sigma^1(\mathcal{M}_\Sigma)$ iff $M^{S5} \in \sigma^0(\mathcal{M}_\Sigma)$ and $Ab(M^{S5}) \in \Phi_\Sigma$. As Σ exists exclusively of Γ_0 and Γ_1, only two selections need to be made, and as Γ_0 is consistent no real selection occurs at this level. This means that the selection entirely concerns Γ_1. As a result the "Minimal abnormal" adaptive models of Σ can be defined as the result of the following two successive selections (with $\mathcal{M}_\Sigma$ the set of **S5**-models of Σ):

$$\sigma^0(\mathcal{M}_\Sigma) =_{df} \mathcal{M}_\Sigma$$

and,

$$\sigma^1(\mathcal{M}_\Sigma) =_{df} \{M^{S5} \in \sigma^0(\mathcal{M}_\Sigma) \mid Ab(M^{S5}) \in \Phi_\Sigma\}$$

A more elaborate description of the "Reliability" and "Minimal Abnormality" strategy can be found in [Batens, 1998].

The "Minimal Abnormality" strategy provides us with a logical tool to handle the parsimonious element incorporated in criterion (d) of definition three.

As each member of Φ_Σ provides us with a hypothesis on what went wrong, (provided we take it for granted that we have reasons to - set theoretically - minimize the number of things that went wrong), all we have to do is select the members of Φ_Σ that are *numerically* smaller than other members. By this means we obtain $\mathbf{D}^{nexp}$, our adaptive logic for non-explanatory diagnosis, as it incorporates principle (e) of definition three. Several members of Φ_Σ may be numerically minimal, and hence all of them have to be treated on a par. Let $\Phi_\Sigma^{\#}$ denote the members of Φ_Σ that have not a larger cardinality than any other members of Φ_Σ. We then select the $\mathbf{D}^{nexp}$-models of Σ as follows:

$$\sigma^0(\mathcal{M}_\Sigma) =_{df} \mathcal{M}_\Sigma$$

and,

$$\sigma^1(\mathcal{M}_\Sigma) =_{df} \{M^{S5} \in \sigma^0(\mathcal{M}_\Sigma) \mid Ab(M^{S5}) \in \Phi_\Sigma^\#\}$$

The $\mathbf{D}^{nexp}$-models of Σ are the members of $\sigma^1(\mathcal{M}_\Sigma)$ and, where $A \in W$, $\Sigma \models_{\mathbf{D}^{nexp}} A$ iff A is verified by all $\mathbf{D}^{nexp}$-models of Σ.

5. The dynamic proof theory of $\mathbf{D}^{nexp}$

As is usual for adaptive logics, the dynamic proof theories for $\mathbf{D}^{nexp}$ are based on a specific relation between the derivability relation of the *upper limit logic* **Triv**, and that of the *lower limit logic* **S5**. This relation is expressed in the following theorem which is proved for the prioritized case in [Batens et al., to appear].

Theorem 1 *If* $A_1, \ldots, A_n \vdash_{\mathbf{Triv}} B$, *then there are* $C_1, \ldots, C_m \in \mathcal{F}^a$ *such that* $A_1, \ldots, A_n \vdash_{\mathbf{S5}} B \vee Dab\{C_1, \ldots, C_m\}$ *(Derivability Adjustment Theorem).*

This theorem warrants that whenever B is **Triv** derivable from $A_1, \ldots, A_n$, there are $C_1, \ldots, C_n$ such that B is **S5**-derivable from $A_1, \ldots, A_n$ or one of the C_i behaves abnormally with respect to $A_1, \ldots, A_n$. This suggests that we derive B from $A_1, \ldots, A_n$, on the condition that none of $C_1, \ldots, C_n$ behaves abnormally.

As is usual for adaptive logics, lines of a proof have five elements: (i) a line number, (ii) the formula A that is derived, (iii) the line numbers of the formulas from which A is derived, (iv) the rule by which A is derived, and (v) a "condition". The condition specifies which formulas have to behave normally in order for A to be so derivable.

We now list the generic rules that govern dynamic proofs from $\Sigma = \langle \Gamma_0, \Gamma_1 \rangle$.

PREM If $A \in \Gamma_i$ (with $i \in \{0, 1\}$, then one may add a line consisting of
     (i) the appropriate line number,
     (ii) $\Diamond^i A$,
     (iii) "$-$",
     (iv) "PREM", and
     (v) $\emptyset$.

RU   If $B_1, \ldots, B_m \vdash_{\mathbf{S5}} A$ and $B_1, \ldots, B_m$ occur in the proof with the conditions $\Delta_1, \ldots \Delta_m$ respectively, then one may add a line consisting of
     (i) the appropriate line number,
     (ii) A,
     (iii) the line numbers of the B_i,

      (iv) "RU", and

      (v) $(\Delta_1 \cup \ldots \cup \Delta_m)$.

RC   If $B_1, \ldots, B_m \vdash_{\mathbf{S5}} A \vee Dab(\Theta)$ and $B_1, \ldots, B_m$ occur in the proof
      with the conditions $\Delta_1, \ldots \Delta_m$ respectively, then one may add
      a line consisting of

      (i) the appropriate line number,

      (ii) A,

      (iii) the line numbers of the B_i,

      (iv) "RC", and

      (v) $(\Theta \cup \Delta_1 \cup \ldots \cup \Delta_m)$.

It is obvious in view of the rules that A is derivable on the condition
Δ in a proof from Σ iff $A \vee Dab(\Delta)$ is **S5**-derivable from Σ. Before
spelling out the marking definition we want to emphasize once more
the distinction between *derivability at a stage* and *final derivability*.
$A \in \mathcal{W}^M$ is *derived at a stage* of a proof from Σ iff A is the second
element of a non-marked line in a proof from Σ. $A \in \mathcal{W}^M$ is *finally
derived* at line i of a proof at a stage from Σ iff A is the second element
of line i, line i is non-marked at the stage, and, whenever line i is marked
in an extension of the proof, there is a further extension in which line
i is not marked.[1] $A \in \mathcal{W}^M$ is *finally derivable* from Σ iff it is finally
derived at some line in a proof at a stage from Σ.

Definition 8 *Where $A \in \mathcal{W}$, $\Sigma \vdash_{\mathbf{D}^{nexp}} A$ iff A is finally derivable
from Σ.*

To obtain the marking definition for $\mathbf{D}^{nexp}$ we need to define which
formulas are reliable with respect to a proof at a stage (which is a concrete matter) rather than with respect to the set of premises (which is an
abstract matter). For that we define $\Phi_s^{\#}(\Sigma)$ from the $\Phi_s(\Sigma)$ that occur
in the proof in the same way as $\Phi_\Sigma^{\#}(\Sigma)$ is defined from the $\Phi_\Sigma(\Sigma)$. This
means that we select the numerically smallest hypotheses from the ones
that tell us what went wrong at level i at a stage of a proof, provided
we already minimized the things that possibly went wrong.

Definition 9 *Marking for $\mathbf{D}^{nexp}$: Line i is marked at stage s iff, where
Δ is its fifth element, $\Delta \cap \Phi_s^{\#}(\Sigma) \neq \emptyset$.*

---

[1] Infinite extensions need to be taken into consideration in accordance with the "Minimal
Abnormality" strategy – see [2].

6.   An illustration of $\mathbf{D}^{nexp}$

To show that $\mathbf{D}^{nexp}$ is adequate to describe the reasoning process that leads to a non-explanatory diagnosis, we will apply it on the circuit that was set out in section 2. For the description of the electric circuit we will use a predicative language in which P_1 is a predicate denoting that input one of gate P has value one. P_0 denotes that the output of gate P has value one and P_2 means the same for input two. The same can be done for gates Q and R, for which we apply the predicates $Q_0, Q_1, Q_2, R_0, R_1, R_2$. By using a predicative logic, we are able to introduce a time factor within the proofs. $P_1 t_1$ then means that the first input of gate P has value one at moment t_1. Whenever a line is marked, its fifth element will be followed by the sign $\sqrt{}$. Applying this formal language, we obtain the following formal representation of the system description SD and the observations OBS:

$$\Gamma_0 = \{R_1 t_1, P_1 t_1, P_2 t_1, R_0 t_1, Q_0 t_1, (\forall x)(P_0 x \equiv R_2 x),$$
$$(\forall x)(P_0 x \equiv Q_1 x), (\forall x)(R_1 x \equiv Q_2 x)\}$$

$$\Gamma_1 = \{(\forall x)(P_0 x \equiv (P_1 x \wedge P_2 x)), (\forall x)(Q_0 x \equiv \neg(Q_1 x \equiv Q_2 x)),$$
$$(\forall x)(R_0 x \equiv \neg(R_1 x \equiv R_2 x))\}$$

Given the information from the system description that is not doubted and the observations in Γ_0 and given the expectancies that P is an AND-gate, Q an XOR-gate and R also an XOR-gate in Γ_1; $\mathbf{D}^{nexp}$ allows to derive the non-explanatory diagnosis "P is not an AND-gate".

1	$R_1 t_1$	-	PREM $\emptyset$
2	$P_1 t_1$	-	PREM $\emptyset$
3	$P_2 t_1$	-	PREM $\emptyset$
4	$R_0 t_1$	-	PREM $\emptyset$
5	$Q_0 t_1$	-	PREM $\emptyset$
6	$(\forall x)(P_0 x \equiv R_2 x)$	-	PREM $\emptyset$
7	$(\forall x)(P_0 x \equiv Q_1 x)$	-	PREM $\emptyset$
8	$(\forall x)(R_1 x \equiv Q_2 x)$	-	PREM $\emptyset$
9	$\Diamond(\forall x)[P_0 x \equiv (P_1 x \wedge P_2 x)]$	-	PREM $\emptyset$
10	$\Diamond(\forall x)[Q_0 x \equiv \neg(Q_1 x \equiv Q_2 x)]$	-	PREM $\emptyset$
11	$\Diamond(\forall x)[R_0 x \equiv \neg(R_1 x \equiv R_2 x)]$	-	PREM $\emptyset$
12	$P_0 t_1 \equiv (P_1 t_1 \wedge P_2 t_1)$	2,3,9	RC

$$\{P_0 x, \neg P_0 x, P_1 x, \neg P_1 x, P_2 x, \neg P_2 x\} \sqrt{}$$

13 $Q_0t_1 \equiv \neg(Q_1t_1 \equiv Q_2t_1)$ 1,5,8,10 RC

$\{Q_0x, \neg Q_0x, Q_1x, \neg Q_1x, Q_2x, \neg Q_2x\}$

14 $R_0t_1 \equiv \neg(R_1t_1 \equiv R_2t_1)$ 1,4,11 RC

$\{R_0x, \neg R_0x, R_1x, \neg R_1x, R_2x, \neg R_2x\}$

15 P_0t_1 2,3,12 RU

$\{P_0x, \neg P_0x, P_1x, \neg P_1x, P_2x, \neg P_2x\}\checkmark$

16 $\neg Q_1t_1$ 1,5,8,13 RU

$\{Q_0x, \neg Q_0x, Q_1x, \neg Q_1x, Q_2x, \neg Q_2x\}$

17 $\neg R_2t_1$ 1,4,14 RU

$\{R_0x, \neg R_0x, R_1x, \neg R_1x, R_2x, \neg R_2x\}$

18 $\exists(\Diamond\neg Q_0x \wedge Q_0x) \vee \exists(\Diamond\neg Q_1x \wedge Q_1x)\vee$

$\exists(\Diamond\neg Q_2x \wedge Q_2x) \vee \exists(\Diamond P_0x \wedge \neg P_0x)\vee$ 1,2,3,5

$\exists(\Diamond\neg P_1x \wedge P_1x) \vee \exists(\Diamond\neg P_2x \wedge P_2x)$ 7,8,9,10 RU $\emptyset$

19 $\exists(\Diamond\neg R_0x \wedge R_0x) \vee \exists(\Diamond\neg R_1x \wedge R_1x)\vee$

$\exists(\Diamond\neg R_2x \wedge R_2x) \vee \exists(\Diamond P_0x \wedge \neg P_0x)\vee$ 1,2,3,4

$\exists(\Diamond\neg P_1x \wedge P_1x) \vee \exists(\Diamond\neg P_2x \wedge P_2x)$ 6,9,11 RU $\emptyset$

20 $(\exists x)\neg(P_0x \equiv (P_1x \wedge P_2x))$ 2,3,6,17 RU

$\{R_0x, \neg R_0x, R_1x, \neg R_1x, R_2x, \neg R_2x\}$

As the two *Dab*-formulas on lines 18 and 19 are unconditionally derived, we know by the "Minimal Abnormality" strategy that either the expectancy "P is an AND-gate" is wrong or that the expectancies "Q is an XOR-gate" and "R is an XOR-gate" are wrong. Of course, as the former is numerically smaller than the latter, we opt to mark the first expectancy (and what is derived from it).

7. Formal analysis of weak explanatory diagnostic reasoning

In this section we discuss the reasoning process that leads to weak explanatory diagnoses. The specifications that were given in section 2 will be extended with the notion of *selective abduction*. In section 8 the strong explanatory diagnostic reasoning process is discussed and we will argue that it is easily obtained whenever a sufficient number of variable observations is available. The adaptive logic $\mathbf{D}^{exp}$ that is defined in sections 9 and 10 is able to formalize both, what will be illustrated in section 11.

If a weak explanatory diagnosis is sought for the determination of a strong explanatory diagnosis, the reasoning process fits the scheme of *abductive hypothesis formation*:

Definition 10

(1) *We observe that Q and want an explanation for this phenomenon.*

(2) *We know that if P would be true, this would (together with the background knowledge R) explain Q.*

(3) *We know that R is true.*

(4) *Because of (1)-(3) we decide to regard P as a hypothesis which deserves further investigation.*

If a weak explanatory diagnosis is sought for its own sake, the abductive reasoning process approximately fits the same scheme. Only the fourth condition is changed in the following way:

Definition 11

...

(4) *Because of (1)-(3), we decide to accept P as true.*

This kind of reasoning process will be called *abductive argumentation*.

Weak explanatory diagnostic reasoning becomes more interesting if we combine it with the notion *selective abduction* as presented in [Magnani, 2001, pp. 72–77]. Although Magnani concentrates on medical diagnosis, this notion can also be applied in the case of diagnosis for faults in systems. The fundamental difference between medical diagnosis (a type of diagnosis for faults in individuals) and the diagnosis for faults in systems is the reference to "background knowledge" (i.e. a non-explanatory diagnosis which is only required whenever a system is divided into components) in the latter case. *Selective abduction* resembles the notion of *abductive hypothesis formation*, but introduces the idea to rank the different hypotheses so as to plan their evaluation. It concerns the search for a set of hypotheses, starting from problem features identified by abstraction [Magnani, 2001, p. 73]. In the case of diagnosis for a fault in a system, the latter information may be understood as the contraction of the original system description by means of the new information we obtained through the non-explanatory diagnosis combined with the *underlying theoretical knowledge*, which can be understood as the expert-knowledge concerning the different parts of the system description. In this way, it it possible to introduce a restriction on the number of conclusions reached by abductive argumentation, or introduce a ranking in the hypotheses considered for strong explanatory diagnosis. according to the degree of expert-knowledge that is available. The ranking of the hypotheses that come forth by *abductive hypothesis formation* can be obtained whenever the *underlying theoretical knowledge* allows for the attribution of different probabilities to the hypotheses that are obtained by the *abductive hypothesis formation*.

Whenever the technical characteristics of the AND-gate are known, some of the weak explanatory diagnoses will be more probable whenever the gate is being diagnosed because of their technical properties. Suppose the AND-gate of our example is of the following type:

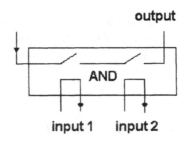

As this AND-gate is constructed by two valves that may interrupt the current in the gate, and as each of them is regulated by one of the inputs, the output of the AND-gate can only have value one when both inputs have value one such that the valves are closed. When the first valve is blocked such that it is always closed, the output of the gate will no longer depend on the value of the first input, such that the output will be one whenever the second input has value one. The five other possible defects can be: the second valve is always closed, both valves are always closed, both valves are always opened and the first or the second valve is always opened (with the latter three giving rise to the same hypothesis). This results in the selection of the following hypothesis: a RIGHT-gate, a LEFT-gate, a TAUT-gate and a CONTR-gate. As the weak explanatory diagnoses stand for the hypotheses that are compatible with the given observations and as these are already known through the process of *abductive hypothesis formation*, we only have to take those hypotheses into consideration that belong to the intersection of the weak explanatory diagnosis and the set of most probable hypotheses. In our example this results in the following hypothesis: a CONTR-gate. Although this will not always be the case, we immediately obtain a strong explanatory diagnosis.

8. Formal analysis of strong explanatory diagnostic reasoning

Strong diagnostic reasoning is a three stage process. First there is the formulation of the weak explanatory hypotheses; then we ask questions and try to answer them; finally, we formulate a strong explanatory diagnosis based on these answers. The first stage was described in the previous section. The second stage consists in the formulation of relevant questions of which at least one possible answer eliminates at least one of the weak explanatory hypotheses. In many cases these questions will be directed toward the generation of new observations such that new selections can be made. In order to formalize this question-answer process, a logic will be required that is both ampliative (because questions must be generated) and dynamic (because questions must be marked whenever an answer is given or whenever a more specific question can be generated.[2]

If the number of variable observations is exhaustive, as is the case for electric gates, and if all of them are known, no questions must be asked and all observations can easily be introduced in the reasoning process. In section 9, we describe a logic that can handle the latter case.

The conclusion we draw after all but one of the weak explanatory hypotheses are eliminated by further experiments, does not follow deductively from our observations. As a consequence, the final argument by which the strong diagnosis is supported has the following format:

Definition 12

(1) *If P, then this explains (together with R) why Q is the case.*

(2) $P \wedge R$ *is better than any other explanation we have of Q.*

(3) *We observe that Q is the case.*

(4) *Because of (1)-(3) we accept that P is true.*

This scheme is the general format for *inference to the best explanation*. There is no general criterion for what "better" is, but here it means *closer to the original system description*.

Although there is a set of most probable hypotheses, there is no absolute guarantee that one of them will be verified. Components can be mixed up in such way that the abnormal behavior of some component is

---

[2]In some contexts it will be impossible to answer all relevant questions. In such cases, a strong explanatory diagnosis is impossible.

not at all caused by a defect in its constructive parts. In our example, there is already a strong explanatory diagnosis with only one observation taken into account. In view of the second observation that is described in section 2, this strong explanatory diagnosis will be falsified such that the remaining weak explanatory hypotheses need to be considered again. When all former hypotheses are falsified, it is highly probable that the diagnosed gate functions in a non-systematic way, which means that it violates the presupposition that all gates function as one of the possible gate-types.

It may also happen that the remaining expectancies give rise to inconsistencies such that no strong explanatory diagnosis is obtainable. In that case the reasoning process for non-explanatory diagnosis has to be continued, including the new observations.

9. An adaptive logic for explanatory diagnostic reasoning

The determination of the set of weak explanatory diagnoses is quite straightforward. When a component is diagnosed, the other possible component-types will become new expectancies. Formally this results in a disjunction of the latter as a new expectancy. When the *underlying theoretical knowledge* allows for a restriction of the number of hypotheses, these will be joined in a separate disjunction which is considered less probable than the general one is, notwithstanding the fact that the separate disjuncts of the former have a higher probability. When a component is diagnosed, it is most probable that it will be one of the remaining component-types or that it is non-systematic. It is less probable that it will be just one of the remaining component-types and supplementary restrictions on the number of disjuncts will cause a further decrease of probability. The decrease of probability comes forth from the inclusion of the latter disjunctions into the former ones. The order of probability of the remaining expectancies will be translated in a ranking according to preference, which allows for a prioritized adaptive logic. A logic is prioritized iff it defines a consequence relation $\Sigma \vdash A$ in which Σ is an n-tuple of sets of closed formulas, $\langle \Gamma_0, \ldots, \Gamma_n \rangle$ and each Γ_i has a different preference ranking. Normality will be understood as the truth of all expectancies. In that case, one of the hypotheses with the highest probability is true.

In our approach *consequences* of members of the Γ_i $(i > 0)$ will or will not be joined to Γ_0. By 'consequences' we shall mean **CL**-consequences. Hence, our consequence relations will be trivial if some member of Γ_i $(i > 0)$ is inconsistent.

The general plot is to interpret expectancies within the modal logic **T** of Feys (which is von Wright's **M**). The essential difference with the application of **S5** is that the accessibility relation R is not transitive in the case of **T**. [3] As there are several versions around of the **T**-semantics, and we need a specific one, we briefly describe it.

Let $\mathcal{L}^M$ the standard modal language with S, P^r, C, and $\mathcal{W}^M$ the sets of sentential letters, predicative letters of rank r, constants, and wffs (closed formulas). To simplify the semantic metalanguage, we introduce a (non-denumerable) set of pseudo-constants $\mathcal{O}$, *requiring that any element of the domain D is named by at least one member of $C \cup \mathcal{O}$*. Let $\mathcal{W}^{M+}$ denote the set of wffs of $\mathcal{L}^{M+}$ (in which $C \cup \mathcal{O}$ plays the role played by C in $\mathcal{L}^M$). The function of $\mathcal{O}$ is to simplify the clauses for the quantifiers.

A **T**-model M is a quintuple $\langle W, w_0, R, D, v \rangle$ in which W is a set of worlds, $w_0 \in W$ the real world, R a binary relation on W, D a non-empty set and v an assignment function. The accessibility relation R is reflexive. The assignment function v is defined by:

C1.1 $v : S \times W \longrightarrow \{0, 1\}$

C1.2 $v : C \cup \mathcal{O} \times W \longrightarrow D$ (where, for all $w \in W$, $\{v(\alpha, w) \mid \alpha \in C \cup \mathcal{O}\} = D$)

C1.3 $v : P^r \times W \longrightarrow \wp(D^r)$ (the power set of the r-th Cartesian product of D)

The valuation function $v_M : \mathcal{W}^{M+} \times W \longrightarrow \{0, 1\}$, determined by the model M is defined by:

C2.1   where $A \in S$, $v_M(A, w) = v(A, w)$

C2.2 $v_M(\pi^r \alpha_1 \ldots \alpha_r, w) = 1$ iff $\langle v(\alpha_1, w), \ldots, v(\alpha_r, w) \rangle \in v(\pi^r, w)$

C2.3 $v_M(\alpha = \beta, w) = 1$ iff $v(\alpha, w) = v(\beta, w)$

C2.4 $v_M(\neg A, w) = 1$ iff $v_M(A, w) = 0$

C2.5 $v_M(A \vee B, w) = 1$ iff $v_M(A, w) = 1$ or $v_M(B, w) = 1$

C2.6 $v_M((\exists \alpha)A(\alpha), w) = 1$ iff $v_M(A(\beta), w) = 1$ for at least one $\beta \in C \cup \mathcal{O}$

C2.7 $v_M(\Diamond A, w) = 1$ iff $v_M(A, w') = 1$ for at least one w' such that Rww'.

A model M verifies $A \in \mathcal{W}^M$ iff $v_M(A, w_0) = 1$. A is valid iff it is verified by all models. Where $\Diamond^i$ abbreviates a sequence of $i \geq 0$ diamonds, M is a **T**-model of $\Sigma = \langle \Gamma_0, \ldots, \Gamma_n \rangle$ iff, for all i ($0 \leq i \leq n$)

---

[3]So, Kripke's **K** would do just as well. However, **T** allows for a simpler formulation of the formal machinery below.

and $A \in \mathcal{L}$, M verifies $\Diamond^i A$ if $A \in \Gamma_i$. We shall write $\Sigma \models_{\mathbf{T}} A$ to denote that all $\mathbf{T}$-models of Σ verify A.

As $\mathbf{T}$ has a decent proof theory, we shall write $\Sigma \vdash_{\mathbf{T}} A$ to denote that $\{\Diamond^i A \mid A \in \Gamma_i\} \vdash_{\mathbf{T}} A$. Remember that each Γ_i is a set of closed formulas of $\mathcal{L}$ (the standard predicative language).

As the accessibility relation of the $\mathbf{T}$-semantics is reflexive, a model that verifies $\Diamond^i A$ also verifies $\Diamond^j A$ for all $j > i$. Thus, some Σ have $\mathbf{T}$-models in which $W = \{w_0\}$ and hence $v_M(A, w_0) = 1$ for all $A \in \Gamma_0 \cup \ldots \cup \Gamma_n$. Such $\mathbf{T}$-models will be called *singleton models*. It is easily seen that Σ has singleton models iff $\Gamma_0 \cup \ldots \cup \Gamma_n$ is consistent.

So, the lower limit logic is $\mathbf{T}$. The upper limit logic, again **Triv**, presupposes the normal situation, viz. the one in which all premises expressing expectancies are compatible with the premises accepted as true (the members of Γ_0). This may be realized by adding to $\mathbf{T}$ the axiom "$\Diamond A \supset A$". A characteristic semantics for **Triv** is obtained by restricting the $\mathbf{T}$-semantics to singleton models.[4] It follows that $\Sigma \vdash_{\mathbf{Triv}} A$ iff $\Gamma_0 \cup \ldots \cup \Gamma_n \vdash_{\mathbf{CL}} A$. In order to define an adaptive logic, we need to figure out the possible abnormalities of an n-tuple of premises. Again, we have to consider the three complications that were described in section 4. A fourth complication is related to the prioritized character of the logic. If $i < j$, then an abnormality of the form $\Diamond^i A \wedge \neg A$ will count as worse than an abnormality of the form $\Diamond^j A \wedge \neg A$. So, here is how we shall handle abnormalities. Let $\mathcal{F}^a$ be the set of atoms (primitive open and closed formulas and their negations) and let $\exists A$ abbreviate A preceded by a sequence of existential quantifiers (in some specified order) over all variables free in A. An abnormality is a formula of the form $\exists(\Diamond^i A \wedge \neg A)$, where $A \in \mathcal{F}^a$. For the semantics, we define, for each $\mathbf{T}$-model M of $\Sigma = \langle \Gamma_0, \ldots, \Gamma_n \rangle$ a set of abnormal parts (where $0 < i \le n$):

Definition 13 $Ab^i(M) =_{df} \{A \in \mathcal{F}^a \mid v_M(\exists(\Diamond^i A \wedge \neg A), w_0) = 1\}$

The adaptive models of Σ will be obtained by making a selection of its $\mathbf{T}$-models, first with respect to the sets $Ab^1(M)$, next with respect to the sets $Ab^2(M)$, etc.

For the proof theory, we need disjunctions of abnormalities. It turns out that we may restrict our attention to disjunctions of formulas of the form $\exists(\Diamond^i A \wedge \neg A)$, in which i is always the same number and all $A \in \mathcal{F}^a$. By $Dab^i(\Delta)$ we denote the disjunction $\bigvee\{\exists(\Diamond^i A \wedge \neg A) \mid A \in \Delta\}$. We

---

[4]A different characterization is obtained by requiring, for all $A \in W$ and for all $w_i, w_j \in W$, that $v_M(A, w_i) = v_M(A, w_j)$.

shall say that $Dab^i(\Delta)$ is a Dab^i-consequence of Σ iff all **T**-models of Σ verify $Dab^i(\Delta)$. A Dab^i-consequence $Dab^i(\Delta)$ of Σ will be called *minimal* iff there is no $\Delta' \subset \Delta$ such that $Dab^i(\Delta')$ is a Dab^i-consequence of Σ.

In order to define the semantics for this logic, we will apply the "Reliability" strategy. It is based on the idea that any n-tuple Σ may define a set of unreliable formulas and that the $\mathbf{D}^{exp}$-models of Σ are the **T**-models of Σ in which only unreliable formulas behave abnormally. The set of unreliable formulas is defined as the set of factors of the minimal Dab-consequences of Σ:

$$U^i(\Sigma) = \bigcup\{\Delta \mid Dab^i(\Delta) \text{ is a minimal } Dab^i\text{-consequence of } \Sigma\}$$

As abnormalities of the form $\diamond^i A \wedge \neg A$ are considered as worse than abnormalities of the form $\diamond^j A \wedge \neg A$ whenever $i < j$, and as only those $Dab^i(\Delta)$ in which i is always the same number are taken into consideration, the $\mathbf{D}^{exp}$-models of an n-tuple Σ will be obtained through the following selection of its T-models:

When $\mathcal{M}_\Sigma$ is the set of all **T**-models of Σ, the $n+1$ selections of $\mathcal{M}_\Sigma$ are defined as follows:

$$\sigma^0(\mathcal{M}_\Sigma) =_{df} \mathcal{M}_\Sigma$$

and, where $0 \leq i < n$,

$$\sigma^{i+1}(\mathcal{M}_\Sigma) =_{df} \{M \in \sigma^i(\mathcal{M}_\Sigma) \mid Ab^{i+1}(M) \subseteq U^{i+1}(\Sigma)\}$$

The $\mathbf{D}^{exp}$-models of Σ are the members of $\sigma^n(\mathcal{M}_\Sigma)$, and where $A \in \mathcal{W}$, $\Sigma \models_{\mathbf{D}^{exp}} A$ iff A is verified by all $\mathbf{D}^{exp}$-models of Σ.

10. The dynamic proof theory of $\mathbf{D}^{exp}$

As for $\mathbf{D}^{nexp}$, the dynamic proof theory is based on a specific relation between the *upper limit logic* and the *lower limit logic*:

Theorem 2 *If $A_1, \ldots, A_n \vdash_{\mathbf{Triv}} B$, then there are $C_1, \ldots, C_m \in \mathcal{F}^a$ and there is an i such that $A_1, \ldots, A_n \vdash_{\mathbf{T}} B \vee Dab^i\{C_1, \ldots, C_m\}$ (Derivability Adjustment Theorem)*[5].

If A is conditionally derived in the proof (that is, on a line the fifth element of which is not $\emptyset$), then the condition will be a couple: a set of formulas and a "level" i ($0 < i \leq n$), indicated by a subscript (as in Δ_i). Sometimes a condition will be compounded from several other

---

[5]This relation is proved in [Batens et al., to appear].

conditions by taking the union of their first members and the maximum of their second members, which will be denoted by max(...).

The generic rules that govern the dynamic proofs in $\mathbf{D}^{exp}$ are:

PREM If $A \in \Gamma^i$, then one may add a line consisting of
   (i) the appropriate line number,
   (ii) $\Diamond^i A$,
   (iii) "$-$",
   (iv) "PREM", and
   (v) $\emptyset$.

RU If $B_1, \ldots, B_m \vdash_{\mathbf{T}} A$ and $B_1, \ldots, B_m$ occur in the proof with the conditions $\Delta^1_{j_1}, \ldots \Delta^m_{j_m}$ respectively, then one may add a line consisting of
   (i) the appropriate line number,
   (ii) A,
   (iii) the line numbers of the B_i,
   (iv) "RU", and
   (v) $(\Delta^1 \cup \ldots \cup \Delta^m)_{\max(j_1, \ldots, j_m)}$.

RC If $B_1, \ldots, B_m \vdash_{\mathbf{T}} A \vee Dab^k(\Theta)$ and $B_1, \ldots, B_m$ occur in the proof with the conditions $\Delta^1_{j_1}, \ldots \Delta^m_{j_m}$ respectively, then one may add a line consisting of
   (i) the appropriate line number,
   (ii) A,
   (iii) the line numbers of the B_i,
   (iv) "RC", and
   (v) $(\Theta \cup \Delta^1 \cup \ldots \cup \Delta^m)_{\max(k, j_1, \ldots, j_m)}$.

It is obvious in view of the rules that A is derivable on the condition Δ_i in a proof from Σ iff $A \vee Dab^i(\Delta)$ is $\mathbf{T}$-derivable from Σ.

Definition 14 *Where $A \in \mathcal{W}$, $\Sigma \vdash_{\mathbf{D}^{exp}} A$ iff A is finally derivable from Σ.*

We now come to the marking definition. For $\mathbf{D}^{exp}$ we need to define which formulas are reliable with respect to a proof at a stage. In other words, we have to define the $U^i_s(\Sigma)$, in which s refers to the stage of the proof. This is a very simple matter: $U^i_s(\Sigma)$ is defined from the $Dab^i(\Delta)$ *that occur in the proof* in precisely the same way as $U^i(\Sigma)$ is defined from the Dab^i-consequences of Δ. This requires only that we locate those $Dab^i(\Delta)$ that are derived in the proof and are *minimal* with respect to the $Dab^i(\Sigma)$ derived in the proof.

Definition 15 *Marking for $\mathbf{D}^{exp}$: line i is marked at stage s iff, where Δ_j is its fifth element, $\Delta_j \cap U^j_s(\Sigma) \neq \emptyset$.*

11. An illustration of $\mathbf{D}^{exp}$

To illustrate $\mathbf{D}^{exp}$, we will formalize the example that was presented in sections 2, 7 and 8. In section 6 we (conditionally) derived the diagnosis "P is not an AND-gate". Now we aim to derive the weak explanatory diagnoses that follows from this non-explanatory diagnosis, while new observations should allow for the derivation of a strong explanatory diagnosis.

For the description of the circuit we will apply the predicative language as it was introduced in section 6. To shorten the disjunctions that are considered as new expectancies, the formalization of the different gate-types will be represented as follows: $P_{\{1111\}}$ represents the formalization of a TAUT-gate (i.e. $(\forall x)P_0 x$), $P_{\{1110\}}$ represents the formalization of an OR-gate (i.e. $(\forall x)(\neg P_0 x \equiv (\neg P_1 x \wedge \neg P_2 x))$), etc.. The predicate N will denote the non-systematic functioning of gate P. The formal representation of the remaining parts of the system description SD, the observations and the new expectancies is as follows:

$\Gamma_0 = \{R_1 t_1, P_1 t_1, P_2 t_1, R_0 t_1, Q_0 t_1, (\forall x)(P_0 x \equiv R_2 x), (\forall x)(P_0 x \equiv Q_1 x),$
$(\forall x)(R_1 x \equiv Q_2 x)\}$
$\Gamma_1 = \{(\forall x)(Q_0 x \equiv \neg(Q_1 x \equiv Q_2 x)), (\forall x)(R_0 x \equiv \neg(R_1 x \equiv R_2 x))\}$
$\Gamma_2 = \{(\forall x)N x \vee P_{\{1111\}} \vee P_{\{1110\}} \vee P_{\{1101\}} \vee P_{\{1011\}} \vee P_{\{0111\}} \vee P_{\{1100\}} \vee$
$P_{\{1010\}} \vee P_{\{1001\}} \vee P_{\{0110\}} \vee P_{\{0101\}} \vee P_{\{0011\}} \vee P_{\{0100\}} \vee P_{\{0010\}} \vee P_{\{0001\}} \vee$
$P_{\{0000\}}\}$
$\Gamma_3 = \{P_{\{1111\}} \vee P_{\{1110\}} \vee P_{\{1101\}} \vee P_{\{1011\}} \vee P_{\{0111\}} \vee P_{\{1100\}} \vee P_{\{1010\}} \vee$
$P_{\{1001\}} \vee P_{\{0110\}} \vee P_{\{0101\}} \vee P_{\{0011\}} \vee P_{\{0100\}} \vee P_{\{0010\}} \vee P_{\{0001\}} \vee P_{\{0000\}}\}$
$\Gamma_4 = \{P_{\{1111\}} \vee P_{\{1100\}} \vee P_{\{1010\}} \vee P_{\{0000\}}\}$

When no supplementary observations are available, the following situation will occur:

1	$R_1 t_1$	-	PREM $\emptyset$
2	$P_1 t_1$	-	PREM $\emptyset$
3	$P_2 t_1$	-	PREM $\emptyset$
4	$R_0 t_1$	-	PREM $\emptyset$
5	$Q_0 t_1$	-	PREM $\emptyset$
6	$(\forall x)(P_0 x \equiv R_2 x)$	-	PREM $\emptyset$
7	$(\forall x)(P_0 x \equiv Q_1 x)$	-	PREM $\emptyset$
8	$(\forall x)(R_1 x \equiv Q_2 x)$	-	PREM $\emptyset$
9	$\Diamond(\forall x)[Q_0 x \equiv \neg(Q_1 x \equiv Q_2 x)]$	-	PREM $\emptyset$
10	$\Diamond(\forall x)[R_0 x \equiv \neg(R_1 x \equiv R_2 x)]$.	-	PREM $\emptyset$
11	$\Diamond^2[(\forall x)N x \vee P_{\{1111\}} \vee P_{\{1110\}} \vee P_{\{1101\}} \vee$		

$P_{\{1011\}} \vee P_{\{0111\}} \vee P_{\{1100\}} \vee P_{\{1010\}} \vee$
$P_{\{1001\}} \vee P_{\{0110\}} \vee P_{\{0101\}} \vee P_{\{0011\}} \vee$
$P_{\{0100\}} \vee P_{\{0010\}} \vee P_{\{0001\}} \vee P_{\{0000\}}]$ -   PREM $\emptyset$

12 $\Diamond^3[P_{\{1111\}} \vee P_{\{1110\}} \vee P_{\{1101\}} \vee P_{\{1011\}} \vee$
$P_{\{1011\}} \vee P_{\{0111\}} \vee P_{\{1100\}} \vee P_{\{1010\}} \vee$
$P_{\{1001\}} \vee P_{\{0110\}} \vee P_{\{0101\}} \vee P_{\{0011\}} \vee$
$P_{\{0100\}} \vee P_{\{0010\}} \vee P_{\{0001\}} \vee P_{\{0000\}}]$ -   PREM $\emptyset$

13 $\Diamond^4[P_{\{1111\}} \vee P_{\{1100\}} \vee P_{\{1010\}} \vee P_{\{0000\}}]$ -   PREM $\emptyset$

14 $Q_0 t_1 \equiv \neg(Q_1 t_1 \equiv Q_2 t_1)$ 1,5,8,9   RC
$\{Q_0 x, \neg Q_0 x, Q_1 x, \neg Q_1 x, Q_2 x, \neg Q_2 x\}_1$

15 $R_0 t_1 \equiv \neg(R_1 t_1 \equiv R_2 t_1)$ 1,4,10   RC
$\{R_0 x, \neg R_0 x, R_1 x, \neg R_1 x, R_2 x, \neg R_2 x\}_1$

16 $\neg P_0 t_1$ 1,4,6,15   RU
$\{R_0 x, \neg R_0 x, R_1 x, \neg R_1 x, R_2 x, \neg R_2 x\}_1$

17 $\neg Q_1 t_1$ 1,5,8,14   RU
$\{Q_0 x, \neg Q_0 x, Q_1 x, \neg Q_1 x, Q_2 x, \neg Q_2 x\}_1$

18 $\neg R_2 t_1$ 1,4,15   RU
$\{R_0 x, \neg R_0 x, R_1 x, \neg R_1 x, R_2 x, \neg R_2 x\}_1$

19 $(\exists x)\neg P_0 x$ 2,3,16   RU
$\{R_0 x, \neg R_0 x, R_1 x, \neg R_1 x, R_2 x, \neg R_2 x\}_1$

20 $(\exists x)\neg(\neg P_0 x \equiv (\neg P_1 x \wedge \neg P_2 x))$ 2,3,16   RU
$\{R_0 x, \neg R_0 x, R_1 x, \neg R_1 x, R_2 x, \neg R_2 x\}_1$

21 $(\exists x)\neg(P_0 x \equiv (\neg P_1 x \wedge P_2 x))$ 2,3,16   RU
$\{R_0 x, \neg R_0 x, R_1 x, \neg R_1 x, R_2 x, \neg R_2 x\}_1$

22 $(\exists x)\neg(P_0 x \equiv (P_1 x \wedge \neg P_2 x))$ 2,3,16   RU
$\{R_0 x, \neg R_0 x, R_1 x, \neg R_1 x, R_2 x, \neg R_2 x\}_1$

23 $(\exists x)\neg(P_0 x \equiv P_1 x)$ 2,3,16   RU
$\{R_0 x, \neg R_0 x, R_1 x, \neg R_1 x, R_2 x, \neg R_2 x\}_1$

24 $(\exists x)\neg(P_0 x \equiv P_2 x)$ 2,3,16   RU
$\{R_0 x, \neg R_0 x, R_1 x, \neg R_1 x, R_2 x, \neg R_2 x\}_1$

25 $(\exists x)\neg(P_0 x \equiv \neg(P_1 x \wedge P_2 x))$ 2,3,16   RU
$\{R_0 x, \neg R_0 x, R_1 x, \neg R_1 x, R_2 x, \neg R_2 x\}_1$

26 $(\forall x)Nx \vee P_{\{0111\}} \vee P_{\{0110\}} \vee P_{\{0101\}} \vee$
$P_{\{0011\}} \vee P_{\{0100\}} \vee P_{\{0010\}} \vee P_{\{0001\}} \vee$ 11,   RC/
$P_{\{0000\}}$ 19-25   RU
$\{P_0 x, \neg P_0 x, P_1 x, \neg P_1 x, P_2 x, \neg P_2 x,$
$R_0 x, \neg R_0 x, R_1 x, \neg R_1 x, R_2 x, \neg R_2 x\}_2$

27 $P_{\{0111\}} \vee P_{\{0110\}} \vee P_{\{0101\}} \vee P_{\{0011\}} \vee$ 12,   RC/
$P_{\{0100\}} \vee P_{\{0010\}} \vee P_{\{0001\}} \vee P_{\{0000\}}$ 19-25   RU
$\{P_0 x, \neg P_0 x, P_1 x, \neg P_1 x, P_2 x, \neg P_2 x,$
$R_0 x, \neg R_0 x, R_1 x, \neg R_1 x, R_2 x, \neg R_2 x\}_3$

28 $P_{\{0000\}}$ 13, RC/
19,23,24 RU
$\{P_0x, \neg P_0x, P_1x, \neg P_1x, P_2x, \neg P_2x,$
$R_0x, \neg R_0x, R_1x, \neg R_1x, R_2x, \neg R_2x\}_4$

The weak explanatory diagnoses, as defined in section 2 are conditionally derived on line 27. The introduction of *underlying theoretical knowledge* is quite advantageous when the former results are compared with the strong explanatory diagnosis that is conditionally derived at line 28. Although the *underlying theoretical knowledge* will not always lead to a strong explanatory diagnosis from the first observation on, it will at least shorten the diagnostic reasoning process. However, one must remind that the latter is an *inference to the best explanation*, such that it can always be overruled by new observations.

Suppose that some new observations are added to Γ_0 such that it becomes:

$$\Gamma_0 = \{R_1t_1, P_1t_1, P_2t_1, R_0t_1, Q_0t_1, \neg R_1t_2, \neg P_1t_2, \neg P_2t_2, R_0t_2, Q_0t_2$$
$$(\forall x)(P_0x \equiv R_2x), (\forall x)(P_0x \equiv Q_1x), (\forall x)(R_1x \equiv Q_2x)\}$$

The proof will continue as follows:

29 $\neg R_1t_2$ - PREM $\emptyset$
30 $\neg P_1t_2$ - PREM $\emptyset$
31 $\neg P_2t_2$ - PREM $\emptyset$
32 R_0t_2 - PREM $\emptyset$
33 Q_0t_2 - PREM $\emptyset$
34 $\neg P_0t_2$ 6,15,29,32 RU
$\{R_0x, \neg R_0x, R_1x, \neg R_1x, R_2x, \neg R_2x\}_1$
35 $(\exists x)\neg P_0x \equiv \neg(P_1x \equiv P_2x)$ 30,31,34 RU
$\{R_0x, \neg R_0x, R_1x, \neg R_1x, R_2x, \neg R_2x\}_1$
36 $(\exists x)\neg P_0x \equiv (P_1x \wedge \neg P_2x)$ 30,31,34 RU
$\{R_0x, \neg R_0x, R_1x, \neg R_1x, R_2x, \neg R_2x\}_1$
37 $(\exists x)\neg P_0x \equiv (\neg P_1x \wedge P_2x)$ 30,31,34 RU
$\{R_0x, \neg R_0x, R_1x, \neg R_1x, R_2x, \neg R_2x\}_1$
38 $(\exists x)P_0x$ 30,31,34 RU
$\{R_0x, \neg R_0x, R_1x, \neg R_1x, R_2x, \neg R_2x\}_1$
39 $\exists(\Diamond^4 R_1x \wedge \neg R_1x) \vee \exists(\Diamond^4 R_2x \wedge \neg R_2x)\vee$
$\exists(\Diamond^4 \neg R_0x \wedge R_0x) \vee \exists(\Diamond^4 \neg P_0x \wedge P_0x)\vee$ 1,2,3,
$\exists(\Diamond^4 \neg R_1x \wedge R_1x) \vee \exists(\Diamond^4 \neg R_2x \wedge R_2x)\vee$ 4,6,10,
$\exists(\Diamond^4 R_0x \wedge \neg R_0x) \vee \exists(\Diamond^4 P_0x \wedge \neg P_0x)\vee$ 13,29,30,

$\exists(\lozenge^4 \neg P_1 x \wedge P_1 x) \vee \exists(\lozenge^4 \neg P_2 x \wedge P_2 x) \vee$ 31,32   RU $\emptyset$

40   $(\forall x) N x \vee P_{\{0111\}} \vee P_{\{0101\}} \vee P_{\{0011\}} \vee$ 11,19-25,   RC/

     $P_{\{0001\}}$ 35-38   RU

             $\{P_0 x, \neg P_0 x, P_1 x, \neg P_1 x, P_2 x, \neg P_2 x,$

                $R_0 x, \neg R_0 x, R_1 x, \neg R_1 x, R_2 x, \neg R_2 x\}_2$

41   $P_{\{0111\}} \vee P_{\{0101\}} \vee P_{\{0011\}} \vee$ 12,19-25,   RC/

     $P_{\{0001\}}$ 35-38   RU

             $\{P_0 x, \neg P_0 x, P_1 x, \neg P_1 x, P_2 x, \neg P_2 x,$

                $R_0 x, \neg R_0 x, R_1 x, \neg R_1 x, R_2 x, \neg R_2 x\}_3$

As the new observations are not compatible with the strong explanatory diagnosis "P is a CONTR-gate", we are able to derive the Dab-formula on line 39, such that line 28 is marked.

Acknowledgments

The research for this paper was financed by the Fund for Scientific Research – Flanders, by the Research Fund of Ghent University, and indirectly by the Flemish Minister responsible for Science and Technology (contract BIL98/73). We want to thank Diderik Batens and Joke Meheus for many discussions on the topic of this paper and their comments on former drafts.

References

Batens, D., 2000, A survey of inconsistency-adaptive logics, in: *Frontiers of Paraconsistent Logic*, D. Batens, C. Mortensen, G. Priest, and J.P. Van Bendegem, eds., Kings College Publications, London, pp. 49–73.

Batens, D., 1998, Inconsistency-adaptive logics, in: *Logic at Work. Essays Dedicated to the Memory of Helena Rasiowa*, E. Orłowska, ed., Springer Verlag, Heidelberg, New York, pp. 445–472.

Batens, D., Meheus J., Provijn D., and Verhoeven L., Some adaptive logics for diagnosis (to appear).

Batens, D. and Meheus, J., 2000, The adaptive logic of compatibility, *Studia Logica* 66:327–348.

Magnani, L., 2001, *Abduction, Reason and Science. Processes of Discovery and Explanation*, Kluwer Academic/Plenum Publishers, New York.

Meheus, J., Erotetic arguments from inconsistent premises (to appear).

Reiter, R., 1980, A logic for default reasoning, *Artificial Intelligence* 13:81–132.

Reiter, R., 1987, A theory of diagnosis first principles, *Artificial Intelligence* 32:57–95.

Weber, E. and De Clercq, K., 2002, Why the logic of explanation is inconsistency-adaptive, in: *Inconsistency in Science*, J. Meheus, ed., Kluwer Academic/Plenum Publishers, Dordrecht, pp. 165–184.

Weber, E. and Provijn, D., A formal analysis of diagnosis and diagnostic reasoning, *Logique & Analyse* (in print).

MODEL-GUIDED PROOF PLANNING

Seungyeob Choi and Manfred Kerber
School of Computer Science, The University of Birmingham
Edgbaston, Birmingham B15 2TT, England
s.choi@cs.bham.ac.uk, m.kerber@cs.bham.ac.uk

Abstract   Proof planning is a form of theorem proving in which the proving procedure is viewed as a planning process. The plan operators in proof planning are called methods. In this paper we propose a strategy for heuristically restricting the set of methods to be applied in proof search. It is based on the idea that the plausibility of a method can be estimated by comparing the model class of proof lines newly generated by the method with that of the assumptions and of the theorem. For instance, in forward reasoning when a method produces a new assumption whose model class is not a superset of the model class of the given premises, the method will lead to a situation which is semantically not justified and will not lead to a valid proof in later stages. A semantic restriction strategy is to reduce the search space by excluding methods whose application results in a semantic mismatch. A semantic selection strategy heuristically chooses the method that is likely to make most progress towards filling the gap between the assumptions and the theorem. Each candidate method is evaluated with respect to the subset and superset relation with the given premises. All models considered are taken from a finite reference subset of the full model class. In this contribution we present the model-guided approach as well as first experiments with it.

1.   Introduction

Although modern computing technologies make it possible to solve many complex mathematical problems, mathematical theorem proving still remains a difficult task for computers as it requires sophisticated reasoning capabilities. When the first theorem proving systems were built there was great optimism that the separation of the syntax from the semantics could lead to a purely syntactical procedure, which – once efficiently implemented – would allow for the automation of difficult

L. Magnani, N.J. Nersessian, and C. Pizzi (eds.),
Logical and Computational Aspects of Model-Based Reasoning, 143–162.
© 2002 Kluwer Academic Publishers. Printed in the Netherlands.

mathematical proofs. This has to a certain degree been realized and become true in so-called machine-oriented theorem provers, but the hyperexponential nature of the search spaces shows that there are severe limitations of this approach. It works well mainly in particular niches of the space of potential theorem proving problems.

The most prominent machine-oriented approaches are based on the resolution [Robinson, 1965] or the tableaux calculus [Smullyan, 1968]. In resolution, the problem that consists of assumptions and a theorem are transformed into a normal form, the so-called clause normal form, and a proof is searched for on this clause level. The approach consists of firstly efficiently searching big search spaces, and secondly making good heuristic choices, since the problem is known to be intractable for hard problems. More and more powerful systems have been built on such paradigms, e.g. MKRP [Eisinger and Ohlbach, 1986], SETHEO [Letz et al., 1992], OTTER [McCune, 1994], SPASS [Weidenbach et al., 1996], and VAMPIRE [Riazanov and Voronkov, 2001].

On the one hand, these systems show remarkable performance on many problems in particular application areas. Even open mathematical problems could be solved with the assistance of automated provers for the first time. Most notably, the problem whether Robbins algebras are Boolean algebras was successfully proved by an automated theorem prover [McCune, 1997]. On the other hand, these methods unfortunately often fail even for problems that mathematicians would consider trivial.

Humans use approaches to theorem proving which drastically differ from the approaches machine-oriented theorem provers use. A deep understanding of what these human methods exactly are is unfortunately still missing and will probably be very difficult to obtain in full generality. However, partial insights can be obtained from the reflections by mathematicians themselves, by which they try to explain the way they find mathematical proofs [Hadamard, 1944; van der Waerden, 1964]. The extensive work by Pólya on teaching mathematics [Pólya, 1945; Pólya, 1954; Pólya, 1962] is a rich source of examples of how mathematicians prove theorems. Furthermore, the field of mathematical reasoning has been investigated in cognitive science [Braine, 1978; Johnson-Laird and Byrne, 1991]. This all led to very valuable insights and significant steps towards a computer model of how humans perform mathematical reasoning. The area in which attempts are made to develop such insights to a computational model of human theorem proving is called human-oriented theorem proving.

Let us look at some differences between the approaches of human-oriented and machine-oriented theorem proving. When humans try to prove theorems, they do not usually transform the whole problem into a normalized form nor do they search for the solution on a logical level. Instead, they search and deduce the conclusion on a more abstract mathematical level. They decide on what proof methodology to use, and as they proceed in a proof attempt, they choose what inference rule to apply. The proof methodologies and inference rules are obtained from either textbooks or previous experience in similar examples. Unlike most machine approaches, these inference rules are not just a set of rules which are fixed once and for all, but any successful proof of a problem can be stored in the mathematicians' memory (in total or in parts) and may be reused as a new inference step for a new problem. Human-oriented theorem proving is an attempt to model (parts of) the human problem solving behavior. Since it makes use of domain specific reasoning and allows for a hierarchical approach, the exponential explosion of the search space which is typical for machine-oriented theorem proving can be avoided for many practical problems. The most prominent approach to human-oriented theorem proving is proof planning [Bundy, 1988]. In proof planning, the search for a proof is considered as a planning process in which existing plan operators, called methods, are combined to form a full plan which can then be (hopefully) carried through.

A major difference between the machine-oriented way of proving theorems and the way how humans do it is the size of the search spaces involved. Machines can traverse search spaces with millions of nodes with reasonable resources. Humans, however, cannot do that, but have to rely on keeping search spaces small by putting enormous effort on adequate representations (and using re-representations) as well as making use of good heuristics that explore the more promising parts of the search space first. A major challenge for the field of human-oriented theorem proving is to understand better how humans arrive at such heuristics. Up to now such knowledge is only patchy. For instance, heuristic information can be encoded in form of specificity which prefers a specific method over a general one, or in form of an explicit ranking which rates methods independently of the concrete situation. None of these approaches provides a satisfying answer to what method to choose.

Semantic information seems, to a large extent, to be used as a form of heuristics in human problem solving. Checking an argument in one concrete model or a couple of models seems to be a very important reasoning pattern. If the argument holds in a couple of key models it has significantly gained in evidence. If it fails, it is rejected and a different argument must be constructed. We believe that a semantic approach will play a very important role in modeling the human reasoning capabilities in mathematics. The main idea of the work presented here is that the plausibility of a method (or a step) in a proof plan can be estimated by means of comparing models taken from a reference class of models.

The remainder of the paper is organized as follows. In the next section we describe proof planning in more depth. We discuss model based reasoning in section 3. The semantic restriction and selection strategies which are the main contributions are presented in section 4. Section 5 discusses the state of our implementation and some initial results. We conclude in section 6.

2.   Proof planning as a way of reasoning

Proof planning [Bundy, 1988] is an approach to human-oriented theorem proving. It is based on the idea that the reasoning process for proving mathematical theorems can be viewed as a planning process which employs abstract plan operators, called *methods*. Assume we want to prove a mathematical theorem (or conclusion) φ from premises $A_1, \ldots, A_n$. The picture of proof planning is that we have a gap between the assumptions and the theorem, which we try to fill by proof steps. We try to do that by systematically reducing gaps by introducing intermediate assertions. This is done by methods which represent on the one hand program fragments which generate specific proof parts, and on the other hand high-level specifications of the program fragments which make planning possible. In each planning stage, the proof planner selects a method which fills in the gap between premises and conclusion. A proof plan is found when there are no more remaining gaps on an abstract level. A proof is found when the proof plan can be successfully executed. Note that proof plans may fail.

2.1   The plan operator: Methods

Methods play the most significant role in proof planning. In Bundy's view [Bundy, 1988] a method consists of a tactic, a precondition, and a postcondition. The tactic is a piece of program code that manipulates the actual proof in a controlled way, and the precondition and the postcondition form a specification of the deductive ability of the tactic. In the ΩMEGA proof planner, a slightly extended view has been taken in order to allow automated adaptation of methods to new problems. [Huang et al., 1992; Huang et al., 1994; Kerber, 1998] proposed an idea of separating the procedural and declarative knowledge in the tactic part of the methods.

An ΩMEGA method has the following slots: declarations, premises, constraints, conclusions, declarative content, and procedural content as displayed in Figure 1. The declarations specify a signature that declares

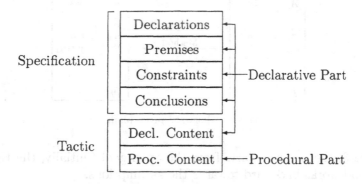

Figure 1. ΩMEGΛ method.

variables of the method. The premises consist of a list of proof line schemata which are used to prove the conclusions. The constraints contain additional restrictions on the premises and the conclusions. The conclusions slot contains the proof lines which are the goal of the proof. The declarative content provides declarative knowledge that is interpreted by the procedural content. And finally, the procedural content is a piece of program code that takes a piece of declarative knowledge and a pointer to the current proof plan tree, and produces a subtree which can be integrated into the current tree.

2.2 Forward and backward proof planning

Mainly two different types of planning strategies, forward and backward planning, are used in proof planning. In forward planning, the planner searches for a method that takes assumptions (axioms, definitions) or already proved intermediate results and produces new facts that can be used as new assertions in the following planning steps. Initially, there is a gap between the initial assumptions $A_1, \ldots, A_n$ and the theorem Th as shown in Figure 2. The planner starts from the assumptions and proceeds toward the theorem by filling in the gaps between assumptions and the theorem. When the line generated by the method is identical to the theorem, a proof plan is found.

Likewise, in backward planning, a method takes a theorem and produces subgoals sufficient to deduce the theorem. Some of the subgoals may be identical to existing assumptions, but typically some others are new and need to be added as open goals to be proved in later steps. In

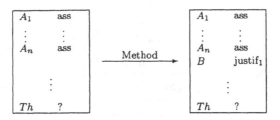

Figure 2.   Forward proof planning.

Figure 3, the planner starts from the open goal (initially, the theorem *Th*) and works backward towards the assumptions.

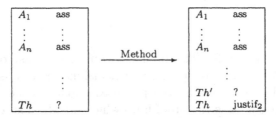

Figure 3.   Backward proof planning.

2.3   An example of a proof plan

The method for mathematical induction is an abstract level method that transforms the problem to a base case and a step case, in which the proof is obtained by recursive applications of induction rules. Note that we describe here only a simplified form. The proof plan for induction was studied in great detail as the first application of proof planning [Bundy et al., 1991].

Figure 4 shows the basic structure of the proof plan for induction.

A proof for the step case consists of two steps, firstly transforming the induction conclusion to make it syntactically similar to the induction hypothesis, and secondly applying the induction hypothesis.

Let us look at a concrete example. With the inductive definition of +:

$$0 + y = y \tag{1}$$
$$s(x) + y = s(x + y) \tag{2}$$

Figure 5 illustrates a concrete inductive proof for the theorem that addition on natural numbers is associative. (We use $x{:}t$ to denote x is of type

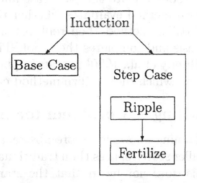

Figure 4. Proof plan for mathematical induction [Ireland and Bundy, 1996].

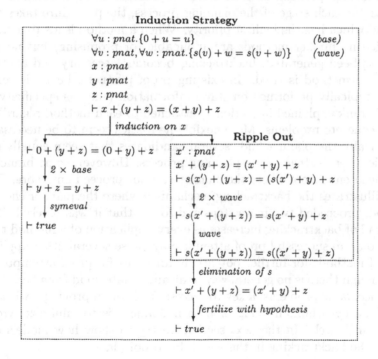

Figure 5. Proof of the associativity [Bundy et al., 1991].

t.) In Figure 5, five methods are used. They are induction, base, wave, fertilize, and symeval (symmetric evaluation). The induction method separates the theorem into two sub-theorems, base case and step case, and replaces x by 0 in the base case and by $s(x')$ in the step case. The base and the wave methods rewrite the base case and the step case, respectively. The symeval method evaluates whether the two expressions are identical. The fertilize method then replaces the expression with the induction hypothesis and completes the proof. The rippling process [Bundy et al., 1993; Bundy et al., 1990] repeats applications of the wave rule to obtain a form to which the fertilize method can be applied.

2.4   Backtracking – a problem for proof planning

Although proof planning allows for more abstract proof search, and typically involves smaller search spaces than traditional machine-oriented theorem proving, this does not mean that the standard problems of search do not constitute a problem. Proof planning can be viewed as a process in which one searches iteratively for a method for the simplification of a given problem until no more subproblems are generated in this process. At each stage of the planning process, the procedure takes one method which has the highest priority. When a method leads to a stage in which no further methods are applicable (or promising) but no full plan has been generated, backtracking becomes necessary and another candidate method is tried. In existing proof planners, heuristic selection is typically performed on static information (such as specificity or explicit rank explained in section 4) attached to each method regardless of the concrete problems. More flexible heuristics seem to be necessary.

Typically the planner finds several candidates that seem equally applicable at each stage of the planning process. However, some branches have dead ends from where the planner can not proceed any more. Figure 6 illustrates the backtracking mechanism where the planner chose a method, proceeded two more steps and found that it was blocked. The overhead of backtracking increases as every application of a method that leads to an unsuccessful proof attempt may cause several other applications of methods for further potentially unsuccessful proof attempts.

Although there is no general way to disambiguate good from bad steps for undecidable problems, a key to successful theorem proving – whether it is done by a human being or by a machine – is to minimize wrong choices in search. In the next section we try to show how models can reduce the backtracking in the search for proof plans.

3.   Model-based reasoning

Humans use models in various forms of reasoning. Models at least contribute to the heuristic information which direction the proof procedure should take. This has been taken into account in several traditional machine-oriented theorem provers [Slaney et al., 1994; Chu and Plaisted,

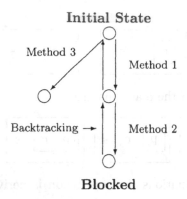

Initial State

Method 3

Method 1

Backtracking →

Method 2

Blocked

Figure 6. Backtracking.

1994; Kerber and Choi, 2000], which have incorporated semantic guidance into their proof search. However, since these provers are based on a low-level refutation procedure, the semantic guidance is restricted to choosing clauses that are most likely to contribute to finding a semantic contradiction. There is a significant difference between machine-oriented and human-oriented attitudes towards how to use models to guide the proof search. Mathematicians use semantics not only to check a contradiction but also to get the implications of a problem and to measure the plausibility of the logical consequences. In this section, we introduce a human-oriented view of model based guidance and present how it works in proof planning.

3.1 Model-based representation of reasoning

A visual representation of reasoning was proposed in mental models [Johnson-Laird and Byrne, 1991], in which deductions are modeled by Venn diagram. For instance, we have two premises: all psychologists are experimenters, and all experimenters are skeptics. The first premise gives rise to two diagrams:

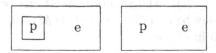

The first diagram illustrates that all psychologists are experimenters and there are also experimenters other than psychologists, and the second diagram shows that all psychologists are experimenters and there are no other experimenters. Likewise, the second premise also gives rise to two diagrams:

The combinations of the diagrams are:

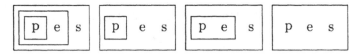

From all the combinations the conclusion is derived: all psychologists are skeptics.

We take up this idea of the mental models to mathematical reasoning by approximating the area of a premise with semantics, and use this information as a heuristic guidance which proof planning steps to perform.

3.2   Models

In order to make use of semantic information we need models. These can be either human generated or machine generated. In our approach we generate a finite set of models from the premises and the theorem with a model generator. Concretely, we use Slaney's FINDER system [Slaney, 1995]. We use the standard definition of models commonly used in classical logic. A domain D is a non-empty set of objects. An interpretation I is a mapping from each constant symbol, variable symbol, and other term to an element of D, and from each formula to a truth value. If a formula F is satisfied in an interpretation I, I is a model of F.

Suppose Γ is a set of assumptions and φ the theorem, (written as $\Gamma \vdash \varphi$). In order to limit the model space, we assume a reference set of interpretations R. We define the model sets $M_\Gamma := \{M \in R \mid M \models \Gamma\}$ and $M_\varphi := \{M \in R \mid M \models \varphi\}$, then it follows from $\Gamma \models \varphi$ that $M_\Gamma \subseteq M_\varphi$. Let Γ consist of assumptions, for instance, $A_1, A_2, \ldots, A_n$, we get model sets $M_1, M_2, \ldots, M_n$ with respect to the model reference class (note, all $M_i \subseteq R$). The semantic relations presented by $A_1 \wedge A_2 \wedge \cdots \wedge A_n \vdash \varphi$ may be described as $(M_1 \cap M_2 \cap \cdots \cap M_n) \subseteq M_\varphi$, as illustrated for $n = 3$ in Figure 7.

3.3   Model-based theorem proving

From the early days of automated theorem proving [Gelernter, 1959] semantics played a significant role in pruning the search for proofs. This

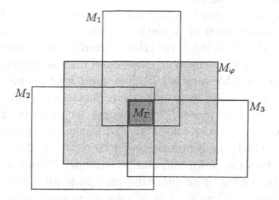

Figure 7. Semantic diagram of assumptions and theorem.

work has been taken up in more recent systems built on traditional the-orem provers such as the SCOTT system [Slaney et al., 1994; Hodgson and Slaney, 2001] and the semantic clause graph procedure [Kerber and Choi, 2000]. Semantic information can indicate what the theory means, in which direction the proof is to proceed, and which one of the candidate steps seems to have the most plausibility to lead to the desired result. In the case of the proof planning approach, in each stage of the above planning procedure we can efficiently reduce the search space if we have a mechanism to rule out less promising methods. The key idea of this research is to use semantics to obtain heuristic knowledge for determining the feasibility of a certain method.

3.4 Model-based proof planning

Models can provide heuristics about whether a formula generated by a method can be used as a new assumption or open goal, and if so, to what extent it fills the gap between the initial assumptions and the theorem. As mathematicians seem to use semantics in order to obtain heuristics to solve problems, it should also be natural that human-oriented automated theorem provers do so. In each step of the theorem proving process, the proof planner chooses a method that fills in the gap between assumptions and the theorem either by adding a new open goal before the theorem or by adding new assumptions after the initial assumptions. The main idea we introduce here is that the plausibility of a method can be estimated by means of comparing the semantics of new lines generated by the method with those of the assumptions and of the theorem. We extend the idea that any model of the assumptions Γ should also be a model of the theorem φ to the models of intermediate results as well. All the newly generated lines must satisfy the semantic rules. In forward proof planning, an application of a method adds new lines and may delete old lines from the planning state (the deleted lines are considered to be replaced by the newly generated ones) and the assumption is updated from the initial set Γ to a new set Γ'. The models of the intersection of all assumptions should also be models of the theorem. That is, $M_{\Gamma'} \subseteq M_\varphi$ must follow from $\Gamma \vdash \varphi \rightsquigarrow \Gamma' \vdash \varphi$, where $\rightsquigarrow$ indicates the transition from one planning state to the next. To put it in different words, any transition for which the semantic restriction $M_{\Gamma'} \subseteq M_\varphi$ does not hold cannot lead to a valid proof and can be excluded from the search space. Likewise, in backward proof planning an application of a method adds a new open goal φ'. The models of new open goals should contain all models of the intersection of all assumptions. $M_\Gamma \subseteq M_{\varphi'}$ must follow from $\Gamma \vdash \varphi \rightsquigarrow \Gamma \vdash \varphi'$; if not it can be excluded from search.

4.   Semantic restriction and selection of methods

Proof planning incorporates formalized proof fragments in methods. However, often existing control rules are not sufficient to heuristically choose one method when two or more methods are applicable. In areas in which strong methods are available – like in the area of mathematical induction – this is no major drawback, since the specification is strong enough to restrict the applicability to one or very few methods. In other areas, however, such heuristic knowledge is not given and existing proof planners can easily fall into dead end situations in which no more methods are applicable. In those cases, backtracking has to be done in order to return to a previous state in which other candidate methods were available. As in any search problem, the efficiency of proof planning depends on exploring the more promising parts of the search space first. We need good heuristics in order to evaluate each branch of the search tree. Some heuristic information can be encoded in form of specificity which prefers the application of a specific method to a general one. Another approach is to explicitly rank methods in plan construction, for instance, by a rating, and to apply the applicable method that has the highest rank (measuring its usefulness) independently of the concrete situation.

In this work we use models in proof planning to exclude implausible candidate methods (semantic restriction) and, where there is still more than one method applicable, to estimate which one would fill the gap better (semantic selection).

4.1 Semantic restriction

A semantic restriction strategy is a strategy that rejects any methods whose applications produce an intermediate result with which the semantic relation discussed in section 3.2 is not satisfied. This simple strategy can avoid many useless steps. For example, the search for a proof plan in ΩMEGA for $\{S \to \forall x Q(x)\} \vdash \forall x(S \to Q(x))$ can be reduced from 10 to only 3 steps with the semantic restriction strategy as we will describe in section 5.

Semantic restriction in forward planning. In some cases, it is possible for the application of a method to produce a new assumption with which the theorem cannot be deduced. As an example, suppose premises and a conclusion that satisfy $A_1, A_2, A_3 \vdash \varphi$ with $M_1 \cap M_2 \cap M_3 \subseteq M_\varphi$. The following method may be applied, which replaces line L1 (indicated by $\ominus$ L1) by line L4 (indicated by $\oplus$ L4). The intermediate result ψ from line L4 will participate in the proof as a new assumption in the next stages.

Method 1	
premises	$\ominus$ L1, L2, L3
conclusions	$\oplus$ L4
proof schema	L1. $H \vdash A_1$
	L2. $H \vdash A_2$
	L3. $H \vdash A_3$
	L4. $H \vdash \psi$

When the range of the model set of each assumption and conclusion is like those illustrated in Figure 8, the set M_φ from the conclusion φ satisfies $M_\Gamma = M_1 \cap M_2 \cap M_3 \subseteq M_\varphi$ but not $M_\Delta = M_2 \cap M_3 \cap M_\psi \subseteq M_\varphi$. This means, from $A_2 \wedge A_3 \wedge \psi$, one cannot derive φ. Therefore, the application of this method should be blocked in the given context.

Semantic restriction in backward planning. Suppose the initial premises are A_1, A_2, A_3 and the theorem is φ. We apply Method 2 that uses premises A_2 and A_3 and adds a new premise A_4 in order to derive φ.

Method 2	
premises	L2, L3, $\oplus$ L4
conclusions	$\ominus$ L5
proof schema	L2. $H \vdash A_2$
	L3. $H \vdash A_3$
	L4. $H \vdash A_4$
	L5. $H \vdash \varphi$

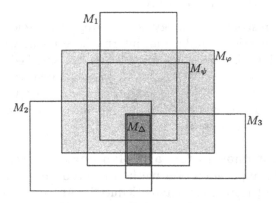

Figure 8.   Wrong conclusion in forward planning.

The range of models of each premise and conclusion is illustrated in Figure 9. The model set from the initial premise set $M_\Gamma := M_1 \cap M_2 \cap M_3$ satisfies $M_\Gamma \subseteq M_\varphi$ and the model set from the premises taken by the method $M_\Delta := M_2 \cap M_3 \cap M_4$ also satisfies $M_\Delta \subseteq M_\varphi$. The proof planner adds the premise A_4 as a new open goal. However, A_4 cannot be deduced from the given premises since $M_\Gamma \not\subseteq M_4$. Although the method

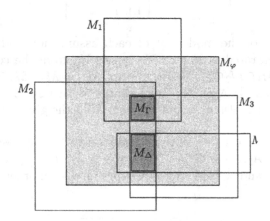

Figure 9.   Wrong assumption in backward planning.

is applicable in the current stage, the newly generated conclusion, A_4, cannot be derived from the initial premises. Therefore, we should not apply this method in this context.

4.2   Semantic selection

A semantic selection strategy does not only restrict the number of methods in a semantic way, but heuristically chooses one that seems to make most progress towards filling the gap. It evaluates the semantics of each method where all methods seem equally applicable if semantics is not taken into consideration. We propose to distinguish more promising methods from the others by estimating how well they semantically match the open goals with respect to the subset and superset relation with the given premises. In a forward reasoning approach, the methods which restrict the model class best in the direction to the theorem are better. Concretely, if the problem is $\Gamma \vdash \varphi$ and two methods μ_1 and μ_2 transform the problem to $\Gamma_1 \vdash \varphi$ and $\Gamma_2 \vdash \varphi$, respectively, with $M_\Gamma \subseteq M_{\Gamma_i}$ and $M_{\Gamma_i} \subseteq M_\varphi$ with $i = 1$ or 2, select M_{Γ_1} rather than M_{Γ_2} if and only if M_{Γ_1} is bigger than M_{Γ_2}. (If the sets have the same cardinality, make a random choice.) Likewise in backwards reasoning, a method that produces the smallest model class should be selected.

Even when two or more candidate methods are applicable without semantic conflict, the plausibility of a method may still be assessed by diagrams. Assume the initial assumption set Γ, assumption set Γ_1 obtained by the application of method μ_1, Γ_2 obtained by the application of μ_2, and the theorem φ. As in Figure 10, if the model set of Γ_2 is

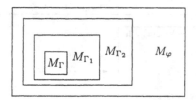

Figure 10.   Semantic selection in proof planning.

larger than that of Γ_1, Γ_2 is likely the one closer to the conclusion and therefore Γ_2 has a greater possibility to deduce the theorem in fewer steps.

The selection strategy in backward proof planning proceeds towards the assumptions. In Figure 10, unlike in forward planning, we choose Γ_1 rather than Γ_2 since it is closer to the initial assumption Γ and it should be easier to be derived from Γ.

5.   Implementation and initial results

We have implemented the semantically restricted proof planner based
on the ΩMEGA system [Benzmüller et al., 1997], which is a multi-initiative
theorem proving environment. We use FINDER [Slaney, 1995] as a
model generator. The semantic proof planner transforms each proof
line to FINDER readable format and generates a finite number of finite
model sets.

We suppose the following problem, which is simple enough to illustrate
the approach:

$$\{S \rightarrow \forall x Q(x)\} \vdash \forall x (S \rightarrow Q(x))$$

In the initial state, there are two methods applicable. The ΩMEGA
proof planner chooses a backward method μ_1 by its existing method
selection strategy. However, the application of μ_1 will give rise to backtracking.

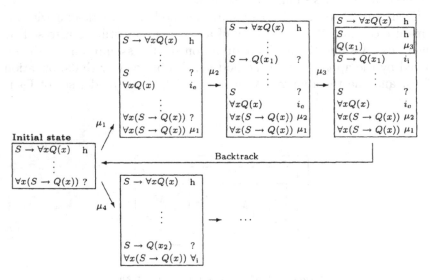

Figure 11.   Proof planning procedure.

In Figure 11, the application of method μ_1 in the initial stage generates two new open lines: S and $\forall x Q(x)$. The new open goal S must
be the logical consequence of the initial assumption $S \rightarrow \forall x Q(x)$. The
theorem $\forall x (S \rightarrow Q(x))$ must be the logical consequence of the new lines
S and $\forall x Q(x)$ along with the initial assumption.

$$\{S \rightarrow \forall x Q(x)\} \vdash S \qquad (3)$$

$$\{(S \rightarrow \forall x Q(x)), S, \forall x Q(x)\} \vdash \forall x (S \rightarrow Q(x)) \qquad (4)$$

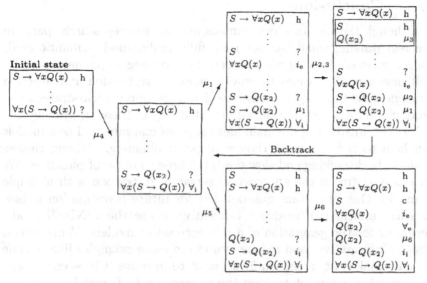

Figure 12. Proof planning procedure (continued).

The relation between models of each line follows:

$$M_{S \to \forall x Q(x)} \subseteq M_S \tag{5}$$

$$M_{\{S \to \forall x Q(x)\}, S, \forall x Q(x)\}} \subseteq M_{\forall x (S \to Q(x))} \tag{6}$$

where M_S means the model set of formula S, and so on.

To verify the first condition, we generate models from the premise $S \to \forall x Q(x)$ with the domain fixed to $D = \{0, 1\}$.

Model	S	Q(0)	Q(1)
m_1	F	F	F
m_2	F	T	F
m_3	F	F	T
m_4	F	T	T
m_5	T	T	T

Among these models, m_5 is the only one that also satisfies S. The other four models are counterexamples. Thus, we see that it is not possible to derive S from the premise. If we do not rule out the method μ_1, the procedure will backtrack after some number of steps.

Similarly, there is another chance to avoid backtracking as shown in Figure 12. As a result of the model-based restriction, the procedure can avoid two backtracking and find the proof in three steps only.

6. Conclusion

Although the way in which humans manage to keep search spaces in theorem proving small has not been fully understood, semantic guidance in form of checking whether certain reasoning steps make sense in well chosen models seems to provide powerful restriction and selection strategies. In this paper we discussed how to apply such strategies to proof planning, which is a human-oriented way of theorem proving.

We have introduced the main ideas of proof planning and how models can help to reduce wrong choices in proof planning. Wrong choices result in backtracking and slow down the process of proof planning. We have demonstrated the approach of semantic restriction with a simple example. One important question left for future investigation is how to select informative models. Currently we use the FINDER model generator for the generation of a reference set of models. While this is already informative (and reduces search) in some examples like the one we have looked at in section 5, we want to improve it in several ways. In particular, we want to keep the reference set of models small and informative at the same time. In order to do so other approaches for using models are necessary. We think that using *typical models* [Kerber et al., 1993] is promising for this purpose.

For the future we plan to deepen our understanding by more experiments. In particular we want to investigate different approaches to semantic selection strategies.

Acknowledgments

We would like to thank Andreas Meier for his generous help to integrate the semantic restriction in the ΩMEGA proof planner. Furthermore we are indebted to Mateja Jamnik, Martin Pollet, and Claus-Peter Wirth for valuable feedback on earlier drafts of this work.

References

Benzmüller, C., Cheikhrouhou, L., Fehrer, D., Fiedler, A., Huang, X., Kerber, M., Kohlhase, M., Konrad, K., Meier, A., Melis, E., Schaarschmidt, W., Siekmann, J., and Sorge, V., 1997, ΩMEGA: Towards a mathematical assistant, in: *Proceedings of the 14th International Conference on Automated Deduction*, W. McCune, ed., Springer, Berlin, pp. 252–255.

Braine, M., 1978, On the relation between the natural logic of reasoning and standard logic, *Psychological Review* 85(1):1–21.

Bundy, A., Stevens, A., van Harmelen, F., Ireland, A., and Smaill, A., Rippling: A heuristic for guiding inductive proofs, *Artificial Intelligence* 62:185–253.

Bundy, A., 1988, The use of explicit plans to guide inductive proofs, in: *Proceedings of the 9th International Conference on Automated Deduction*, E. Lusk and R. Overbeek, eds., Springer, Berlin, pp. 111–120.

Bundy, A., van Harmelen, F., Hesketh, J., and Smaill, A., 1991, Experiments with proof plans for induction, *JAR* 7(3):303–324.

Bundy, A., van Harmelen, F., Smaill, A., and Ireland, A., 1990, Extensions to the rippling-out tactic for guiding inductive proofs, in: *Proceedings of the 10th International Conference on Automated Deduction*, M.E. Stickel, ed., Springer, Berlin, pp. 132–146.

Chu, H. and Plaisted, D. A., 1994, Semantically guided first-order theorem proving using hyper-linking, in: *Proceedings of the 12th International Conference on Automated Deduction*, A. Bundy, ed., Springer, Berlin, pp. 192–206.

Eisinger, N. and Ohlbach, H. J., 1986, The Markgraf Karl Refutation Procedure (MKRP), in: *Proceedings of the 8th International Conference on Automated Deduction*, J. Siekmann, ed., Springer, Berlin, pp. 681–682.

Gelernter, H., 1959, Realization of a geometry theorem-proving machine, in: *Proceedings of the International Conference on Information Processing*, UNESCO.

Hadamard, H., 1944, *The Psychology of Invention in the Mathematical Field*, Dover Publications, New York.

Hodgson, K. and Slaney, J., 2001, Development of a semantically guided theorem prover, in: *Proceedings of the International Joint Conference on Automated Reasoning*, Springer, Berlin, pp. 443–447.

Huang, X., Kerber, M., and Kohlhase, M., 1992, Methods – the basic units for planning and verifying proofs, SEKI Report, SR-92-20, Fachbereich Informatik, Universität des Saarlandes, Saarbrücken, Germany.

Huang, X., Kerber, M., Richts, J., and Sehn, A., 1994, Planning mathematical proofs with methods, *Journal of Information Processing and Cybernetics, EIK* 30(5-6):277–291.

Ireland, A. and Bundy, A., 1996, Productive use of failure in inductive proof, *JAR* 16(1–2):79–111.

Johnson-Laird, P.N. and Byrne, R., 1991, *Deduction*, Lawrence Erlbaum Associates Publishers, Hove.

Kerber, M., Melis, E., and Siekmann, J., 1993, Analogical reasoning based on typical instances, in: *Proceedings of the IJCAI-Workshop on Principles of Hybrid Reasoning and Representation*, Chambéry, France, pp. 85–95.

Kerber, M. and Choi, S., 2000, The semantic clause graph procedure, in: *Proceedings of CADE-17 Workshop on Model Computation – Principles, Algorithms, and Applications*, P. Baumgartner, C. Fermüller, N. Peltier and H. Zhang, eds., pp. 29–37.

Kerber, M., 1998, Proof planning – a practical approach to mechanised reasoning in mathematics, in: *Automated Deduction – A Basis for Applications*, W. Bibel and P. H. Schmitt, eds., vol. III, Kluwer Academic Publishers, pp. 77–95.

Letz, R., Schumann, J., Bayerl, S., and Bibel, W., 1992, SETHEO: A high-performance theorem prover, *JAR* 8:183–212.

McCune, W., 1994, *OTTER 3.0 Reference Manual and Guide*, Mathematics and Computer Science Division, Argonne National Laboratory, Argonne, IL.

McCune, W., 1997, Solution of the Robbins problem, *JAR* 19:263–276.

Pólya, G., 1945, *How to Solve It*, Princeton University Press, Princeton, NJ.

Pólya, G., 1954, *Mathematics and Plausible Reasoning*, Princeton University Press, Princeton, NJ.

Pólya, G., 1962, *Mathematical Discovery – On Understanding, Learning, and Teaching Problem Solving*, Princeton University Press, Princeton.

Robinson, J.A., 1965, A machine-oriented logic based on the resolution principle, *JACM* 12(1):23–41.

Riazanov, A. and Voronkov, A., 2001, Vampire 1.1, in: *Proceedings of the International Joint Conference on Automated Reasoning*, R. Goré, A. Leitsch, T. Nipkow, eds., Springer, Berlin, pp. 376–380.

Slaney, J., 1995, *FINDER – Finite Domain Enumerator Version 3.0 Notes and Guide*, Centre for Information Science Research, Australian National University, Canberra, Australia.

Slaney, J., Lusk, E., and McCune, W., 1994, SCOTT: Semantically Constrained Otter, in: *Proceedings of the 12th International Conference on Automated Deduction*, A. Bundy, ed., Springer, Berlin, pp. 764–768.

Smullyan, R.M., 1968, *First-Order Logic*, Springer, Berlin.

van der Waerden, B., 1964, Wie der Beweis der Vermutung von Baudet gefunden wurde, *Abh. Math. Sem. Univ. Hamburg* 28:6–15.

Weidenbach, C., Bernd, G., and Georg, R., 1996, SPASS & FLOTTER, Version 0.42, in: *Proceedings of the 13th International Conference on Automated Deduction*, M.A. McRobbie and J.K. Slaney, eds., Springer, Berlin, pp. 141–145.

DEGREES OF ABDUCTIVE BOLDNESS

Isabella C. Burger

Department of Mathematics, Rand Afrikaans University, P.O. Box 524, Auckland Park, 2006, South Africa

icb@na.rau.ac.za

Johannes Heidema

Department of Mathematics, University of South Africa, P.O. Box 392, Unisa, 0003, South Africa

heidej@unisa.ac.za

Abstract   Underlying most processes and algorithms of deduction and abduction we find an entailment relation between propositions. This consequence relation may be classical, modulo a theory, nonmonotonic and defeasible, statistical or probabilistic, fuzzy, etc. We define and study relations between propositions, some of which employ information to boldly go beyond the cautiousness of conventional inference. They do this by merging an inferential and a conjectural relation to a single new relation, but with varying relative strengths of the two components in the blend, endowing the new relation with varying degrees of abductive boldness. They may lead to algorithms for new deductive and abductive processes.

1.   Introduction: Cautious inference and bold conjecture

Abduction is ubiquitous. Roughly speaking, it is the inverse of deduction. In various contexts it coincides with, or relates to, conjecture; diagnosis [Hamscher et al., 1992; Peng and Reggia, 1990]; induction [Flach and Kakas, 2000]; inference to the best explanation; hypothesis formulation; interpretation and disambiguation [Hobbs et al., 1993]; learning [Gabbay and Smets, 2000]; pattern recognition; etc. [Aliseda, 1997; Hintikka, 1998; Paul, 1993]. Some of its contexts are rather remote from logic, for instance disambiguation in perception and in the

L. Magnani, N.J. Nersessian, and C. Pizzi (eds.),
Logical and Computational Aspects of Model-Based Reasoning, 163-180.
© 2002 Kluwer Academic Publishers. Printed in the Netherlands.

understanding of natural language [Josephson and Josephson, 1994, ch. 10], [Marr, 1982]. Closer to a formal approach to abduction are works exhibiting its relevance to the philosophy of science [Kuipers, 1999; Magnani, 2001; Magnani et al., 1999]. We are concerned with *logical* aspects of abduction and in particular with the characterization of degrees of boldness of the relation between propositions which underlies a rational abductive process. This relation is usually taken to be classical entailment or a somewhat bolder nonmonotonic or defeasible entailment. We suggest that a whole spectrum of relations, ranging between classical entailment and its inverse, is relevant for deduction and abduction and their interrelationships.

At the heart of logic lie the relations of *inference* (entailment, consequence) and its inverse, *conjecture* (abduction, antecedence) – whether in their classical or newer non-classical guises. Intuitively we may think of these relations on propositions as inducing dynamic processes: inference is a *cautious* process, since the truth of the premisses ensures the truth – or at least the plausibility, i.e. the truth in all the most normally occurring of the relevant states – of the conclusion.

Conjecture, on the other hand, is a *bold* process; it moves from given sentences to a logically stronger one, which may be false. Exactly how bold depends i.a. on the extra information available beyond the given sentences. Some extra information is needed to justify the conjectural move as rational.

Defeasible inference as in nonmonotonic logic, involves a mildly daring form of abduction – it takes an initial abductive jump, and from there on the flow of the process is according to classical entailment. We sketch a spectrum of relations on propositions of varying degrees of boldness which may be employed in new ways in deduction and abduction. They are obtained by merging an inferential and a conjectural relation to a single new relation, but with varying relative strengths of the two components in the blend. In this way very bold abductions may still be rationally justified. The extra information needed to construct these relations in a judicious way will be in the form of orderings on the set of all states.

To keep the exposition simple, we restrict the logic to a propositional language L generated by finitely many atoms, with all the usual propositional connectives and the usual semantics of *states* (possible worlds, interpretations, truth valuations). Illustrative examples employ the language generated by p and q with its four states 11, 10, 01 and 00 (where $1 = $ true, $0 = $ false, and the first bit indicates the valuation of p, the second that of q).

A state which satisfies a sentence, i.e. makes it true, is a *model* of that sentence. The term *proposition* will, with benign ambiguity, indicate any of the following: a logical equivalence class of sentences; a representative sentence from such a class; the set of models of a (class of) sentence(s); the corresponding element of the Lindenbaum-Tarski algebra of the language.

2. Abduction as defeasible inference

The archetypal case where from proposition X we cautiously infer (or deduce) proposition Y has the underlying relation "X semantically entails Y", i.e. $X \models Y$ or $Mod(X) \subseteq Mod(Y)$ (where $Mod(X)$ denotes the set of models of proposition X, maybe restricted to models of some background theory). The entailment relation $\models$ is cautious in the sense that it preserves truth, but may lose logical information. Inversely, we may also consider the case that from X we may daringly conjecture (or abduct) Y, when $Mod(X) \supseteq Mod(Y)$, in which case truth may be lost, but logical information is preserved. Here inference and conjecture are simply inverses of each other. To illustrate, consider the propositional language generated by the two atomic symbols p and q. Entailment is depicted in the Lindenbaum-Tarski algebra $\mathcal{B} = (B, \models, \wedge, \vee, \neg, \perp, \top)$ of this language (Figure 1) in which $X \models Y$ if X lies below Y in the diagram and X and Y are connected by a line (either directly or transitively). The class of tautologies (logical truths) is the top element $\top$ in $\mathcal{B}$; the class of contradictions is the bottom element $\perp$. Note that $p + q$ is logically equivalent to $(p \wedge \neg q) \vee (\neg p \wedge q)$ and that $p \leftrightarrow q$ is logically equivalent to $(p \wedge q) \vee (\neg p \wedge \neg q)$. The conjectural relation is obtained by just turning the picture upside-down. These two relations are completely independent of any (extra) data or information we may have available – they are the completely ignorant person's way of founding the processes of inference and conjecture.

Defeasible inference describes the process of drawing plausible conclusions which cannot be guaranteed absolutely on the basis of the available knowledge, i.e. inference where one needs to go beyond the definite knowledge, but without making blind guesses. This may be done by employing a "default rule", justifying shrinking the set of models of X to that of X' – a proposition logically stronger that X – in strengthening the relation $X \models Y$. In other words: $X \mid\sim Y$ (read as 'X defeasibly entails Y') if and only if $X' \models Y$ (where $X' \models X$). For the purposes of this paper, we shall require that a default rule should be expressible as an *epistemic state*, i.e. a total preorder (reflexive, transitive, connecting relation) $\leq_{ep}$ on the set $\mathbf{W}$ of all possible states. An epistemic state

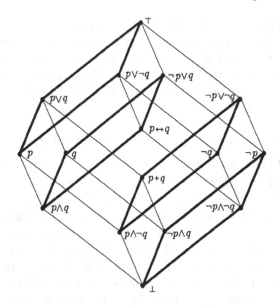

Figure 1.   The Lindenbaum-Tarski algebra B generated by $\{p, q\}$.

partitions the set of states into equivalence classes on which a linear order is induced. The higher up a state lies in the ordering $\leq_{ep}$, the higher its degree of preference (or, normality). Within this framework of preferential model semantics [Shoham, 1998], $Mod(X')$ is taken as the set of *maximal* models of X in $\leq_{ep}$:

Definition 1 *For any* $X, Y \in B$,

$$X \mid\sim Y :\Longleftrightarrow \max_{\leq_{ep}} \mathrm{Mod}(X) \subseteq \mathrm{Mod}(Y).$$

An element $x \in \mathbf{W}$ is *maximal* in $\mathbf{X} \subseteq \mathbf{W}$ iff $x \in \mathbf{X}$ and there is no $y \in \mathbf{X}$ such that $x \leq_{ep} y$ and $y \not\leq_{ep} x$.

Jumping from X to the logically stronger X' requires a (judicious) conjectural action, but that is where the bold part of this type of abduction ends – when X' is obtained, we proceed with classical inference. Referring the reader again to Figure 1, finding the defeasible consequences of X, means going down in the diagram to a proposition X' (according to an epistemic state $\leq_{ep}$), and then taking the filter above X', i.e. taking the deductive closure of X'. Nayak and Foo [Nayak and Foo, 1999] employ epistemic states in a slightly different way to obtain abduction in the sense of defeasible inference.

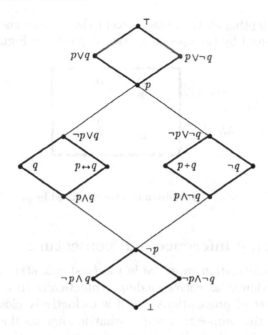

Figure 2. B ordered by $\models_{T,\models}$ with $T = p$.

The entailment relation $\models$ is the special case of $|\sim$ when the epistemic state $\leq_{ep}$ consists of only one partition class – that of all possible states. An epistemic state with two classes dichotomizes the states: "normal" versus "abnormal"; "good" versus "bad". In this case, if we take the top class as the set of models of a proposition T, then $|\sim$ is a refinement of the usual *T-expanded entailment relation*: $X \models_T Y :\Longleftrightarrow X \wedge T \models Y$. The refinement occurs only on those propositions which are inconsistent with T – they are not all $|\sim$-equivalent, but are ordered by $\models$ in $|\sim$. If we would refine $\models_T$ by 'blowing up' the equivalence classes in $\models_T$ by means of $\models$, but retaining the $\models_T$-ordering between the classes, that would give us a relation $\models_{T,\models}$ which is a refinement of any T-induced *epistemic entrenchment ordering* (cf. [Antoniou, 1997, p. 192] for a definition) on the propositions in B. In general, the more partition classes there are in the epistemic state (and hence, roughly speaking, the fewer states there are in a class), the higher the degree of boldness in the conjectural jump of $|\sim$. In Figure 2 we illustrate $\models_{T,\models}$ for the p, q language where T is the proposition p. The equivalence classes in $\models_T$ are obtained by collapsing each of the four classes drawn with thick lines, but retaining the ordering between the classes (indicated by the thin lines). Collapsing all the classes except the bottom one gives us the relation $|\sim$ induced by the epistemic state depicted in Figure 3.

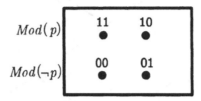

Figure 3.   An epistemic state – induced by p.

3.   Merging inference and conjecture

The semantic information captured in the epistemic state given in Figure 3 may be viewed as corresponding syntactically to a *belief set* or *theory* (i.e. a set of propositions which is deductively closed) which is represented by the proposition p. (In what follows we shall think of a belief set as the single logically strongest proposition in the set that represents the set.) In general, every belief set T sorts the states in $\mathbf{W}$ into the *accepted set* $Mod(T)$ (the top class in the epistemic state) and the *rejected set* $\mathbf{W} - Mod(T) = Mod(\neg T)$ (the bottom class). If now X and Y are any two propositions (each representing a belief set), we want to compare them as to how well they sort the possible states into two classes, given the T-sorting as a norm (a fixed bench-mark). Looking at Figure 4, it seems natural to say that 'X *sorts at best as well as Y does (relative to the T-sorting as norm)*' if and only if the following two conditions hold:

$$Mod(X) \cap Mod(T) \subseteq Mod(Y) \cap Mod(T) \text{ and}$$
$$Mod(Y) \cap Mod(\neg T) \subseteq Mod(X) \cap Mod(\neg T),$$

in which case we write $X \sqsubseteq_T Y$.

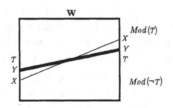

Figure 4. Y sorts closer to T than X.

Syntactically $\sqsubseteq_T$ is defined as follows:

Definition 2 *For any* $X, Y, T \in B$,

$$X \sqsubseteq_T Y :\Longleftrightarrow (X \wedge T) \models (Y \wedge T) \text{ and } (Y \wedge \neg T) \models (X \wedge \neg T).$$

This definition has previously been proposed by David Miller [Miller, 1978] and Theo Kuipers [Kuipers, 1987] in the context of the study of verisimilitude [Brink, 1989; Niiniluoto, 1998; Zwart, 1998], and studied in detail by [Burger and Heidema].

Note that $X \sqsubseteq_T Y \Longleftrightarrow [(X \leftrightarrow T) \models (Y \leftrightarrow T)]$. The relation $\sqsubseteq_T$ is a Boolean ordering, and hence induces a Boolean algebra $\mathcal{B}_T = (B, \sqsubseteq_T, \sqcap_T, \sqcup_T, \neg, \neg T, T)$, which has T as its top, $\neg T$ as its bottom, the same complementation $\neg$ (negation) as $\mathcal{B}$, and meet and join operations which are related to those in $\mathcal{B}$ ($\wedge$ and $\vee$) as follows:

$$X \sqcap_T Y = (X \wedge Y) \vee [(X \vee Y) \wedge \neg T]$$
$$X \sqcup_T Y = (X \vee Y) \wedge [(X \wedge Y) \vee T].$$

The original Lindenbaum-Tarski algebra $\mathcal{B}$ (e.g. Figure 1) and the new *T-modulated* Boolean algebra $\mathcal{B}_T$ are isomorphic under the isomorphism $m_T : B \longrightarrow B$, $m_T(X) := (X \leftrightarrow T)$, which is its own inverse, and hence also establishes an isomorphism in the opposite direction. To illustrate, we show $\mathcal{B}_p$ for the p, q language in Figure 5.

When for a pair of propositions (X, Y), $X, Y \in B$, $Mod(X)$ and $Mod(Y)$ differ by a single state, we say that (X, Y) is a *step*. If (X, Y) is a step and $X \models Y$, then (X, Y) is an *inferential step*, while in the opposite case, when $Y \models X$, (X, Y) is a *conjectural step*. The 32 (pairs of) steps of the p, q language are represented by the 32 lines in Figure 1 and by the same 32 lines in Figure 5, although now arranged differently. Whereas in Figure 1 every step upward is inferential and every step downward is conjectural, it is no longer the case in $\mathcal{B}_T$. In Figure 5, depicting $\sqsubseteq_p$, we call a step upward *p-positive*, since it goes from a sorting to another one which is more similar to the p-sorting. A step downward in $\sqsubseteq_p$ is *p-negative*. Some p-positive steps are inferential, i.e. they go up in both $\models$ and $\sqsubseteq_p$. These are represented by the *thin* lines going in the same direction in $\models$ and $\sqsubseteq_p$. The other p-positive steps are conjectural, i.e. they go up in $\sqsubseteq_p$ but down in $\models$. They are represented by the *thick* lines, which go in opposite directions in $\models$ and $\sqsubseteq_p$. Thus, $\sqsubseteq_p$ is a merging of inference and conjecture; of cautiousness and boldness.

This mixed nature of $\sqsubseteq_T$ can also be seen in its definition, which consists of two parts, equitably combined by "and":

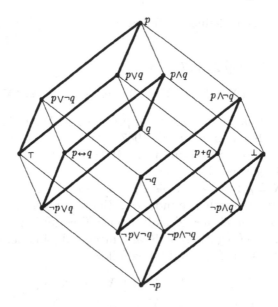

Figure 5.   The p-modulated Boolean algebra $\mathcal{B}_p$ generated by $\{p, q\}$.

$$X \sqsubseteq_T Y :\Longleftrightarrow (X \wedge T) \models (Y \wedge T) \text{ and } (Y \wedge \neg T) \models (X \wedge \neg T).$$

The first part, $X \wedge T \models Y \wedge T$, or equivalently $X \wedge T \models Y$, holds of course when $X \models Y$. The second part, $Y \wedge \neg T \models X$, holds when $Y \models X$. So the first part expands entailment, while the second part expands inverse entailment. If $T = \top$, then $X \sqsubseteq_T Y$ iff $X \models Y$ – in which case every T-positive step is inferential. If $T = \bot$, then $X \sqsubseteq_T Y$ iff $Y \models X$, i.e. every T-positive step is conjectural. In general, the logically stronger T becomes, the higher $\bot$ moves up in $\sqsubseteq_T$, and the closer the relation $\sqsubseteq_T$ comes to the inverse of entailment.

The very special case when we know exactly which world is the actual world, i.e. the case when T has one model only, yields a diagram of $\sqsubseteq_T$ in which T is only one inferential step upwards from $\bot$.

Should we daringly want to give, e.g., the conjectural component of $\sqsubseteq_T$ precedence over its inferential one, we could combine the two parts lexicographically rather than by conjunction. This would mean that we firstly order the propositions by employing only the conjectural part of the definition of $\sqsubseteq_T$, and if necessary, i.e. when two propositions are equivalent in terms of the conjectural part, we employ the inferential component to establish a refinement amongst those propositions which are equivalent with respect to the conjectural component. Favoring the conjectural component of $\sqsubseteq_T$ yields an relation $\sqsubseteq_T^d$ which orders the propositions much more daringly than $\sqsubseteq_T$ (that explains the "d" in the notation). The diagram on the left in Figure 6 depicts $\sqsubseteq^d_T$ for $T = p$ in our illustrative language.

The counterpart of $\sqsubseteq^d_T$ is the relation $\sqsubseteq^c_T$ (see Figure 6, diagram on the right, for an illustration) in which we cautiously favor the inferential component of $\sqsubseteq_T$ (in a lexicographical way) above the conjectural component.

Formally:

Definition 3 *For any* $X, Y, T \in B$,

$$X \sqsubseteq^d_T Y : \quad \Longleftrightarrow \quad (Y \wedge \neg T) \models (X \wedge \neg T)$$
$$\text{and if } (Y \wedge \neg T) = (X \wedge \neg T), \text{ then } (X \wedge T) \models (Y \wedge T);$$
$$X \sqsubseteq^c_T Y : \quad \Longleftrightarrow \quad (X \wedge T) \models (Y \wedge T)$$
$$\text{and if } (X \wedge T) = (Y \wedge T), \text{ then } (Y \wedge \neg T) \models (X \wedge \neg T).$$

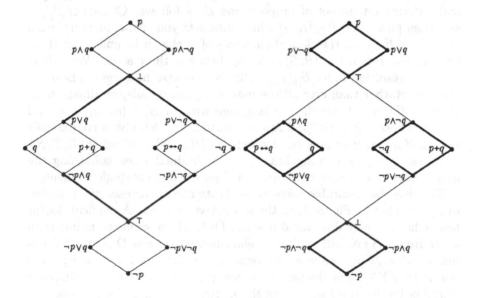

Figure 6. B ordered by $\sqsubseteq^d_p$ (left) and $\sqsubseteq^c_p$ (right).

If we now consider an epistemic state $\leq_{ep}$ with more than two classes, we can distinguish logically weaker – and hence more secure – *knowledge* from logically stronger – and hence bolder – *beliefs*. We now call the (non-empty) top equivalence class the set of *accepted* states, and the bottom class the set of *rejected* states. The union of the classes between the top and the bottom class will be the set of *undecided* states. Given an epistemic state, the proposition R corresponding to the accepted class of states is called its *belief (set)*, while the proposition S corresponding

to the set **W**− *(rejected class)* of states is called its *knowledge (set)*.
Note that $R \models S$. We can now generalize the $\sqsubseteq_T$-relation by replacing
the role of T by a similar role, but this time played by the interval
$[R, S] = \{U \in B \mid R \models U \text{ and } U \models S\}$ in B between belief and knowledge
[Burger and Heidema, 2000]. The latter paper also explains how the new
$\sqsubseteq_{[R,S]}$-relation may be employed in belief revision, a topic which is also
discussed by Aliseda [Aliseda, 1997], Boutilier and Becher [Boutilier and
Becher, 1995], and Pagnucco [Pagnucco, 1996].

Propositions whose model sets differ only on the set of undecided
states are equivalent in the new ordering. The equivalence classes are
also intervals in B and they become the elements of the new Boolean
algebra, say $B_{[R,S]}$, which has $[R, S]$ as top and $[\neg S, \neg R]$ as bottom. To
come again to a classification of (at least some) inferential and conjectural steps as being either positive or negative, we define a lexicographical ordering on the set of propositions B as follows: Construct $B_{[R,S]}$
and then pick some $T \in [R, S]$ which interests you. Now 'blow up' each
element of $B_{[R,S]}$ to the class of elements of B that it is, and order them
by $\sqsubseteq_T$, but retain the $[R, S]$-ordering between the classes. We call B
with the resulting order $B_{[R,T,S]}$. This is of course no longer a Boolean
algebra. Rather than give all the mathematical details, we illustrate in
Figure 7 the results for the p, q language with $[R, S] = [p \wedge q, p \vee q]$ and
$T = p$. Those inferential and conjectural steps which give a comparable
pair in $B_{[R,T,S]}$, may now be classified as $[R, T, S]$-*positive* or $[R, T, S]$-
negative. The Boolean algebra $B_{[R,S]}$ is obtained when collapsing the
four classes drawn with thick lines in Figure 7 to four single elements.

This may be a suitable stage to illustrate one of the uses of our orderings, e.g. that in Figure 7, in the abductive process. A medical doctor
researches a newly discovered disease, DIS, which seems to manifest in
a syndrome, SYN, although the relationship between DIS and SYN is
not yet clear. Suppose $p =$ *"the patient has DIS"* and $q =$ *"the patient
manifests SYN"*. The doctor sees a new patient and after a preliminary
examination is pretty sure – has the knowledge – that $S = p \vee q =$ *"the
patient has DIS, or at least shows SYN"*, while his bold belief is that
$R = p \wedge q =$ *"the patient has DIS and shows SYN"*, and the crucial part
of this belief is of course his theory that $T = p =$ *"the patient has DIS"*.

The lattice of Figure 7 is one rational way to represent the epistemic
ranking that the doctor imposes on propositions at this stage. Further
tests now conclusively establish as empirical evidence that $E = q =$ *"the
patient shows SYN"*.

The doctor now confronts the following typical abductive problem
situation: Given his current theory T and the observation E that has to
be explained, he must find one or more suitable hypotheses H such that

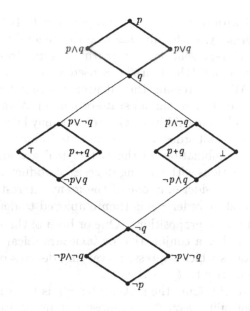

Figure 7. $\mathcal{B}_{[R,T,S]}$ with $R = p \wedge q$, $T = p$, $S = p \vee q$.

T together with H entails E: $T \wedge H \models E$. Here we usually demand that
(i) H is consistent with T;
(ii) T does not entail E; and
(iii) H does not entail E.

In our example (and Figure 1 may help to see this) there are only two candidates satisfying these conditions, namely $H_1 = p \leftrightarrow q$, and $H_2 = \neg p \vee q = p \rightarrow q$.

H_2 is the logically weakest candidate, and if logical caution were the only consideration, it should be preferred. But look at Figure 7: H_1 ranks above H_2, being closer to what the doctor knows and believes than H_2.

Given the whole epistemic situation with its ranking of propositions, the doctor should dare to hypothesize H_1. This relative ranking of H_1 and H_2 is stable across different orderings (on the propositions) which incorporate a p-endorsed abductive component: in all the Figures 5, 6, and looking ahead, 10, H_1 ranks above H_2.

4.　Abduction via power relations

Another way of merging inference and conjecture is by employing the mathematical notions of *power set* and *power ordering* to order the propositions in B with respect to some proposition T. Before we discuss these relations formally, we briefly describe the general idea.

A proposition T (representing "the truth" or some (possibly) incomplete information about "the truth") is used to induce partial orders on the states in $\mathbf{W}$ – expressing the "degree of compatibility with the models of T". Some of these partial state-orders on $\mathbf{W}$ may represent a daring approach to "compatibility", while others may be more cautiously orientated. These partial orders are then used to induce power orderings on sets of states. Combinations of the latter yield a whole spectrum of orderings on propositions with varying degrees of abductive boldness. In this paper we shall consider only one of the many interesting T-induced state-orders. This state-order $\preceq_T$ is then employed to define two power relations $\preceq_T^{\downarrow}$ and $\preceq_T^{\uparrow}$ on propositions. One or both of these two relations are then used in either a conjunctive or lexicographical way to obtain orderings on theories which express, in varying degrees of cautiousness and boldness, similarity to T.

The first step in obtaining the state-order $\preceq_T$ is to introduce the notion of a "compatibility vector". Suppose our propositional language has n atoms and that $\mathbf{2}$ is the two-element Boolean algebra $\{0, 1\}$ with $0 < 1$. Every state may then be seen as an n-tuple of zeros and ones. We now want to think of the elements of $\mathbf{W}$ also in another way, namely as *compatibility* or *comparison* vectors – expressing the degree of compatibility or agreement or similarity of two states of the world:

Definition 4 *Given any two elements t and x of $\mathbf{W}$, we define $t(x) \in \mathbf{W}$ by*

$$[t(x)]_i := \begin{cases} 1 & \text{if } t_i = x_i, \\ 0 & \text{if } t_i \neq x_i, \end{cases} \text{ for any } i \in \{1, 2, ..., n\}.$$

The vector $t(x)$ $(= x(t))$ expresses the compatibility or structural similarity of two states x and t: the more 1's in $t(x)$, the closer state t and x resemble each other. If $t = x$, then $[t(x)]_i = 1$ for every i and one may schematically think of $t(x)$ as the vector $11...1$ – expressing *maximum compatibility* between the vectors t and x.

On the other hand, $t(x) = 00...0$ expresses *maximum incompatibility* – the case when $t_i \neq x_i$ for every i. Every other vector in between these extreme cases represents a partial compatibility between t and x. In other words: the higher up $t(x)$ lies within $(\mathbf{W}, \leq_{pr})$, where $\leq_{pr}$ (the product order on $\mathbf{W} = \mathbf{2}^n$) compares the compatibility vectors componentwise, the better the concordance of t and x. Hence, if $t(x) \leq_{pr} t(y)$, then x *is at most as compatible with t as y is*. The order $\leq_{pr}$ on $\mathbf{W}$ is defined formally as follows:

Definition 5 *For every* $x, y \in \mathbf{W}$,

$$x \leq_{pr} y :\Longleftrightarrow x_i \leq y_i \ (in \ \mathbf{2}) \ for \ all \ i \in \{1, 2, ..., n\}.$$

The idea of compatibility between two possible states can now be expanded to the notion of compatibility between a possible state and a set of states.

A convention on notation: In what follows we shall denote the set of models of a theory X by the corresponding bold letter $\mathbf{X}$.

Definition 6 *Let* $\mathbf{T} = Mod(T)$. *For every* $x \in \mathbf{W}$ *we define*

$$\mathbf{T}(x) := \{t(x) \mid t \in \mathbf{T}\}.$$

The set $\mathbf{T}(x)$ of comparison vectors represents x in so far as compatibility with the data T is concerned. Roughly speaking, the higher the set $\mathbf{T}(x)$ lies within $(\mathbf{W}, \leq_{pr})$, the higher the compatibility of the state x with the data T or the set of states $\mathbf{T}$.

In Table 1 we illustrate the set $\mathbf{T}(x)$ and the corresponding proposition $T(x)$ for every $x \in \mathbf{W} = \{11, 10, 01, 00\}$ when $\mathbf{T} = \{11, 10\}$, i.e. when $T = p$. It is easily seen that $11...1 \in \mathbf{T}(x)$ if and only if $x \in \mathbf{T}$ – meaning that x is maximally compatible with one model of T.

x:	11	10	01	00
$\mathbf{T}(x)$:	$\{11, 10\}$	$\{10, 11\}$	$\{01, 00\}$	$\{00, 01\}$
$T(x)$:	p	p	$\neg p$	$\neg p$

Table 1. $\mathbf{T}(x)$ and $T(x)$ when $T=p$.

The definition of the state-order $\preceq_T$ is built on the following notion which is defined in terms of the product order $\leq_{pr}$:

Definition 7 *For any set of states* $\mathbf{X} \subseteq \mathbf{W}$, *we define*

$$\triangle\mathbf{X} := \{w \in \mathbf{W} \mid (\exists x \in \mathbf{X})(x \leq_{pr} w)\}, \ the \ \text{upset of } \mathbf{X}.$$

(We may dually define the notion *downset of* $\mathbf{X}$. For a more detailed discussion on "upsets" and "downsets" and their relationships to "positive" and "negative" sentences respectively, we refer the reader to [Burger and Heidema, 1994].)

Let us go back to the idea of viewing the elements of $(\mathbf{W}, \leq_{pr})$ as comparison vectors. For a given proposition T, every $x \in \mathbf{W}$ is represented by the configuration $\mathbf{T}(x) \subseteq \mathbf{W}$ of comparison vectors. To every one of these sets we may then apply $\triangle$ – defined in terms of the product ordering $\leq_{pr}$ on $\mathbf{W}$. We then obtain the upsets $\triangle \mathbf{T}(x)$. In agreement with our previous conventions on notation, we shall denote the corresponding propositions by $\triangle T(x)$. Table 2 shows the propositions $\triangle T(x)$ corresponding to the upsets $\triangle \mathbf{T}(x)$ of the sets $\mathbf{T}(x)$ listed in Table 1. These sets were obtained by means of the product ordering depicted in Figure 8.

x:	11	10	01	00
$\triangle T(x)$:	p	p	$\top$	$\top$

Table 2. $\triangle T(x)$ when $T = p$.

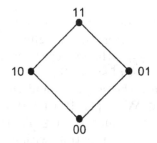

Figure 8. The product ordering, $\leq_{pr}$ on $\mathbf{W} = \{11, 10, 01, 00\}$.

We now employ the sets $\triangle \mathbf{T}(x)$ to define $\preceq_T$:

Definition 8 *Let* $\mathbf{T} = Mod(T)$, *where* T *is any proposition. For all* $x, y \in \mathbf{W}$ *we define*

$$x \preceq_T y :\Longleftrightarrow \triangle \mathbf{T}(y) \subset \triangle \mathbf{T}(x) \text{ or } x = y.$$

If $x \preceq_T y$, we say that x *is at most as compatible with* T *as* y *is.* (By employing downsets instead of upsets, we obtain a totally different state-order which compares states in a more cautious way with respect to the data T; and which induces power orders that are in general more cautiously orientated than those which we shall consider – induced by $\preceq_T$.)

If T is either tautological or contradictory, $\preceq_T$ is the identity relation on $\mathbf{W}$. The ordering $\preceq_T$ is different from the product ordering $\leq_{pr}$ except in the very special case $\mathbf{T} = \{11...1\}$ – the case when $\mathbf{T}(x) = \{x\}$ for all $x \in \mathbf{W}$ and hence $\Delta\mathbf{T}(y) \subset \Delta\mathbf{T}(x)$ if and only if $x <_{pr} y$. Figure 9 illustrates $\preceq_T$ when $T = p$.

We now consider two relations on propositions which are defined via power relations (induced by $\preceq_T$) on the corresponding sets of models.

Definition 9 *For any $X, Y, T \in B$,*

$$X \preceq_T^{\downarrow} Y : \iff (\forall x \in \mathbf{X})(\exists y \in \mathbf{Y})[x \preceq_T y];$$
$$X \preceq_T^{\uparrow} Y : \iff (\forall y \in \mathbf{Y})(\exists x \in \mathbf{X})[x \preceq_T y].$$

If T is either tautological or contradictory, $X \preceq_T^{\downarrow} Y \iff X \models Y$ and $X \preceq_T^{\uparrow} Y \iff Y \models X$. Thus, $\preceq_T^{\downarrow}$ generalizes the entailment relation, while $\preceq_T^{\uparrow}$ does the same with respect to the inverse of entailment. The way in which the "inferential" relation $\preceq_T^{\downarrow}$ and the "conjectural" relation $\preceq_T^{\uparrow}$ are combined in a new relation $\ll_T$ gives rise to a whole spectrum of mergings which ranges from cautious to daring when $\ll_T$ is defined in five different ways as follows:

Definition 10 *For any $X, Y, T \in B$,*

(i) $X \ll_T Y :\iff X \preceq_T^{\downarrow} Y$;

(ii) $X \ll_T Y :\iff X \preceq_T^{\downarrow} Y$ *and if also* $Y \preceq_T^{\downarrow} X$, *then* $X \preceq_T^{\uparrow} Y$;

(iii) $X \ll_T Y :\iff X \preceq_T^{\downarrow} Y$ *and* $X \preceq_T^{\uparrow} Y$;

(iv) $X \ll_T Y :\iff X \preceq_T^{\uparrow} Y$ *and if also* $Y \preceq_T^{\uparrow} X$, *then* $X \preceq_T^{\downarrow} Y$;

(v) $X \ll_T Y :\iff X \preceq_T^{\uparrow} Y$.

Brink and Heidema [Brink and Heidema, 1987] and Burger and Heidema [Burger and Heidema, 1994] discuss the relation $\ll_T$ as defined in

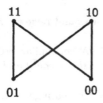

Figure 9. $\preceq_T$ on $\mathbf{W} = \{11, 10, 01, 00\}$ for $T = p$.

(iii) for the case when T represents "the truth". Figure 10 illustrates $\ll_T$ as defined in (iv) for $T = p$. One senses that this relation, while staying rational in the sense of being judiciously induced by the state-order $\preceq_T$ representing some form of compatibility with the theory T, is bolder than anything which could be considered to be some type of entailment.

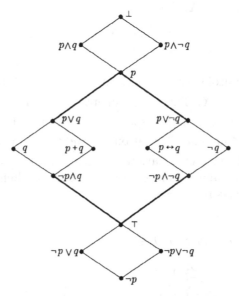

Figure 10. B ordered by $\ll_T$ (version (iv)) with $T = p$.

References

Aliseda, A., 1997, Seeking Explanations: Abductions in Logic, Philosophy of Science and Artificial Intelligence (ILLC Dissertation Series 1997-04), University of Amsterdam, Amsterdam.

Antoniou, G., 1997, *Nonmonotonic Reasoning*, MIT Press, Cambridge, MA.

Boutilier, C. and Becher, V., 1995, Abduction as belief revision, *Artificial Intelligence* 77:43-94.

Brink, C., 1989, Verisimilitude: Views and reviews, *History and Philosophy of Logic* 10:181–201.

Brink, C. and Heidema, J., 1987, A verisimilar ordering of theories phrased in a propositional language, *The British Journal for the Philosophy of Science* 38:533–549.

Burger, I.C. and Heidema, J., 1994, Comparing theories by their positive and negative contents, *The British Journal for the Philosophy of Science* 45:605-630.

Burger, I.C. and Heidema, J., 2000, Epistemic states guiding the rational dynamics of information, in: *PRICAI 2000: Topics in Artificial Intelligence* (Proceedings of the 6th Pacific Rim International Conference on Artificial Intelligence, Melbourne, Australia, August/September 2000, LNAI 1886), R. Mizoguchi and J. Slaney, eds., Springer-Verlag, Berlin, pp. 275–285.

Burger, I.C. and Heidema, J., Merging inference and conjecture by information, *Synthese*, forthcoming.

Flach, P.A. and Kakas, A.C., eds., 2000, *Abduction and Induction. Essays on their Relation and Integration* (Applied Logic Series, Vol. 18), Kluwer Academic Publishers, Dordrecht.

Gabbay, D.M. and Smets, P., eds., 2000, *Handbook of Defeasible Reasoning and Uncertainty Management Systems Volume 4: Abductive Reasoning and Learning*, Kluwer Academic Publishers, Dordrecht.

Hamscher, W., Console, L., and De Kleer, J., 1992, *Readings in Model-Based Diagnosis*, Morgan Kaufmann Publishers, San Mateo, CA.

Hintikka, J., 1998, What is abduction? The fundamental problem of contemporary epistemology, *Transactions of the Charles S. Peirce Society* 34(3):503–533.

Hobbs, J., Stickel, M., Appelt, D., and Martin, P., 1993, Interpretation as abduction, *Artificial Intelligence* 63:69–142.

Josephson, J.R. and Josephson, S.G., 1994, *Abductive Inference*, Cambridge University Press, Cambridge.

Kuipers, T.A.F., 1987, A structural approach to truthlikeness, in: *What is Closer-to-the-Truth? A Parade of Approaches to Truthlikeness* (Poznan Studies in the Philosophy of Science, Vol.10), T.A.F. Kuipers, ed., Editions Rodopi, Amsterdam, pp. 79–99.

Kuipers, T.A.F., 1999, Abduction aiming at empirical progress or even truth approximation, *Foundations of Science* 4.3:307–323.

Magnani, L., 2001, *Abduction, Reason, and Science Processes of Discovery and Explanation*, Kluwer Academic/Plenum Publishers, Dordrecht.

Magnani, L., Nersessian, N.J., and Thagard, P., eds., 1999, *Model-Based Reasoning in Scientific Discovery*, (Proceedings of the International Conference on Model-Based Reasoning in Scientific Discovery, Pavia, Italy, 1998), Kluwer Academic/Plenum Publishers, New York.

Marr, D., 1982, *Vision*, W.H. Freeman, San Francisco.

Miller, D., 1978, On distance from the truth as a true distance, in: *Essays in Mathematical and Philosophical Logic*, J. Hintikka, I. Niiniluoto, and E. Saarinen, eds., D. Reidel, Dordrecht, pp. 415–435.

Nayak, A.C. and Foo, N.Y., 1999, Abduction without minimality, in: *Advanced Topics in Artificial Intelligence* (Proceedings of the 12th Australian Joint Conference on Artificial Intelligence, AI'99, LNAI 1747), N. Foo, ed., Springer-Verlag, Berlin, pp. 365–377.

Niiniluoto, I., 1998, Verisimilitude: The third period, *The British Journal for the Philosophy of Science* 49:1–29.

Pagnucco, M., 1996, The Role of Abductive Reasoning within the Process of Belief Revision, PhD thesis, University of Sydney, Sydney.

Paul, G., 1993, Approaches to abductive reasoning: an overview, *Artificial Intelligence Review* 7:109–152.

Peng, Y. and Reggia, J.A., 1990, *Abductive Inference Models for Diagnostic Problem-Solving*, Springer-Verlag, Berlin.

Shoham, Y., 1988, *Reasoning about Change: Time and Causation from the Standpoint of Artificial Intelligence*, MIT Press, Cambridge, MA.

Zwart, S.D., 1998, Approach to the truth: Verisimilitude and truthlikeness, *(ILLC Dissertation Series 1998-02)*, University of Amsterdam, Amsterdam.

SCIENTIFIC EXPLANATION AND MODIFIED SEMANTIC TABLEAUX

Angel Nepomuceno-Fernández

Unidad de Lógica, Lenguaje e Información, Dpto. de Filosofía y Lógica,

Universidad de Sevilla, calle Camilo Jose Cela s/n, 41018 Sevilla, España

nepomuce@us.es

Abstract   Standard semantic tableaux method and a modification of its δ-rule, due to [Boolos, 1984] and [Díaz, 1993], allows us to obtain δ'-tableaux and apply that to abduction problem, paying attention to some versions given in [Aliseda, 1997]. Our approach of abduction in semantic tableaux faces up to the problem of the existence of infinite branches. Defined Cn, a basic logical operation, a new operation Cn* is obtained. An abduction problem $\langle \theta, \varphi \rangle$ can be seen as the problem of choosing the appropriated sentence of the set $\mathrm{Ab}(\langle \theta, \varphi \rangle)$, defined from Cn, or, taking into account Cn*, $\mathrm{Ab}(\langle \theta, \varphi \rangle)$. The main results are: (i) a (finite) set Γ of L-sentences is n-satisfiable iff the δ'-tableau of Γ has an open branch in which only n constants occur; (ii) if a finite set Γ of L-sentences is satisfiable, then the δ'-tableau provides a minimal interpretation that satisfies Γ; (iii) for any abduction problem $\langle \theta, \varphi \rangle$ such that θ is n-satisfiable, if there is a δ'-tableau of $\theta \cup \{\neg\varphi\}$ different from its standard tableau, then there is a solution α for $\langle \theta, \varphi \rangle$; (iv) given an abduction problem $\langle \theta, \varphi \rangle$ such that θ is n-satisfiable, if α is a solution and $\theta \cup \{\alpha\}$ is consistent, then $\alpha \in \mathrm{Ab}^*(\langle \theta, \varphi \rangle)$ and it is an explanatory solution (with respect to Ab*); (v) given an abduction problem $\langle \theta, \varphi \rangle$ such that θ is n-satisfiable, if α is an explanatory solution, then $\varphi \in \mathrm{Th}^*(\theta \cup \{\alpha\})$. They are Theorem 3, Corollary 4, Theorems 14 and 15 and Corollary 17, respectively. Several examples to illustrate that are given.

1.   Introduction

Semantic tableaux are a successful tool for drawing some scientific practices and can be applied to explanation problems. More specifically, an approach of abduction in semantic tableaux is possible as has been done recently [Aliseda, 1997; Mayer and Pirri, 1993]. Given an abduction problem $\langle \Theta, \varphi \rangle$, where Θ is a (finite) set of sentences of a language and

L. Magnani, N.J. Nersessian, and C. Pizzi (eds.),

Logical and Computational Aspects of Model-Based Reasoning, 181–198.

φ a sentence of the same, since φ is not logical consequence of Θ, the semantic tableau of $\Theta \cup \{\neg\varphi\}$ has at least an open branch. A solution for such problem can be obtained by means of sentences that, when added, close the open branches. Of course, to close any of that by adding certain sentences (φ, for example) should be avoided. Definitely, the semantic tableaux method can be applied to the search for solutions to an abduction problem.

Though the method is very effective for classical propositional logic, it is well known that an important problem raises when used for first order predicate logic, namely the existence of infinite branches. This represents a serious difficulty in order to achieve a system of reasoning for calculating solutions to explanation problems. However, some infinite tableaux may be transformed into others with completed branches [Díaz, 1993; Boolos, 1984]; those new finite branches provide information that permits the construction of a type of explanation.

In this paper the main aim is to study the problem of the existence of infinite branches when semantic tableaux are applied to a systematic search of a solution to an abductive problem. We propose the use of semantic tableaux by modifying a rule (the so called δ-rule) in a way that if a theory is finitely satisfiable, then the new semantic tableau presents the same advantages than the standard one to obtain certain explanatory facts, not only providing a solution to the abductive problem but giving a deductive justification for the fact that has to be explained. This is a first approximation that may be developed further. It should not be seen as opposite to others but as a complementary way of tackling a complex problem that is interesting from a philosophical, logical and computational points of view.

To abbreviate, in definitions and related contexts, we write "iff" instead of "if and only if". In the first section the main theorem of Beth's semantic tableaux is presented, though previous necessary notions are subjected to the appendix, where some of them are given in a summarized form. The new rule that substitutes the δ-rule is presented, then, taking the abduction problem as a prototype of an explanation problem, it is applied to define modified semantic tableaux from which provisional solutions are defined. Some examples to illustrate the process are given.

2. Modified semantic tableaux

To prove the main result in semantic tableaux we need some notions that are developed through the final appendix, so that we can pay attention to the most essential, in despite of consulting them when necessary.

Theorem 1 *Any finite set of sentences is unsatisfiable iff its semantic tableau is closed.*

Proof. Because of two lemmas given in appendix, a finite set of sentences is satisfiable iff its semantic tableau is open, by appropriated negations, the theorem is proved. □

Corollary 2 *For any finite set of sentences Γ and a sentence φ, $\Gamma \models \varphi$ iff the semantic tableau of $\Gamma \cup \{\neg\varphi\}$ is closed.*

Proof. $\Gamma \models \varphi$ iff $\Gamma \cup \{\neg\varphi\}$ is unsatisfiable, which (by the above theorem and proposition 27 –appendix–) is equivalent to the corresponding semantic tableau is being closed. □

This result remains though the δ-rule is modified as follows. Taking into account some interesting remarks [Díaz, 1993; Boolos, 1984] concerning infinite branches in some tableaux, we can adopt a variant of δ-rule, named δ'-rule (the resultant tableau is called δ'-tableau):

$$\frac{\exists x \varphi}{\varphi(b_1/x) \mid ... \mid \varphi(b_{n+1}/x)};$$

if $b_1, ..., b_n$ are the constants that occur in sentences of the branch. The branch Φ is divided:

$$\Phi + \varphi(b_1/x); \ ...; \ \Phi + \varphi(b_{n+1}/x),$$

$\varphi(b_i/x)$ is a δ'-sentence, for every $i \leq n+1$.

Theorem 3 *A (finite) set Γ of L-sentences is n-satisfiable, for $n \geq 1$, iff the δ'-tableau of Γ has an open branch in which only n constants occur.*

Proof. Suppose that Γ is n-satisfiable. Induction on the levels of application of the rules: let Γ' be 1-satisfiable, when the first constant appears, because of a γ-, δ- or δ'-rule, corresponding γ-, δ- or δ'-sentences are 1-satisfiable, because these rules are not applied further, and by applications of α- or β-rules if necessary, the branch becomes completed and open. Taking that for $k \geq 1$, the result can be achieved for $k + 1$. On the other hand, suppose the δ'-tableau of Γ has an open branch in which $n \geq 1$ constants occur, then a Herbrand interpretation that satisfies Γ can be defined, but the Herbrand universe must have n constants. □

Corollary 4 *If a finite set Γ of L-sentences is satisfiable, then the δ'-tableau provides a minimal Herbrand interpretation that satisfies Γ.*

Proof. If Γ is satisfiable but not finitely satisfiable, then the Herbrand universe is infinite and denumerable. If Γ is finitely satisfiable, then there is a minimal natural number $m \geq 1$ such that Γ is m-satisfiable, so its δ'-tableau has an open branch with only m constants and the corresponding Herbrand universe has m elements. $\square$

An example of a sentence 1-satisfiable is $\forall x \exists y Rxy$. If we apply Beth's method to the sentence, an infinite branch is generated, but working with δ'-tableau makes it more manageable:

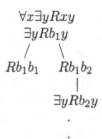

As can be seen, the branch on the left is open, though the one on the right is not completed. In fact, it grows infinitely on the right, since (it should be noted) it is the same branch as the one obtained by Beth's method. However, we have the following Herbrand interpretation:

$$D_H = \{b_1\}, \Im^H(R) = \{\langle b_1, b_1 \rangle\},$$

and, therefore $\langle D_H, \Im^H \rangle \models \forall x \exists y Rxy$ obviously.

3.   Scientific explanation

3.1   Preliminaries

Despite the controversial aspects of the notion of explanation, as a first approach, it is a form of inference that achieves premises from conclusions, so that from the theory and the explanation the conclusion could be deduced. Among the four kinds of explanations [Nagel, 1979], one is viewed approached from a logical point of view: The Hempelian or deductive-nomological model. We shall not study it in detail, but it is interesting to consider a form of explanation that, when produced, gives a formal reasoning, which (from any theory and the explanation) infers the phenomenon that had to be explained. This is the case of *abduction*.

A theory can be seen as a set Θ of L-sentences, so if φ is a L-sentence that formalizes the statement to be explained, the aim is to find another L-sentence α such that $\Theta \cup \{\alpha\} \models \varphi$.

3.2   Versions of abductions

Taking *abduction* as scientific explanation par excellence, any explanation problem can be reduced to an *abduction problem*, which is given by a theory $\Theta \subseteq L$ and a sentence $\varphi \in L$ such that $\Theta \not\models \varphi$ and $\Theta \not\models \neg\varphi$ [Mayer and Pirri, 1993], in symbols $\langle \Theta, \varphi \rangle$.

Definition 5 $\text{Ab}(\langle \Theta, \varphi \rangle) = \{\psi \in L : \Theta \cup \{\psi\} \models \varphi\}$. *Any sentence* $\alpha \in L$ *is a solution of the abduction problem* $\langle \Theta, \varphi \rangle$ *iff* $\alpha \in \text{Ab}(\langle \Theta, \varphi \rangle)$.

Taking into account the above notation, if $\varphi \in \text{Cn}(\Theta \cup \{\alpha\})$, then $\alpha \in \text{Ab}(\langle \Theta, \varphi \rangle)$. However, not all elements of $\text{Ab}(\langle \Theta, \varphi \rangle)$ are good explanations. In fact, a trivial solution is φ, since $\varphi \in \text{Ab}(\langle \Theta, \varphi \rangle)$, but it is not a satisfactory scientific explanation. As shown by Aliseda [Aliseda, 1997], we can consider five versions of scientific explanation (more specifically, abduction), though we shall pay attention to the following three:

1 Plain. $\alpha \in \text{Ab}(\langle \Theta, \varphi \rangle)$, that is to say $\Theta \cup \{\alpha\} \models \varphi$. This is the most basic one.

2 Consistent. $\alpha \in \text{Ab}(\langle \Theta, \varphi \rangle)$, that is to say $\Theta \cup \{\alpha\} \models \varphi$, and $\Theta \cup \{\alpha\}$ is consistent.

3 Explanatory. $\alpha \in \text{Ab}(\langle \Theta, \varphi \rangle)$, $\varphi \in \text{Cn}(\Theta \cup \{\alpha\})$, $\varphi \notin \text{Cn}(\Theta)$, and $\varphi \notin \text{Cn}(\{\alpha\})$, that is to say $\Theta \cup \{\alpha\} \models \varphi$, $\Theta \not\models \varphi$ and $\alpha \not\models \varphi$.

The definition of Ab rests on a logical operation ($\models$ or $\vdash$) that may be the classical one, characteristically deductive, which verifies several minimal conditions (see appendix, where Cn is defined from $\mathcal{P}(L)$ to $\mathcal{P}(L)$ by means of four clauses), but it is not a necessary requirement, since abduction and deduction are different after all. In fact, Ab is not a basic logical operation as the following proposition shows that one of the conditions fails.

Proposition 6 *Given an abduction problem* $\langle \Theta, \varphi \rangle$, $\text{Ab}(\langle \Theta, \varphi \rangle)$ *is such that*

$$\Theta \cap \text{Ab}(\langle \Theta, \varphi \rangle) \neq \Theta.$$

If $\Theta \cap \text{Ab}(\langle \Theta, \varphi \rangle) = \Theta$, then $\Theta \subseteq \text{Ab}(\langle \Theta, \varphi \rangle)$, but for any $\psi \in \Theta$, $\Theta \cup \{\psi\} = \Theta$ and by definition $\Theta \not\models \varphi$. $\square$

Semantic tableaux can be one of the most successful ingredients to work in abduction as a prototype of scientific explanation. As it is well known, for any (finite) set of L-sentences Θ, and L-sentences α and φ, $\Theta \cup \{\alpha\} \models \varphi$ iff $\Theta \cup \{\neg\varphi\} \models \neg\alpha$. The semantic tableau of $\Theta \cup \{\neg\varphi\}$

must be open, since the semantic tableau of $\Theta \cup \{\neg\varphi\} \cup \{\alpha\}$ must be closed, that is to say, by adding α to open branches of semantic tableau of $\Theta \cup \{\neg\varphi\}$, the closure of that should be achieved. In despite of any refinement of the tableaux method, δ'-tableaux may be another tool for obtaining explanations systematically in first order logic.

Definition 7 *Given an open branch of any semantic tableau Φ, a set Λ of non complementary literals is a closing set for Φ iff for every $l \in \Lambda$, $\Phi + l$ is closed.*

Definition 8 *Given an abduction problem $\langle \Theta, \varphi \rangle$ such that the theory Θ is n-satisfiable, $n \geq 1$, and Φ, a completed open branch of δ'-tableau of $\Theta \cup \{\neg\varphi\}$; if there is $\alpha = l_1 \wedge \dots \wedge l_m$, $m \geq 1$, such that $l_i \neq \neg\varphi$, $l_i \in \Lambda$, and $\Phi + l_i$ is closed, for every $i \leq m$, then α is a solution.*

Example 9 $\Theta = \{\forall x \exists y Rxy\}$, and φ is Ra_1a_2, a part of the δ'-tableau of $\{\forall x \exists y Rxy, \neg Ra_1a_2\}$ is

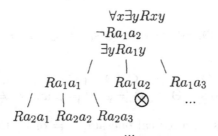

there are two finished open branches (another is closed, and two more are not finished):

$$\langle \forall x \exists y Rxy; \neg Ra_1a_2; \exists y Ra_1y; Ra_1a_1; Ra_2a_1 \rangle$$

and

$$\langle \forall x \exists y Rxy; \neg Ra_1a_2; \exists y Ra_1y; Ra_1a_1; Ra_2a_2 \rangle$$

so

$$\{Ra_1a_2; \neg Ra_1a_1; \neg Ra_2a_1\}$$

and

$$\{Ra_1a_2; \neg Ra_1a_1; \neg Ra_2a_2\}$$

are its closing set respectively, then

$$\neg Ra_1a_1 \wedge \neg Ra_2a_1$$

and
$$\neg Ra_1a_1 \wedge \neg Ra_2a_2$$
are provisional solutions of that abduction problem.

Example 10 $\Theta = \{\forall x \exists y (Rxy \wedge \neg Ryx)\}$ and φ is Ra_1a_3. One finished open branch of the δ'-tableau of $\Theta \cup \{\neg Ra_1a_3\}$ (whose developed presentation we omit to abbreviate) consists of the following sentences:

$$\forall x \exists y (Rxy \wedge \neg Ryx), \neg Ra_1a_3, \exists y (Ra_1y \wedge \neg Rya_1),$$

$$\exists y (Ra_2y \wedge \neg Rya_2), \exists y (Ra_3y \wedge \neg Rya_3),$$

$$Ra_1a_2 \wedge Ra_2a_1, Ra_1a_2, \neg Ra_2a_1,$$

$$Ra_2a_3 \wedge Ra_3a_2, Ra_2a_3, \neg Ra_3a_2,$$

and its closing set is

$$\{Ra_1a_3, \neg Ra_1a_2, Ra_2a_1, \neg Ra_2a_3, Ra_3a_2\},$$

according to definition,

$$\neg Ra_1a_2 \wedge Ra_2a_1 \wedge \neg Ra_2a_3 \wedge Ra_3a_2$$

is a provisional solution of that problem. More precisely,

Theorem 11 *For any abduction problem* $\langle \Theta, \varphi \rangle$ *such that* Θ *is n-satisfiable,* $n \geq 1$, *if there is a δ'-tableau of* $\Theta \cup \{\neg\varphi\}$ *(different from its standard tableau), then there is a solution α for* $\langle \Theta, \varphi \rangle$.

Proof. As $\varphi \notin \mathrm{Cn}(\Theta)$, δ'-tableau of $\Theta \cup \{\neg\varphi\}$ is open, there is a branch Φ in which m constants occur, $m \geq n$. Let Λ be the closing set for Φ, and $l_{i_1}, l_{i_2}, ..., l_{i_r} \in \Lambda$ such that $l_{i_j} \neq \neg\varphi$ for every $j \leq r$, then α is $l_{i_1} \wedge l_{i_2} \wedge ... \wedge l_{i_r}$. $\square$

Definition 12 Cn^* *is defined from* Cn *and the following rule: If* Γ *is a set of L-sentences,* $\exists x\beta$ *is a L-sentence and* $\Gamma \cup \{\exists x\beta\}$ *is n-satisfiable,* $n \geq 1$, *then*

$$\frac{\Gamma, \exists x\beta \models_n \psi}{\Gamma, \beta(a_1/x) \vee ... \vee \beta(a_n/x) \models_n \psi}.$$

So that for any set of axioms

$$\mathrm{Cn}^*(X) = \mathrm{Cn}(X) \cup \{\psi \in L : X \models_n \psi\}.$$

Theorem 13 Cn^* *is a basic logical operation.*

Proof. For any set of axioms X,

1 $X \subseteq \mathrm{Cn}(X) \subseteq \mathrm{Cn}^*(X)$, so that $X \subseteq \mathrm{Cn}^*(X)$,

2 if $X' = \{\psi \in L : X \models_n \psi\}$, then

$$\mathrm{Cn}^*(\mathrm{Cn}^*(X)) = \mathrm{Cn}^*(\mathrm{Cn}(X) \cup X');$$

but

$$\mathrm{Cn}^*(\mathrm{Cn}(X) \cup X') = \mathrm{Cn}^*(\mathrm{Cn}(X)) \cup \mathrm{Cn}^*(X'),$$

and

$$\mathrm{Cn}^*(\mathrm{Cn}(X)) \cup \mathrm{Cn}^*(X') = \mathrm{Cn}^*(X),$$

then

$$\mathrm{Cn}^*(\mathrm{Cn}^*(X)) = \mathrm{Cn}^*(X),$$

3 if $X \subseteq Y$, then $\mathrm{Cn}^*(X) \subseteq \mathrm{Cn}^*(Y)$,

4 if $\eta \in \mathrm{Cn}^*(Y)$, then $\eta \in \mathrm{Cn}^*(X)$, for some finite $X \subseteq Y$. $\square$

Definition 14 *Given an abduction problem $\langle \Theta, \varphi \rangle$ such that Θ is n-satisfiable, $n \geq 1$:*

$$\mathrm{Ab}^*(\langle \Theta, \varphi \rangle) = \{\psi \in L : \varphi \in \mathrm{Cn}^*(\Theta \cup \{\psi\})\}.$$

Theorem 15 *Given an abduction problem $\langle \Theta, \varphi \rangle$ such that Θ is n-satisfiable, $n \geq 1$, if α is a solution and $\Theta \cup \{\alpha\}$ is consistent, then $\alpha \in \mathrm{Ab}^*(\langle \Theta, \varphi \rangle)$ and it is an explanatory solution (with respect to Ab^*).*

Proof. Let $\langle \Theta, \varphi \rangle$ be an abduction problem such that Θ is n-satisfiable, $n \geq 1$, and α a provisional solution, such that $\Theta \cup \{\alpha\}$ is consistent. Let α be a provisional solution, then $\Theta \cup \{\neg\varphi\} \models_n \neg\alpha$, because of which $\Theta \cup \{\alpha\} \models_n \varphi$, that is to say $\varphi \in \mathrm{Cn}^*(\Theta \cup \{\alpha\})$, so that $\alpha \in \mathrm{Ab}^*(\langle \Theta, \varphi \rangle)$; $\varphi \notin \mathrm{Cn}(\Theta)$ and $\mathrm{Cn}(\Theta) \subseteq \mathrm{Cn}^*(\Theta)$, then $\varphi \notin \mathrm{Cn}^*(\Theta)$. On the other hand, $\varphi \notin \mathrm{Cn}^*(\{\alpha\})$. Therefore, α is an explanatory solution. $\square$

Definition 16 *Given an abduction problem $\langle \Theta, \varphi \rangle$ such that Θ is n-satisfiable, $n \geq 1$ and α is a provisional solution, the operation Th^* is obtained from Th by adding the following new rule (n constants occur in sentences of Γ; it is called $n\exists$-elimination)*

$$\frac{\Gamma, \exists x \beta \vdash_n \psi}{\Gamma, \beta(a_1/x) \vee \ldots \vee \beta(a_n/x) \vdash_n \psi}.$$

Corollary 17 *Given an abduction problem $\langle \Theta, \varphi \rangle$ such that Θ is n-satisfiable, $n \geq 1$, if α is an explanatory solution, then $\varphi \in \mathrm{Th}^*(\Theta \cup \{\alpha\})$.*

Proof. In general, $\text{Th}^*(X) = \text{Th}(X) \cup \{\psi \in L : X \vdash_n \psi\}$. $\text{Th}(X) = \text{Cn}(X)$; we can easily see that for any L-sentence η, $X \vdash_n \eta$ iff $X \models_n \eta$ and, as its consequence,

$$\{\psi \in L : X \vdash_n \psi\} = \{\psi \in L : X \models_n \psi\},$$

therefore, $\text{Th}^*(X) = \text{Cn}^*(X)$. As $\text{Th}^*(\Theta \cup \{\alpha\}) = \text{Cn}^*(\Theta \cup \{\alpha\})$ and (by that theorem) $\varphi \in \text{Cn}^*(\Theta \cup \{\alpha\})$, so that $\varphi \in \text{Th}^*(\Theta \cup \{\alpha\})$. $\square$

Remark 1 *As already pointed out (corollary 2), given a finite set of sentences Γ and a sentence φ, if the semantic tableau of $\Gamma \cup \{\neg\varphi\}$ is closed, then $\Gamma \models \varphi$ and, by completeness of first order logic, $\Gamma \vdash \varphi$. In this case, the tableau has provided a refutation for $\neg\varphi$, by means of which a derivation of φ from Γ can be obtained. On the other hand, according to previous corollary, given an abduction problem $\langle \Theta, \varphi \rangle$ such that Θ is n-satisfiable, $n \geq 1$, α is a solution and defined Th^*, $\Theta \cup \{\alpha\} \vdash_n \varphi$, that is to say, we can obtain a derivation of φ from $\Theta \cup \{\alpha\}$. Let us see the studied examples:*

(A) $\Theta = \{\forall x \exists y Rxy\}$; $\varphi = Ra_1a_2$ and

$$\alpha = \neg Ra_1a_1 \wedge \neg Ra_2a_2.$$

A derivation is presented as follows:

1 $\forall x \exists y Rxy$,   (assumption),

2 $\neg Ra_1a_1 \wedge \neg Ra_2a_2$,   (assumption),

3 $\exists y Ra_1y$,   ($\forall$-elimination 1),

4 $\exists y Ra_2y$,   ($\forall$-elimination 1),

5 $Ra_1a_1 \vee Ra_1a_2$,   ($2\exists$-elimination 3),

6 $\neg Ra_1a_1$,   ($\wedge$-elimination 2),

7 Ra_1a_2,   (disjunctive syllogism 5,6).

(B) $\Theta = \{\forall x \exists y (Rxy \wedge \neg Ryx)\}$; $\varphi = Ra_1a_3$; as solution α we obtained

$$\neg Ra_1a_2 \wedge Ra_2a_1 \wedge \neg Ra_2a_3 \wedge Ra_3a_2.$$

A derivation is presented as follows:

1 $\forall x \exists y (Rxy \wedge \neg Ryx)$,   (assumption),

2 $\neg Ra_1a_2 \wedge Ra_2a_1 \wedge \neg Ra_2a_3 \wedge Ra_3a_2$,   (assumption),

3 $\exists y(Ra_1y \wedge \neg Rya_1)$,   ($\forall$-elimination 1),

4 $\exists y(Ra_2y \wedge \neg Rya_2)$,   ($\forall$-elimination 1),

5 $\exists y(Ra_3y \wedge \neg Rya_3)$,   ($\forall$-elimination 1),

6 $(Ra_1a_1 \wedge \neg Ra_1a_1) \vee (Ra_1a_2 \wedge \neg Ra_2a_1) \vee (Ra_1a_3 \wedge \neg Ra_3a_1)$,
   ($3\exists$-elimination 3),

7 $\neg(Ra_1a_1 \wedge \neg Ra_1a_1)$,   (negation of contradiction),

8 $(Ra_1a_2 \wedge \neg Ra_2a_1) \vee (Ra_1a_3 \wedge \neg Ra_3a_1)$, (disjunctive syllogism 6,7),

9 $\neg Ra_1a_2$,   ($\wedge$-elimination 2),

10 $\neg Ra_1a_2 \vee Ra_2a_1$,   ($\vee$-introduction 9),

11 $\neg(Ra_1a_2 \wedge \neg Ra_2a_1)$,   (de Morgan 10),

12 $Ra_1a_3 \wedge \neg Ra_3a_1$,   (disjunctive syllogism 8,11),

13 Ra_1a_3,   ($\wedge$-elimination 12).

4.   Conclusions

As already said, the semantic tableaux method can be taken to deal with abduction in classical logic. In the propositional case it runs without problems, but in the first order case, though defining first order abductive systems is possible, an important problem arises, namely the existence of infinite tableaux. In this work we have tackled an approach of abduction in semantic tableaux in a new direction. Other approaches take standard semantic tableaux whereas our proposal takes the δ'-tableaux, which have the so called δ'-rule instead of the well known δ-rule. So we have been able to achieve some results: in short, given an abduction problem $\langle \Theta, \varphi \rangle$, when Θ is n-satisfiable (so $\Theta \cup \{\neg\varphi\}$ is n-satisfiable), $n \geq 1$:

- There is at least a completed open branch.

- From a such branch a closing set of literals is defined.

- If the closing set of literals is $\{\alpha_1, ..., \alpha_m\}$, then $\alpha = \alpha_1 \wedge ... \wedge \alpha_m$, $m \geq 1$, is a solution to the proposed problem.

- Defined new logical operations, taking into account the (minimal) metalogical properties, an effective derivation of φ from $\Theta \cup \{\alpha\}$ can be obtained.

Our proposal has only shown a part of a broader field, so we can set some points, among to other possibilities, for further research:

- To improve our presentation to check what requirements for the different versions of abduction are verified by the solution that a δ'-tableau provides. A logical structural analysis could be advisable to study the basic logical operations and a complete characterization of abduction.

- According to previous suggestion, to develop a complete proof theory of generated solutions by δ'-tableaux.

- To explore this form of providing solutions to abduction problems as computation, making comparison with the approaches in Artificial Intelligence and some related areas.

5. Appendix

5.1 Semantic tableaux

Preliminaries.   Let L be a first order formal language that is defined over a vocabulary consisting of a set of variables, a set of constants, a set of predicate or relational symbols, all with an associated arity $1, 2, ...,$ (these three sets may be denumerable infinite), a set of auxiliary symbols (parentheses, points, etc.) and the set of logical symbols $\{\neg, \wedge, \vee, \rightarrow, \exists, \forall\}$. The union set of variables and constants is called the *set of terms*.

Definition 18 *The set of L-formulae is the smallest that verifies*

1 if R is a predicate symbol of arity $n \geq 1$, and $t_1, t_2, ..., t_n$ are terms (repetition is allowed), then $Rt_1, t_2, ..., t_n$ is a L-formula that is called atomic formula.

2 If φ is a L-formula, then $\neg\varphi$ is a L-formula,

3 If φ, ψ are L-formulae, then $\varphi \vee \psi$, $\varphi \wedge \psi$ and $\varphi \rightarrow \psi$ are L-formulae,

4 If φ is a L-formula and x is a variable, then $\exists x\varphi$ and $\forall x\varphi$ are L-formulae.

Definition 19 *the set of literals is the set of all atomic formulae and all negations of atomic formulae. If l and l' are two literals such that one is the negation of the other, then l is the complement of l', and l' is the complement of l, that is to say, they are complementary.*

Subformula, free and bound occurrences of variables, substitution, and *L*-sentence are as usual. $\varphi(t/x)$ represents the formula obtained from φ by replacing all free occurrences of x by the term t.

Definition 20 *An L-structure $M = \langle D, \Im \rangle$, where $D \neq \emptyset$, and $\Im$ is the interpretation function defined from constants and relational symbols of L to D and predicates and relations over D, such that*

1 *If b is a constant, then $\Im(b) \in D$,*

2 *If R is a predicate symbol of arity $n \geq 1$, $\Im(R) \subseteq D^n$, or, what is the same, $\Im(R) \in \mathcal{P}(D^n)$.*

Definition 21 *Given an L-structure $M = \langle D, \Im \rangle$ and an L-sentence φ, the relation of satisfaction (M satisfies φ, in symbols $M \models \varphi$) is defined according to the following clauses:*

1 *if φ is $Rb_1, ..., b_n$, $n \geq 1$, $M \models \varphi$ iff $\langle \Im(b_1), ..., \Im(b_n) \rangle \in \Im(R)$,*

2 *if φ is $\neg\psi$, $M \models \varphi$ iff it is not the case that $M \models \psi$ ($M \not\models \psi$),*

3 *if φ is $\psi \vee \eta$, $M \models \varphi$ iff $M \models \psi$ or $M \models \eta$,*

4 *if if φ is $\psi \wedge \eta$, $M \models \varphi$ iff $M \models \psi$ and $M \models \eta$,*

5 *if φ is $\psi \rightarrow \eta$, $M \models \varphi$ iff $M \models \neg\psi$ or $M \models \eta$,*

6 *if φ is $\exists x \psi$, $M \models \varphi$ iff there is an L-structure M', that may differ from M at most with respect to the interpretation of a (new) constant b –so, for every constant $c \neq b$, $\Im(c) = \Im'(c)$, though $\Im(b) = \Im'(b)$ or $\Im(b) \neq \Im'(b)$. In short, $M' =_b M$– such that $M' \models \psi(b/x)$,*

7 *if φ is $\forall x \psi$, $M \models \varphi$ iff for every $M' =_b M$, $M' \models \psi(b/x)$.*

From this point of view satisfaction is a relation between *L*-structures and *L*-sentences (not formulae) that is in accordance with some applications of semantic tableaux, as we shall see below. To indicate that M satisfies every sentence of a set of *L*-sentences Γ, we shall write $M \models \Gamma$. The class of all *L*-structures is referred as $MOD(L)$, and the class of *L*-structures that satisfy any set of sentences Γ is naturally defined as

$$MOD(\Gamma) = \{M \in MOD(L) : M \models \Gamma\}.$$

Definition 22 *An L-sentence φ is satisfiable iff there exists $M \in MOD(L)$ such that $M \models \varphi$ (otherwise, φ is unsatisfiable).*

Definition 23 *For every* $n \geq 1$, *an L-sentence* φ *is n-satisfiable iff there exists* $M \in MOD(L)$, $M = \langle D, \Im \rangle$ *and* $\mid D \mid = n$, *such that* $M \models \varphi$.

Theorem 24 *If an L-sentence* φ *is n-satisfiable,* $n \geq 1$, *then it is m-satisfiable for every* $m \geq n$.

Proof. Let D be a domain such that $\mid D \mid = n \geq 1$, $M = \langle D, \Im \rangle$ and, for an L-sentence φ, $M \models \varphi$. Let D^* be another domain whose cardinal is $m \geq 1$, such that $\mid D \mid \leq \mid D^* \mid$, and D' such that $D' \subseteq D^*$ and $\mid D \mid = \mid D' \mid$. Let f be a bijection between D' and D and s an distinguished element of D'. Define a function g from D^* to D such that the following conditions are met for every $r \in D^*$:

1 if $r \in D'$, then $g(r) = f(r)$,

2 $g(r) = g(s)$, otherwise.

Then define a function h from relational symbols R of L, for every arity $k \geq 1$, to $\mathcal{P}(D^{*^k})$, such that

$$h(R) = \{\langle s_1, ..., s_k \rangle \in D^{*^k} : \langle g(s_1), ..., g(s_k) \rangle \in \Im(R)\}.$$

From that, by induction over the complexity of φ (which we omit), it is proved that $\Im^*$ is definable in such a way that

$$\Im^*(R) = h(R),$$

$$M^* = \langle D^*, \Im^* \rangle$$

and $M^* \models \varphi$ iff $M \models \varphi$. $\square$

Given a set of L-sentences Γ, we can pay attention to a special subclass of $MOD(\Gamma)$.

Definition 25 *Let* Γ *be a set of L-sentences.* D_Γ, *the Herbrand universe of* Γ, *is the set of constants that occur in sentences of* Γ. *If there are no sentences of* Γ *in which a constant occurs, then* $D_\Gamma = \{b_1\}$.

Definition 26 *A Herbrand interpretation for* Γ *is an L-structure*

$$H = \langle D_\Gamma, \Im_H \rangle$$

such that, for every $b \in D_\Gamma$, $\Im_H(b) = b$.

Theorem 27 *If* Γ *is a set of L-sentences and* Γ *is satisfiable, then there is a Herbrand interpretation that satisfies* Γ.

Proof. Suppose $M = \langle D, \Im \rangle$ such that $M \models \Gamma$. Let D_H be the Herbrand universe of Γ, then we define a mapping from D_H to D: for every $b \in D_H$, $g(b) = \Im(b)$ and for every relational symbol R of arity $n \geq 1$ that occurs in sentences of Γ,

$$\langle b_1, ..., b_n \rangle \in \Im_H(R) \text{ iff } \langle g(b_1), ..., g(b_n) \rangle \in \Im(R).$$

Let φ be any sentence of Γ and $M \models \Gamma$, the proof continues by induction on the complexity of φ, the base case: if φ is $Rb_1, ..., b_n$, then, as $M \models Rb_1, ..., b_n$,

$$\langle \Im(b_1), ..., \Im(b_n) \rangle \in \Im(R),$$

but $\Im(b_i) = g(b_i)$ for all $i \leq n$, so

$$\langle g(b_1), ..., g(b_n) \rangle \in \Im(R),$$

that is to say $\langle D_\Gamma, \Im_H \rangle \models \varphi$ (induction steps follow from the definition of satisfiability). $\square$

Other semantic notions as *logical consequence* or *semantic entailment*, *independence, validity*, etc. are taken as usual. An important result to be used in the context is given in the following:

Proposition 28 *Let Γ be a set of L-sentences an φ a L-sentence. $\Gamma \models \varphi$ (Γ entails φ, or φ is logical consequence of Γ) iff $\Gamma \cup \{\neg\varphi\}$ is unsatisfiable.*

Beth's method. Beth developed a semantic tableaux method that has been also elaborated by Hintikka and Smullyan [Letz, 1999]. The most extended approach is due to the last author, we consider the semantic tableaux obtained by applying some rules that are grouped into four types, named as usual though including more forms of sentences in the ¬-rule in order to reduce the cases of γ- and δ-rules.

Definition 29 *Given Γ, a non empty (finite) set of L-sentences, which can be regarded as premises or assumptions, the semantic tableau is the set of its branches,*

$$T(\Gamma) = \{\Gamma_1, ...\Gamma_n\},$$

$n \geq 1$. *Γ_i is a branch, for every $i \leq n$, obtained by applications of certain rules (explained below) to each non signed sentence. The process of building a branch continues until a literal, which is complement of another one, appears or, otherwise, every non literal sentence is signed.*

Definition 30 *A branch is (atomically) closed when two complementary literals occur in it. If a branch is not closed, then it is open.*

Definition 31 *A branch is completed when it is closed or when every non literal sentence is signed. A semantic tableau is closed when all its branches are closed.*

To sign a non-literal sentence, it should be noted, is to apply a rule. If Φ is the part of any branch, an application of any rule gives a new part (it may be divided) by adding new sentences in accordance with the rules. To indicate that, we shall write $\Phi + \eta$ and identify η by means of the name of the corresponding rule. The rules are the following:

1 $\neg$-rules:

$$\frac{\neg\neg\varphi}{\varphi}; \quad \frac{\neg(\varphi \vee \psi)}{\neg\varphi \wedge \neg\psi}; \quad \frac{\neg(\varphi \wedge \psi)}{\neg\varphi \vee \neg\psi}; \quad \frac{\neg(\varphi \rightarrow \psi)}{\varphi \wedge \neg\psi}; \quad \frac{\neg\exists x\varphi}{\forall x\neg\varphi}; \quad \frac{\neg\forall x\varphi}{\exists x\neg\varphi};$$

from Φ, $\Phi + \eta$ is obtained, where η is a $\neg$-sentence, one of the sentences φ, $\neg\varphi \wedge \neg\psi$, etc.,

2 α-rule:

$$\frac{\alpha_1 \wedge \alpha_2}{\begin{array}{c}\alpha_1 \\ \alpha_2\end{array}};$$

the new part of the branch is $\Phi + \alpha_1 + \alpha_2$ (α_1 and α_2 are α-sentences),

3 β-rule:

$$\frac{\beta_1 \vee \beta_2}{\beta_1 \mid \beta_2};$$

then Φ is divided in two new branches: $\Phi + \beta_1$ and $\Phi + \beta_2$ (β_1 and β_2 are β-sentences),

4 γ-rule:

$$\frac{\forall x\varphi}{\begin{array}{c}\varphi(b_1/x) \\ \varphi(b_2/x) \\ \cdots \\ \varphi(b_n/x)\end{array}}$$

$b_1, ..., b_n$ are all constants that occur in sentences of the branch, $n \geq 1$ (if no one occurs, then $n = 1$), the new part of the branch is

$$\Phi + \varphi(b_1/x) + ... + \varphi(b_n/x)$$

($\varphi(b_i/x)$ is a γ-sentence, for all $i \leq n$)

5 δ-rule:

$$\frac{\exists x \varphi}{\varphi(b/x)}$$

where b is a new constant that does not occur in any sentence of the branch, the new part of the branch is $\Phi + \varphi(b/x)$ and this is a δ-sentence.

Lemma 32 *If the semantic tableau of a finite set of L-sentences has an open branch, then that set is satisfiable.*

Proof. Let Γ be the (finite) set of sentences and Φ_Γ an open branch of its semantic tableaux. Let $\Gamma_0 = \{l_1, l_2, ..., l_n\}$ be all literals of Φ_Γ. For every $m \geq 1$, Γ_{m+1} is the smallest set verifying:

1. if $\varphi \in \Gamma_m$, then $\neg\neg\varphi \in \Gamma_{m+1}$,

2. if $\varphi, \psi \in \Gamma_{m+1}$, $\varphi \wedge \psi \in \Gamma_{m+1}$,

3. if $\varphi \in \Gamma_m$, then $\varphi \vee \eta \in \Gamma_{m+1}$,

4. if $\neg\varphi \wedge \neg\psi$, $\neg\varphi \vee \neg\psi$, $\varphi \wedge \neg\psi$, $\exists x \neg\varphi$ or $\forall x \neg\varphi \in \Gamma_m$, then $\neg(\varphi \vee \psi)$, $\neg(\varphi \wedge \psi)$, $\neg(\varphi \rightarrow \psi)$, $\neg\forall x\varphi$ or $\neg\exists x\varphi \in \Gamma_{m+1}$, respectively,

5. if $\varphi(b/x) \in \Gamma_m$, for a constant b that occurs in any sentence of Γ, then $\exists x\varphi \in \Gamma_{m+1}$,

6. if $\varphi(b/x) \in \Gamma_m$, for every constant b that occur in any sentence of Γ, then $\forall x\varphi \in \Gamma_{m+1}$.

So $\Phi_\Gamma = \bigcup_{m \geq 0} \Gamma_m$. An interpretation is defined as follows:

$$M = \langle D_\Gamma, \Im_H \rangle.$$

D_Γ is the Herbrand universe set of constants that occur in literals of Γ_0. Besides, for every $b \in D_\Gamma$, $\Im(b) = b$, and if R is a predicate symbol of arity $n \geq 1$, for any

$$\langle b_1, ..., b_n \rangle \in D_\Gamma^n,$$

$$\langle b_1, ..., b_n \rangle \in \Im(R) \text{ iff } Rb_1, ..., b_n \in \Gamma_0.$$

A consequence of this definition is: if $l \in \Gamma_0$, then

a) when l is $Rb_1, ..., b_n$, since $\langle b_1, ..., b_n \rangle \in \Im(R)$, $M \models l$,

b) when l is $\neg Rb_1, ..., b_n$, since $\langle b_1, ..., b_n \rangle \notin \Im(R)$,

$$M \not\models Rb_1, ..., b_n,$$

so that

$$M \models \neg Rb_1, ..., b_n;$$

whatever the case may be, $M \models l$, therefore, $M \models \Gamma_0$. By induction, taking as hypothesis that $M \models \Gamma_m$ (we omit to abbreviate), $M \models \Gamma_{m+1}$ is proved and, as it is a consequence, $M \models \Gamma$. $\square$

Lemma 33 *If a finite set of L-sentences is satisfiable, then there exists a semantic tableau with such set as premises that has an open branch.*

Proof. Suppose that the set is Γ and there is an interpretation M such that $M \models \Gamma$. Induction by the number of applications of rules, the base case is obvious. Let $n \geq 1$ be the number of applications of rules, giving an open branch Φ. After the $n+1$-application, if $M \models \Phi$, then by examining the rules we can prove that M satisfies any $\neg$-sentence, the two α-sentences, one of the β-sentences, all γ-sentences or the δ-sentence, doing so until all branches are completed, one of them must be open. $\square$

5.2   Basic logical operation

Let Cn be an operation defined from $\mathcal{P}(L)$ to $\mathcal{P}(L)$:

Definition 34 Cn *is a basic logical operation defined from $\mathcal{P}(L)$ to $\mathcal{P}(L)$. For any set of axioms X, $\mathrm{Cn}(X) = \{\psi \in L : X \models \psi\}$, and it is verified that:*

1 $X \subseteq \mathrm{Cn}(X)$,

2 $\mathrm{Cn}(\mathrm{Cn}(X)) = \mathrm{Cn}(X)$,

3 if $X \subseteq Y$, then $\mathrm{Cn}(X) \subseteq \mathrm{Cn}(Y)$,

4 if $\psi \in \mathrm{Cn}(Y)$, then $\psi \in \mathrm{Cn}(X)$, for some finite $X \subseteq Y$ [Wolenski, 1999].

On the other hand, let Th be another operation (its domain and range are the same), $\mathrm{Th}(X) = \{\psi \in L : X \vdash \psi\}$. Because of the *first order thesis*, first order languages are the most suitable for many mathematical proposals, so that, as $\langle L, \vdash \rangle$ is sound and complete, that is to say, for every set of axioms $X \subseteq L$, it is verified that

$$\{\psi \in L : X \models \psi\} = \{\psi \in L : X \vdash \psi\}.$$

So $\models$ and $\vdash$ are interchangeable and the former schema is equivalent to $\Theta \cup \{\alpha\} \vdash \varphi$. A deductive theory can be identified with the set of its consequences: Θ is a deductive theory iff $\Theta = \mathrm{Cn}(\Theta)$ (or $\mathrm{Th}(\Theta)$).

In general, if $\Theta \cup \{\alpha\} \models \varphi$ (equivalently, $\Theta \cup \{\alpha\} \vdash \varphi$), that is to say $\varphi \in \mathrm{Cn}(\Theta \cup \{\alpha\})$ (equivalently, $\varphi \in \mathrm{Th}(\Theta \cup \{\alpha\})$), then the semantic tableau of $\Theta \cup \{\alpha\} \cup \{\neg\varphi\}$ is closed.

References

Aliseda, A., 1997, Seeking Explanations: Abduction in Logic, Philosophy od Science and Artificial Intelligence, Institute for Logic, Language and Computation, Amsterdam.

Barwise, J., 1985, Model-theoretic logics: Background and aims, in: *Model-Theoretic Logics*, J. Barwise and S. Feferman, eds., Springer-Verlag, Berlin, pp. 3–23.

Beth, E.W., 1969, Semantic entailment and formal derivability, in: *The Philosophy of Mathematics*, Hintikka, J., ed., Oxford University Press, London, pp. 9–41.

Boolos, G.S., 1984, Trees and finite satisfactibility, *Notre Dame Journal of Formal Logic* 25:110–115.

Díaz, E., 1993, Arboles semánticos y modelos mínimos, in: *Actas del I Congreso de la Sociedad de Lógica, Metodología y Filosofía de la Ciencia en España*, Pérez, E., ed., Universidad Complutense, Madrid, pp. 40–43.

Letz, R., 1999, First-order tableau methods, in: *Handbook of Tableau Methods*, M. D'Agostino, D.M. Gabbay, R. Hahnle, and J. Posegga, eds., Kluwer Academic Publisher, Dordrecht, pp. 125–196.

Mayer, M.C. and Pirri, F., 1993, First order abduction via tableau and sequent calculi, *Bulletin of The I.G.P.L.*, pp. 99–117.

Nagel, E., 1979, *The Structure of Science. Problems in the Logic of Scientific Explanation*, Harcourt, Brace & World, New York.

Wolenski, J., 1999, Logic from a metalogical point of view, in *Logic at Work. Essays Dedicated to the Memory of Helena Rasiowa*, Orlowska, E., ed., Springer-Verlag, Heidelberg, pp. 25–35.

COMPUTATIONAL ASPECTS OF MODEL-BASED REASONING

COMPUTATIONAL DISCOVERY OF COMMUNICABLE SCIENTIFIC KNOWLEDGE

Pat Langley and Jeff Shrager
Institute for the Study of Learning and Expertise
2164 Staunton Court, Palo Alto, CA 94306, USA
{langley,shrager}@isle.org

Kazumi Saito
NTT Communication Science Laboratories
2-4 Hikaridai, Seika, Soraku, Kyoto 619-0237 Japan
saito@cslab.kecl.ntt.co.jp

Abstract   In this paper we distinguish between two computational paradigms for knowledge discovery that share the notion of heuristic search, but differ in the importance they place on using scientific formalisms to state discovered knowledge. We also report progress on computational methods for discovering such communicable knowledge in two domains, one involving the regulation of photosynthesis in phytoplankton and the other involving carbon production by vegetation in the Earth ecosystem. In each case, we describe a representation for models, methods for using data to revise existing models, and some initial results. In closing, we discuss related work on the computational discovery of communicable scientific knowledge and outline directions for future research.

1.   Introduction

Scientific discovery is generally viewed as one of the most complex human creative activities. As such, it seems worth understanding for both theoretical and practical reasons. One powerful metaphor treats the discovery process as a form of computation, and in fact work that adopts this metaphor has a long history that dates back over two decades (e.g., [Langley, 1979; Lenat, 1977; Lindsay et al., 1980]). Research within

L. Magnani, N.J. Nersessian, and C. Pizzi (eds.),
Logical and Computational Aspects of Model-Based Reasoning, 201–225.
© 2002 Kluwer Academic Publishers. Printed in the Netherlands.

this framework has advanced steadily until, in recent years, it has led
to new discoveries deemed worth publication in the scientific literature
(e.g., see [Langley, 2000]). However, despite this progress, work on the
topic remains subject to important limitations.

In this paper, we describe a new computational approach to discovery of scientific knowledge and illustrate its application to two domains.
The first focuses on constructing regulatory models for photosynthesis in
phytoplankton using data from DNA microarrays. The second involves
finding a quantitative model of the Earth ecosystem that fits environmental data obtained from satellites and ground stations. In both cases,
we report our formalism for representing models, a computational technique for producing them from observations, and initial results with
actual data.

Although these two applications differ on many dimensions, they also
share a reliance on three concerns: the discovered knowledge must be
communicable to domain scientists; the new model must be linked to
previous domain knowledge; and the model must move beyond a descriptive summary to explain the observations. We should also note
that our long-term goal is not to automate the discovery process, but
instead to provide interactive tools that scientists can direct and use to
aid their model development.

After describing our approaches to discovery in microbiology and
Earth science, we discuss related work on computational discovery and
outline some likely directions for future research. However, before pre-
senting our computational framework and its application, we must first
place it in a broader historical context of work on knowledge discovery.

2.   Paradigms for computational discovery

As Kuhn [Kuhn, 1962] has noted, the *paradigm* within which scientific
research occurs has a major impact on both its content and its method,
and computational research on knowledge discovery is no exception. For
this reason, we should review the two major frameworks for studying
the discovery process in computational terms. These two paradigms
hold some important assumptions in common, but they diverge on a key
issue.

2.1   The data mining paradigm

A number of developments have made possible the progress on computational approaches to knowledge discovery. The most recent break-
through, which we may call the *data* revolution, came from the insight that one can benefit by collecting and storing, automatically, vast

amounts of data that describe natural, engineering, and social domains of interest. These abilities have been made practical by the availability of inexpensive computer memory storage, the advent of new measurement techniques that ease data acquisition, and the introduction of communication infrastructure (e.g., the Internet) that supports rapid transfer of data. We can set the date for this revolution around 1995, when these technologies became common, but awareness of the coming situation was widespread five years earlier. Naturally, the access to electronic data sets holds great potential to support knowledge discovery, and many scientists, engineers, and businessmen have focused their energies on fulfilling that potential.

A somewhat earlier development, which we may call the *search* revolution, resulted from the insight that computers are general symbol manipulators and that one can view many tasks which require intelligence as involving search through a space of symbolic structures. This ability became practical with the introduction of computer programming languages that could represent and manipulate symbolic structures, as well as algorithms for carrying out heuristically-guided search through a space of such structures. We can date this revolution to the middle 1950s, when Newell and Simon [Newell and Simon, 1956] created the first list-processing language and used it to automate search for proofs of logical theorems. Notions of heuristic search preceded this achievement, but computationalists began to apply the idea in earnest only after this proof of concept. Simon [Simon, 1966] was also one of the first authors to view the discovery process in terms of search.

In recent years, these two insights have been combined by researchers and developers in a paradigm known as *data mining* or *knowledge discovery in databases*. Work in this arena emphasizes the availability and potential of large, electronic data sets, as well as computational techniques that can represent and search for knowledge implicit in those data. The data mining community has inherited its key techniques from two parent disciplines – machine learning and databases – that have focused historically on computational processing of data. This approach has become especially popular in the commercial sector, where it has been applied successfully to manufacturing, marketing, and finance, but it has also been put to good effect in a variety of scientific fields.

However, despite its impressive track record, the data mining framework has an important drawback related to its emphasis on the discovery of knowledge in understandable forms. In principle, this concern is perfectly legitimate, since we typically assume that knowledge can be represented explicitly and communicated among humans. Yet the data mining community's efforts along these lines have focused on particular formalisms it has inherited from its parent disciplines, notably decision trees, logical rules, and Bayesian networks. Researchers regularly take positions about the understandability of such representations, but their stances are based more on popular myths than on careful reasoning or empirical evidence.

One such myth concerns the claim that univariate decision trees, with their logical semantics, are inherently easier to understand than alternative notations, like probabilistic classifiers, that involve numeric weights and degrees of match. Yet Igor Kononenko [personal communication, 1993], who originally believed this intuition, found that medical doctors felt a naive Bayesian classifier, which computes probabilistic summaries, was easier to comprehend than decision trees induced from the same patient data. Presumably, this was because the physicians had more exposure to probability theory than to nonparametric schemes like decision trees. We can draw a tentative conclusion from this result: knowledge is more understandable when cast in a formalism familiar to the recipient.

A similar myth involves the claim that computational methods like backpropagation, which learns weights in a multilayer neural network, produce results that are inherently opaque. Yet Saito and Nakano [Saito and Nakano, 1997] have shown that, by carefully structuring the network architecture, one can use backpropagation to discover numeric equations like those central to physics and other sciences, and which, presumably, are interpretable by experts in those domains. We can draw another plausible lesson from this result: whether the discovered knowledge is understandable depends far less on the search algorithm than on the manner in which one uses that algorithm.

2.2   Computational scientific discovery

These observations suggest the relevance of a third, much older, historical development, the *scientific* revolution, which introduced not only the idea of evaluating laws and theories in terms of their ability to fit observations, but also emphasized the casting of such knowledge in some formal notation. We can date this insight to around 1700, when Newton's theory of gravitation became widely accepted, though it was predated by similar formal statements like Kepler's laws. Over the past 300 years, scientists and engineers have developed a variety of formalisms to represent knowledge that bear little resemblance to the notations which dominate the data mining community. We hold that such formalisms from science and engineering are more appropriate targets for knowledge discovery, at least in such domains, than data mining notations.

In fact, there exists an alternative computational paradigm, predating the data mining framework, that combines the representational insights of the scientific revolution with the notion of heuristic search. We will refer to this framework as *computational scientific discovery*, since its primary focus has been finding laws and theories in scientific domains. This paradigm also assumes the presence of data or observations, but emphasizes their role less than the search metaphor and scientific notations. Research in this area addressed originally the rediscovery of knowledge from the history of science (e.g., [Langley et al., 1987; Shrager and Langley, 1990], but the last decade has seen numerous examples of novel discoveries that have led to publications in the relevant scientific literature [Langley, 2000]. We maintain that this approach is more appropriate for the discovery of *communicable* knowledge than the data mining framework precisely because it utilizes formalisms already familiar to domain experts.

Note that there has been considerable work within the data mining tradition on scientific domains. Much has involved applications to molecular biology, such as learning predictors for protein folding, but Fayyad et al. [Fayyad et al., 1996] review similar efforts in astronomy, such as distinguishing stars from galaxies, and planetology, such as detecting volcanoes on Venus. This work has proven valuable to the disciplines involved, but we hold that the knowledge discovered in these cases is not communicable in the same sense as described above. The learned predictors, whether stated as decision trees, neural networks, or probabilistic classifiers, are unlikely to appear as knowledge themselves in scientific papers, and thus would not be *communicated*. Rather, they play the role of measuring instruments, which are essential to scientific progress but which constitute *tacit* knowledge [Polanyi, 1958] rather than the communicable variety.

By this point, we hope to have convinced readers that the task of communicable knowledge discovery differs in important ways from the problems typically pursued in the data mining community, and that this task deserves significantly increased attention among knowledge discovery researchers. For despite the success stories to date, there remain many open problems that require additional effort. For instance, most research on computational scientific discovery has focused on finding knowledge from scratch, but scientists are typically concerned with revising and improving existing theories. Researchers in the field have also concentrated primarily on discovery of descriptive regularities, but scientists often aim for models that explain observed phenomena in terms of unobserved variables and processes. Finally, most work on computational discovery has emphasized automating this activity, but scientists would benefit more from interactive tools that assist them in their efforts rather than ones that aim to replace them.

In the sections that follow, we report progress on these issues in the context of two scientific domains. In both cases, we review an existing explanatory model that accounts partially for some phenomena, describe a computational system that revises this model to fit these data better, and present some initial results of such improvement. Our research on interactive tools has advanced less, but we have designed our revision techniques to support such a capability. As in other work on computational scientific discovery, the systems cast their discovered knowledge in a familiar scientific notation to ensure communicability.

3.   Revising regulatory models in microbiology

Despite biologists' basic understanding of the mechanisms through which genes produce biochemical behavior, their elucidation of the regulatory networks that control gene expression remains very incomplete. Nevertheless, for specific organisms under particular conditions, they have devised partials accounts of gene regulation. Many of their inferences along these lines come from the observation of gene expression levels in terms of RNA concentration, either through Northern blots or cDNA microarrays.

The most popular computational approach to processing such expression data – clustering genes into coregulated classes – is a clear example of the data mining paradigm. This knowledge-lean approach reduces the dimensions of microarray data to a more reasonable level, but the results take the form of descriptive summaries rather than explanations. An alternative paradigm, often utilized by traditional biologists, tests specific hypotheses about regulatory pathways with focused measurements of expression levels. This knowledge-rich strategy lets the scientist evaluate their explanatory models, but it does not, by itself, suggest revised models that improve on the initial hypotheses.

In this section, we describe an approach that combines knowledge with data to revise an initial biological model. We focus on the regulation of photosynthesis in Cyanobacteria, an area for which we have both a model proposed by domain scientists and microarray data collected to evaluate this model. As outlined above, our goal is to develop computational methods that can utilize data to improve such a model while retaining its communicability and its links to existing biological knowledge.

3.1   Representing models of gene regulation

Any computational method designed to improve regulatory models must first have some representation for those models. As we have noted, most work in machine learning and data mining relies on representational formalisms that were invented by artificial intelligence researchers and that, as a result, make little contact with notations commonly used by scientists. In contrast, our approach represents models of gene regulation in terms that biologists will find familiar and comprehensible.

Figure 1 presents a regulatory model for Cyanobacteria, obtained from a plant biologist, that attempts to explain the fact that this organism bleaches when exposed to bright light. The model denotes observable and theoretical variables by nodes, whereas it indicates biological processes, in which one variable influences another, by directed links. Processes internal to the organism are shown by solid lines and processes that connect to the environment by dashed lines.

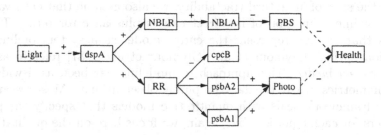

Figure 1.   An initial model for regulation of photosynthesis in Cyanobacteria.

The model states that an increase in light leads to heightened activity of dspA, a protein that the biologist believes to act as a sensor. This protein influences the expression of NBLR and thus up regulates NBLA, which in turn decreases the number of phycobilisome (PBS) rods, measurable from the organism's greenness, that are responsible for absorbing light. The PBS reduction decreases the amount of light the organism absorbs, which protects its health, as measured by culture density, because the bright light would damage it otherwise. The protein dspA also influences the organism's health by another pathway that operates through RR, a hypothesized response regulator that down regulates expression of three additional gene products. Decreases in two of these – psbA1 and psbA2 – in turn cause an increase in photosynthetic activity (Photo), which would reduce organism health if left unaltered under high light conditions.

Although this model incorporates quantitative variables, it specifies only the directions of influence and not their specific form or their parameters. AI research in qualitative physics (e.g., [Forbus, 1984]) has used similar notations to support common sense reasoning. We have focused on such qualitative causal models, as opposed to quantitative ones, because biologists usually reason and communicate in these terms, and we want our computational methods to support their typical operating style.

The example model is also partial and abstract, in that the biologist who proposed it clearly viewed it as a working hypothesis. Some causal links are abstract in that they denote entire chains of subprocesses. In particular, the link from dspA to NBLR denotes a complex signaling pathway for which the details are unknown or irrelevant at this level of analysis. The model also includes abstract variables like RR, which refers to an unspecified gene or set of genes that mediates between dspA and other variables. Thus, our formalism can express partial, abstract, and qualitative models like those often used by biologists.

For the sake of analytical tractability, we also assume that each variable is a linear function of its direct causes plus an error term. This means that we can represent the entire model as a system of linear equations, which Glymour et al. [Glymour et al., 1987] refer to as a *linear causal model*. This approach to modeling has been used widely in econometrics, where the data are purely observational. Most research in this framework deals with quantitative models that specify the parameters for each equation, but, again, we focus here on the qualitative version.

3.2   Utilizing, evaluating, and revising models

Since our models are qualitative, they cannot predict directly the continuous expression levels one can observe for genes, but they do imply certain relations among variables. In particular, they predict which variables should be correlated and the direction of those relationships. If two variables are connected directly, then we expect their correlation to have the same sign as that on their link. If they are connected indirectly, we multiply the signs on the path that connects them. For instance, the model in Figure 1 predicts that NBLA and cpcB will be negatively correlated, even though neither has a direct causal influence on the other and the path connecting them passes through RR, an unobservable variable.

In some cases, there exist multiple paths between a pair of variables. When the predicted sign for all paths between these nodes agree, the system simply makes that prediction. However, when two or more paths disagree, we assume the model includes an annotation that indicates either the positive or negative paths are dominant, which gives an unambiguous prediction. This extended formalism lets a qualitative model predict a positive or negative correlation for each pair of observed variables, even without information about the quantity of each link's effect.

In addition, casting our regulatory structures as linear causal models lets us make other important predictions about *partial correlations*, which describe the relationship between two variables once the effects of other terms have been factored out. For instance, the partial correlation $\rho_{12.3}$ denotes the correlation between X_1 and X_2 when controlling for X_3. Simon [Simon, 1954] has shown that a zero partial correlation $\rho_{12.3}$ implies that X_1 and X_2 are connected through X_3. In contrast, a nonzero partial correlation implies that X_1 and X_2 are connected through paths that do not involve X_3. Thus, the model in Figure 1 predicts that the partial correlation of dspA and PBS given NBLA will be zero, because the variable NBLA lies along the path between them. Glymour et al. have generalized these conditions for more complicated models, but the intution remains the same.

Our approach evaluates a candidate regulatory model by predicting, for each set of three variables, which partial correlations should occur and which ones should not. The system then calculates these partial correlations from the data and determines, for each one, whether it differs significantly from zero. Upon comparing the predicted partial correlations with those supported by the data, it obtains the number of true positives (tp), true negatives (tn), false positives (fp), and false negatives (fn). The system combines these counts using

$$score = fp + fn - tp - tn \; ,$$

which provides an overall measure of the model's qualitative fit to the observations. Because most linear causal models imply different partial correlations, this metric lets it discriminate among many alternative regulatory structures.

To revise its model of gene regulation, the system carries out a two-stage heuristic search through a space of candidate models. The first stage, which focuses on the causal structure, starts from the initial model proposed by biologists with the signs on links removed. The operators for generating alternative models include adding a link between variables, removing an existing link, and reversing the direction of a link.[1] The

---

[1]These operators are constrained by biological knowledge. For instance, the system knows that stimulus variables like Light must serve as causal influences to gene variables, and that behavioral variables like Photo must be caused by the latter.

system invokes the *score* metric described above to select among models, and it carries out hill-climbing search through the model space, on each step selecting the revision that most improves the evaluation metric. The search halts after a prespecified number of revision steps.

Because experiments that measure gene expression typically collect few samples, this approach is unstable in that small changes to the data can produce very different models. To offset this, the system generates 20 different training sets by sampling with replacement from the original data, then runs its revision algorithm to generate 20 new models. The program then counts how many times each revision occurs in these models and retains only those that appear in at least 75 percent of them.

Once the system has induced the model's causal structure, the second stage carries out another search to determine the signs on links. In this case, the evaluation function measures instead the number of correlations for which the predicted and observed signs agree. If the model involves only a few links, the system considers exhaustively all possible assignments of pluses and minuses on the links, then selects the best-scoring assignment. Otherwise, it resorts to hill climbing through the space of assignments, starting from those in the initial model and halting when no further improvement occurs.

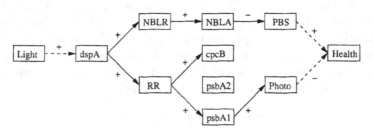

Figure 2.   A revised model for regulation of photosynthesis in wild Cyanobacteria.

3.3   Initial results on photosynthetic regulation

We applied our revision method to data for the wild type Cyanobacteria and a mutant that does not bleach under high light conditions. We have data from cDNA microarrays about the expression levels for approximately 300 genes believed to play a role in photosynthesis. For the initial analysis, we focused on genes in the initial model shown in Figure 1 and did not consider links to other genes. The microarray data, which reflects the concentration of mRNA for each gene relative to that in a control condition, were measured at 0, 30, 60, 120, and 360 minutes after high light was introduced, with four replicated measurements at each time point. We treated the data as independent samples, ignoring their temporal aspects and dependencies among the replicates.

Figure 2 shows the revised model that the system produced from these data. There are five differences from the initial regulatory account. Two changes, removal of the links to and from psbA2, involve the model structure. The other three revisions concern changes of signs, in particular for the links from RR to psbA1, from RR to cpcB, and from PBS to Health. Discussions with the biologist who proposed the original model indicate a strong belief that RR influences Photo, but uncertainty about the exact pathways. This means that the changes which involve RR are not problematic, since the presence of one gene product (psbA1) is enough to regulate the photosynthetic center (Photo). However, the reversed sign on the link from PBS to Health raises a problem, since the belief that excessive light causes damage means this link should be positive. We hypothesize that, in this study, the light exposure was not enough to overcome benefits from the energy it provides, which the model omits.

We also tested the system on expression data for a mutant of Cyanobacteria that does not bleach under high light conditions. Presumably, such a mutant differs genetically from the wild organism in only a few ways, so we started search from the same model as in our first study. In this case, the system removed the link from dspA to RR, but made no other revisions. This is a plausible change, since the mutation involved removal of the dspA gene from the organism. However, the new model does not explain why the mutant fails to bleach when exposed to high light. One possibility is that the 20 samples did not provide enough statistical power to let the system remove the link from dspA to NBLR, which would produce the desired effect. Although these initial results are encouraging, it seems clear that we can still improve our approach to revising qualitative models of gene regulation. Elsewhere [Shrager et al., 2002] we discuss some directions for future research along these lines.

4.　　Revising quantitative models in Earth science

Earth scientists have reached a broad enough understanding of ecosystem processes to develop models for the entire biosphere. These differ from the microbiological models we considered in the last section in that they are primarily quantitative rather than qualitative. Ecosystem models can also be quite complex, containing tens of equations, many theoretical variables, and parameters for each grid cell, which can number in the thousands. Such models are consistent with high-level ecosystem phenomena, but the availability of new data from satellites and other sources provides the opportunity to refine them further.

One such model, Potter and Klooster's [1997, 1998] CASA, predicts, with reasonable accuracy, the global production and absorption of biogenic trace gases in the Earth's atmosphere, as well as explaining changes in the geographic vegetation patterns on the land. The model's predictive variables include surface temperature, moisture levels, and soil properties, along with global satellite observations of the land surface. CASA incorporates both instantaneous and difference equations that describe changes over time due to the terrestrial carbon cycle and processes that mineralize nitrogen and control vegetation type. The model operates on gridded input, with typical usage involving grid cells that are eight kilometers square, since this matches the resolution for land surface observations obtained from satellites.

Although CASA has been quite successful at modeling Earth's ecosystem, its predictions still differ from observations in certain ways, and in this section we describe a computational approach to improving its fit to the data available. As before, the result is a revised model, cast in the same notation as the original one, that incorporates changes that are scientifically plausible and, we hope, interesting to Earth scientists.

4.1   A portion of the CASA model

Rather than attempting to refine the complete CASA model, which is quite complex, we decided to focus on a submodel near the 'top' that leads directly to the main dependent variable, NPPc, which denotes the net production of carbon. Table 1 lists the variables that occur in this submodel and summarizes the quantities they represent, whereas Table 2 shows the equations that relate these variables, with indentation reflecting the submodel's logical structure.

The first equation in Table 2 states that NPPc is the product of two unobservable variables, the photosynthetic efficiency at a site, E, and the solar energy intercepted at that site, IPAR. Photosynthetic efficiency is calculated in turn as the product of three stress factors that reduce the maximum efficiency (0.56). The first stress term, T2, involves the difference between the actual temperature for a site, Tempc, and the optimum temperature for that site, Topt. The second factor, T1, takes into account how close Topt comes to the global optimum for all sites, thus incorporating the idea that plants which have adapted to harsh temperatures are less efficient overall.

The third term that influences photosynthetic efficiency, W, denotes stress that results from absence of moisture as determined by EET, the estimated water loss from evaporation and transpiration, and by PET, the water loss that would result from these two processes if the

NPPc is the net plant production of carbon at a site during the year.

E is the photosynthetic efficiency at a site after factoring various sources of stress.

T2 is a temperature stress factor ($0 < T2 < 1$), nearly Gaussian in form but falling off more quickly at higher temperatures.

T1 is a temperature stress factor ($0 < T1 < 1$) for cold weather.

W is a water stress factor ($0.5 < W < 1$) for dry regions.

Topt is the average temperature for the month at which MON-FAS-NDVI takes on its maximum value at a site.

Tempc is the average temperature at a site for a given month.

EET is the estimated evapotranspiration (water loss due to evaporation and transpiration) at a site.

PET is the potential evapotranspiration (water loss due to evaporation and transpiration given an unlimited water supply) at a site.

PET-TW-M is a component of potential evapotranspiration that takes into account the latitude, time of year, and days in the month.

A is a polynomial function of the annual heat index at a site.

AHI is the annual heat index for a given site.

MON-FAS-NDVI is the relative vegetation greenness for a given month as measured from space.

IPAR is the energy from the sun that is intercepted by vegetation after factoring in time of year and days in the month.

FPAR-FAS is the fraction of energy intercepted from the sun that is absorbed photosynthetically after factoring in vegetation type.

MONTHLY-SOLAR is the average solar irradiance for a given month at a site.

SOL-CONVER is 0.0864 times the number of days in each month.

UMD-VEG is the type of ground cover (vegetation) at a site.

Table 1. Variables used in the NPPc portion of the CASA ecosystem model.

water supply were unlimited. The potential evapotranspiration PET is influenced in turn by its two components, the annual heat index AHI and PET-TW-M.

The model predicts IPAR, the energy that a site intercepts from the sun, as a product of three terms: the fraction of energy absorbed at the site through photosynthesis (FPAR-FAS), the average radiation that occurs during a given month (MONTHLY-SOLAR), and the number of days in that month (SOL-CONVER). The variable FPAR-FAS is in turn influenced by the relative greenness at a site as observed from space (MON-FAS-NDVI) and by the intrinsic property SRDIFF, which takes

$$\mathrm{NPPc} = \sum_{month} \max(\mathrm{E} \cdot \mathrm{IPAR}, 0)$$

$$\mathrm{E} = 0.56 \cdot \mathrm{T1} \cdot \mathrm{T2} \cdot \mathrm{W}$$

$$\mathrm{T1} = 0.8 + 0.02 \cdot \mathrm{Topt} - 0.0005 \cdot \mathrm{Topt}^2$$

$$\mathrm{T2} = 1.18/[(1 + e^{0.2 \cdot (\mathrm{Topt}-\mathrm{Tempc}-10)}) \cdot (1 + e^{0.3 \cdot (\mathrm{Tempc}-\mathrm{Topt}-10)})]$$

$$\mathrm{W} = 0.5 + 0.5 \cdot \mathrm{EET}/\mathrm{PET}$$

$$\mathrm{PET} = 1.6 \cdot (10 \cdot \mathrm{Tempc}/\mathrm{AHI})^A \cdot \mathrm{PET\text{-}TW\text{-}M} \text{ if } \mathrm{Tempc} > 0$$

$$\mathrm{PET} = 0 \text{ if } \mathrm{Tempc} \leq 0$$

$$\mathrm{A} = 0.000000675 \cdot \mathrm{AHI}^3 - 0.0000771 \cdot \mathrm{AHI}^2 + 0.01792 \cdot \mathrm{AHI} + 0.49239$$

$$\mathrm{IPAR} = 0.5 \cdot \mathrm{FPAR\text{-}FAS} \cdot \mathrm{MONTHLY\text{-}SOLAR} \cdot \mathrm{SOL\text{-}CONVER}$$

$$\mathrm{FPAR\text{-}FAS} = \min((\mathrm{SR\text{-}FAS} - 1.08)/\mathrm{SRDIFF}(\mathrm{UMD\text{-}VEG}), 0.95)$$

$$\mathrm{SR\text{-}FAS} = -(\mathrm{MON\text{-}FAS\text{-}NDVI} + 1000)/(\mathrm{MON\text{-}FAS\text{-}NDVI} - 1000)$$

Table 2. Equations used in the NPPc portion of the CASA ecosystem model.

on different numeric values for different values of UMD-VEG, a discrete variable that specifies the vegetation type at a site.

Making predictions from this submodel is a straightforward process, in that one simply starts from the observable[2] input variables – Tempc, MONTHLY-SOLAR, SOL-CONVER, MON-FAS-NDVI, UMD-VEG, EET, PET-TW-M, and AHI – and calculates values for the variables that depend on them. The resulting quantities are then passed to other equations that compute values for other terms, with this continuing until a value for NPPc is predicted. One repeats this process with each grid cell for which observations are available.

4.2 An approach to quantitative model revision

As before, our approach to scientific discovery involves refining a model like that in Table 2 rather than constructing one from scratch. Thus, this initial model constitutes the starting point for heuristic search through a space of models, with the search process directed by candidates' ability to fit the data. However, in this case our models are quantitative rather than qualitative and, as such, require different operators and a different evaluation function to direct search.

---

[2] Actually, the variables EET, PET-TW-M, and AHI are unobservable terms defined elsewhere in the model. To make the revision task more tractable, we assumed their definitions were correct and treated them as observables, using the model to compute their values.

To this end, we assume that the overall structure of the model is correct, but that the specific equations and their parameters can be improved. For example, after the revision process, NPPc would still be defined in terms of E and IPAR, but the functional form of this definition may no longer be NPPc = E · IPAR. Moreover, we can utilize parameter revision to mimic revision of equation forms by encoding each expression in the initial model as a multivariate polynomial equation of the form

$$y \;=\; w_0 + \sum_{j=1}^{J} w_j \prod_{k=1}^{K} X_k^{w_{jk}},$$

where y is a continuous variable that depends on continuous variables $X_1, \ldots, X_K$. For example, the equation W = 0.5 + 0.5 · EET/PET in this scheme becomes W = 0.5 + 0.5 · $\text{EET}^{1.0}$· $\text{PET}^{-1.0}$. Such functional relations subsume many of the numeric laws found by earlier quantitative discovery systems like BACON [Langley, 1979] and FAHRENHEIT [Żytkow et al., 1990], as well as the expressions in Table 2.

This encoding transforms our set of equations into the equivalent of a multilayer neural network, with one subnetwork for each relationship in the model. More specifically, each equation becomes a two-layer network with product units in the first level, to encode multiplicative terms, and additive units in the second level, to encode their weighted summation. This transformation maps the set of possible models into a weight space. By adapting Saito and Nakano's [Saito and Nakano, 1997] BPQ algorithm for discovering numeric equations, we can implement a gradient descent search through this space. Briefly, this method incorporates a second-order learning technique that calculates both the descent direction and the step size automatically. The search process halts when it finds a set of weights that minimize the squared error on the dependent variable y. The method then transforms the resulting network back into a set of polynomial equations, with weights on product units becoming exponents and weights on linear units becoming coefficients.

We can see readily how this approach can improve the parameters for an equation. Although the NPPc submodel contains some parameterized equations that our Earth science collaborators believe are reliable, like that for computing the variable A from the annual heat index AHI, it also includes equations with parameters about which there is less certainty, like the expression that predicts the temperature stress factor T2 from Tempc and Topt. By fixing the weights that correspond to reliable parameters, as well as the weights that encode exponents, the BPQ algorithm searches through the weight space associated with the other parameters to find settings that reduce predictive error. We can use the same mechanism to revise the form of an equation by specifying that the weights for exponents should not be fixed.

We must extend the approach slightly to support revision of values for an intrinsic property (e.g., SRDIFF) that the model associates with the discrete values for some nominal variable (e.g., the vegetation type UMD-VEG). In such cases, we encode each nominal term as a set of dummy variables, one for each discrete value, setting the dummy variable equal to one if the discrete value occurs and zero otherwise. We introduce one hidden unit for the intrinsic property, with links from each dummy variable and weights that correspond to the intrinsic value associated with each discrete value. We then utilize Saito and Nakano's BPQ algorithm to search the weight space that corresponds to alternative sets of intrinsic values, using the original model to initialize weights.

Although this approach to model refinement can modify more than one equation or intrinsic property at a time, the results we report in the next section assume that the user focuses the system's attention on one portion of the model. We envision an interactive mode in which the scientist identifies a portion of the model that he thinks could be better, runs the revision method to improve its fit to the data, and repeats this process until he is satisfied.

4.3   Initial results on ecosystem model revision

Our Earth science collaborators provided us with data related to the NPPc submodel that let us evaluate our approach to model revision. These data came from 303 ground sites for which the variables Tempc, MON-FAS-NDVI, MONTHLY-SOLAR, SOL-CONVER, and UMD-VEG had been measured for each month of the year. We used other portions of the CASA model to calculate variables for the three additional variables AHI, EET, and PET-TW-M. We utilized the resulting 303 training cases to carry out initial tests of our revision methods, focusing on portions of the NPPc submodel recommended by the Earth scientists.

Discussions with our Earth science collaborators suggested the expression for T2, one of the temperature stress variables, as a likely candidate for revision. As we saw in Table 2, the initial equation for this term was

$$T2 \;=\; 1.8/[(1 + e^{0.2(Topt - Tempc - 10)})(1 + e^{-0.3(Tempc - Topt - 10)})]\,,$$

which generates the curve shown in Figure 3. The Gaussian-like, but slightly assymetrical, form encodes the idea that photosynthetic efficiency will decrease as the actual temperature (Tempc) moves further from the optimal temperature (Topt) in either direction. When we asked our system to improve the parameters in this expression but to retain

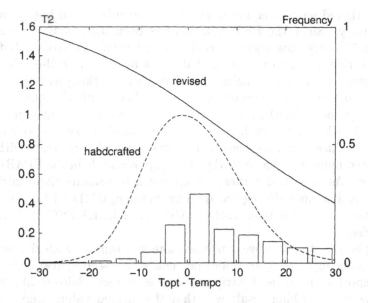

Figure 3. Behavior of handcrafted and revised equations for the stress variable T2.

its original form, it produced

$$T2 = 1.80/[(1 + e^{0.05(Topt-Tempc-10.8)})(1 + e^{-0.03(Tempc-Topt-90.33)})] ,$$

which has fairly similar values to the initial ones for some parameters but quite different values for others. The root mean squared error for the revised model was 461.466, as measured by leave-one-out cross validation, which was only one percent better than the 467.910 error for the original model.

Although this result seems disappointing at first glance, the curves in Figure 3 reveal a more interesting picture. The revised equation for the temperature stress factor T2 still produces a Gaussian-like curve when plotted as a function of the difference Topt − Tempc, but its values decrease monotonically within the effective range from −30 to 30 Celsius. The Earth scientists found this revision interesting, although counterintuitive, because it suggests that this stress term has little effect on carbon production. Because the original equation for T2 was not based on principles of plant physiology, such an observation is beneficial to the modeling enterprise even when the empirical improvement is small.

Our second attempt at parameter revision involved the equation for PET, which calculates the potential loss of water due to evaporation and transpiration if unlimited water were available in the atmosphere. In this case, the revised parameter values were all very similar to those in the original model's equation and led to no substantial improvement in accuracy. Since the PET equation has been used in Earth science for over 50 years, this negative result was not overly surprising. Indeed, the fact that our revision method did not alter such well-established parameters was an encouraging sign that it was working as intended.

We also tested our revision method's ability to alter values for the intrinsic property SRDIFF that are associated with different vegetation types UMD-VEG. For each site, the latter variable takes on one of 11 nominal values, such as grasslands, forest, and desert, each with an associated numeric value for SRDIFF that plays a role in the FPAR-FAS equation. As described earlier, our approach to revising these intrinsic values involved introducing one dummy variable, $UMD\text{-}VEG_k$, for each vegetation type such that $UMD\text{-}VEG_k = 1$ if $UMD\text{-}VEG = k$ and 0 otherwise.

In this case, the improvement was more substantial, with the revised model reducing error by over four percent, which seems substantial. We have reported the revised intrinsic values elsewhere [Saito et al., 2001], but the most striking result was that the altered values were nearly always lower than the initial values. This result is certainly interesting from an Earth science viewpoint. Our domain experts suspect that measurements of NPPc and related variables from a wider range of sites would produce intrinsic values closer to those in the original model, but such a test must await additional observations.

Because the original NPPc submodel grouped the 11 intrinsic values into four distinct values, we also applied a clustering algorithm to group the revised values in the same way. We specified different numbers of target clusters, ranging from one to five, and examined the effect on error rate. As expected, the training error decreased monotonically with the number of clusters, but the cross-validation error was lowest for three clusters. The error for this revised model was better than that for the model with 11 distinct values, but only slightly. Again, the clustered values were nearly always lower than the initial ones.

As we noted earlier, our system can also revise the functional forms in a quantitative model. One candidate for such revision was the equation for photosynthetic efficiency, E, which is calculated as a product of three stress terms in

$$E \;=\; 0.56 \cdot T1 \cdot T2 \cdot W \;.$$

Multiplying the stress terms has the effect of reducing photosynthetic efficiency below the maximum 0.56 possible [Potter and Klooster, 1998], since each factor takes on a value less than one.

In this case, a natural extension was to consider the space of equations that included exponents on the stress terms, which we initialized to 1.0, as in the original model, and constrained to be positive. This time, the system produced the revised equation

$$E = 0.521 \cdot T1^{0.00} \cdot T2^{0.03} \cdot W^{0.00},$$

which reduced error over the original model by almost five percent. The new equation has a similar coefficient, but it also has a small exponent for T2 and zero exponents for T1 and W. These results were very interesting to our Earth science collaborators, as they suggest that the T1 and W stress terms are not needed for predicting NPPc. One explanation is that the influence of these factors is already being captured by the NDVI measure available from space, for which the signal-to-noise ratio has been steadily improving since CASA was first developed. They are also consistent with our results with the T2 equation, which revealed monotonically changing values for this variable over the relevant range.

5.   Related research on computational discovery

As we noted earlier, there is a substantial literature on the computational discovery of communicable scientific knowledge (e.g., [Langley et al., 1987; Džeroski and Todorovski, 1993; Washio and Motoda, 1998]), but most of this research has focused on the construction of laws and models, rather than on their revision. There also exists a nearly disjoint literature on the computational revision of knowledge bases cast in non-scientific formalisms, most often using Horn clauses and related logical notations (e.g., [Ourston and Mooney, 1990]). However, there has been some work on the revision of scientific theories, which we should review here briefly.

One body of related research has involved revision of structural models from the history of chemistry and physics. For example, Żytkow and Simon's [Żytkow and Simon, 1986] STAHL detected inconsistencies in chemical reactions and revised its componential models by adding or removing constituents. Rose and Langley's [Rose and Langley, 1986] STAHLp improved on this approach and applied it to additional historical episodes. Kocabas' [Kocabas, 1991] BR-3 system extended this framework to include detection of incomplete theories and postulation of new properties to explain the absence of reactions, then applied these strategies to the history of particle physics. Finally, O'Rorke et al. [O'Rorke et al., 1990] developed AbE, an abductive system for model revision which they used to model the shift from the phlogiston to the oxygen theory.

Other work on the revision of qualitative scientific theories, more akin to our own, has focused on process models that explain causal events.

Rajamoney's [Rajamoney, 1990] COAST system incorporated ideas from qualitative physics to represent and revise models about fluid and heat flow, whereas Karp's [Karp, 1990] HYPGENE used a qualitative biochemical notation to support revision of models about attenuation in gene regulation. Kulkarni and Simon [Kulkarni and Simon, 1990] describe KEKADA, a system that reproduced many steps in Krebs' discovery of the urea cycle. All three systems augmented the revision process with methods for experiment design that aimed to distinguish among competing hypotheses.

There exists less research on the revision of quantitative scientific models. Chown and Dietterich [Chown and Dietterich, 2000] report an approach that improves an existing ecosystem model's fit to continuous data, but their method only alters parameter values and does not revise equation structure. Džeroski and Todorovski [Todorovski and Džeroski, 2001] present LAGRAMGE, a system that revises both the structure of a model's equations and their parameters, using a grammatical formalism to specify domain constraints on acceptable models. They have applied this approach to the same portion of the CASA ecosystem model as we have addressed and obtained similar improvements. Early research by Glymour et al. [Glymour et al., 1987] addressed revision of linear causal models that took a quantitative form, but their methods are more closely related to those we have used for qualitative model revision.

Our vision for an interactive discovery environment directly derives from Mitchell et al. [Mitchell et al., 1997], who developed a similar environment to support discovery in metallurgy. Their DAVICCAND system let users select pairs of numeric variables to relate, specify qualitative conditions that focus attention on subsets of the data, and find numeric laws that relate variables within a given region. The program also included mechanisms for identifying outliers that violate these numeric laws and for using the laws to infer the values of intrinsic properties. DAVICCAND presented its results using graphical displays and functional forms that were familiar to metallurgists.

We should note that the notion of communicable knowledge discovery is not limited to scientific domains. Another example comes from Rogers et al. [Rogers et al., 1999], who developed methods for revising the contents of digital maps based on traces from a global positioning system. Their system improved estimates of center lines for road segments, inferred the number of lanes associated with each segment, and added content about the type of traffic signal at intersections. The revised knowledge took the same form as the initial digital map, letting it be displayed in a graphical format familiar to mapmakers and drivers while increasing the map's overall accuracy and detail.

6.   Concluding remarks

In this paper, we distinguished between two broad computational approaches to discovery: the paradigm of data mining, which emphasizes the availability of large data sets to drive the search process, and computational scientific discovery, which takes advantage of established scientific formalisms to state the resulting knowledge in a communicable fashion. We argued that the latter is more appropriate for aiding discovery in scientific disciplines, but we also noted the need for more research in this promising framework.

In response, we reported progress on the discovery of communicable scientific knowledge in two domains, one involving gene regulation of photosynthesis in Cyanobacteria, and the other involving carbon production by vegetation as a function of environmental factors. In both cases, we developed algorithms that discovered knowledge in the same formalisms as utilized by domain scientists. Our methods also reflected two additional concerns that have received little attention in the discovery literature: the revision of initial models, rather than their generation from scratch, and the development of explanatory models, with theoretical variables and processes, rather than purely descriptive summaries. We showed that our discovery methods, one designed for qualitative models and the other for quantitative, led to improvements over existing models in terms of their fit to available data.

Although our results to date are encouraging, we must extend our computational discovery techniques in a number of directions before they become useful tools for scientists. For example, both discovery algorithms we presented can alter an initial model's relations among variables, but they cannot introduce new variables during the revision process. Another shared limitation is the methods' support for models with instantaneous relationships among variables but not ones that involve change over time. We should augment both discovery algorithms to consider additional variables during the revision process and to support models that express temporal relations. For quantitative models like CASA, we envision using ordinary differential equations and drawing on methods like Džeroski and Todorovski's [Todorovski and Džeroski, 2001] LAGRAMGE for revision; for qualitative models, we will borrow formalisms developed in the qualitative physics community (e.g., [Forbus, 1984]).

Clearly, such additions will increase the search space that our revision methods must explore, which in turn suggests the need for domain constraints to direct the process. To this end, we intend to introduce a taxonomy of variables and an analogous taxonomy of processes, with the latter making reference to the former. For instance, regarding biochemical models, one might know that metabolic processes are influenced by a certain class of genes and that they involve instantaneous relations, whereas transcription processes are controlled by another class and involve a time delay. Knowledge of this sort can constrain significantly the number of models that are included in the search space, and thus increase the chances of finding a candidate that scientists find acceptable. Analogous knowledge about which types of variables can occur in which types of equations can place similar constraints on the search for quantitative models.

Another challenge that we have encountered in our research has been the need to translate existing models into a declarative form that our discovery methods can manipulate. In response, we have started to develop a modeling language in which scientists can cast their initial models and carry out simulations, but that can also serve as the declarative representation for our discovery methods. The ability to automatically revise models places novel constraints on such a language. We envision this software developing into an interactive discovery aide that lets a scientist specify initial models, focus the system's attention on particular data sets and on parts of the model it should attempt to improve, and generally control high-level aspects of the discovery process. Thus, future versions will need a graphical interface for creating models, editing them, and marking fragments that can be revised, as well as tools for displaying matches to data, linking to other knowledge bases, and tracking changes in models over time. Taken together, these extensions should produce a valuable aide for practicing scientists.

Naturally, we also hope to evaluate our approach to model revision on other aspects of photosynthesis regulation and other portions of the CASA model as additional data become available. A more serious test of generality would be application of the same methods to other scientific domains in which there already exist formal models that can be revised. In the longer term, we should evaluate our interactive system not only in its ability to increase the predictive accuracy of an existing model, but in terms of the satisfaction the system provides to scientists who use it for model development.

Acknowledgments

This work was supported by Grants NCC 2-1202, NCC 2-5471, and NCC 2-1335 from NASA Ames Research Center, and by NTT Communication Science Laboratories, Nippon Telegraph and Telephone Corporation. We thank Arthur Grossman and C. J. Tu for the initial model of photosynthesis regulation and associated microarray data, Stephen Bay for implementing the system that analyzed these data, and Christopher Potter, Alicia Torregrosa, and Steven Klooster for access to their CASA model and related ecosystem data. Earlier versions of material in this paper have appeared in *Proceedings of the Fourth International Conference on Discovery Science* and *Proceedings of the Pacific Symposium on Biocomputing*.

References

Chown, E. and Dietterich, T.G., 2000, A divide and conquer approach to learning from prior knowledge, in: *Proceedings of the Seventeenth International Conference on Machine Learning*, Morgan Kaufmann, San Francisco, CA, pp. 143–150.

Džeroski, S., and Todorovski, L., 1993, Discovering dynamics, in: *Proceedings of the Tenth International Conference on Machine Learning*, Morgan Kaufmann, San Francisco, CA, pp. 97–103.

Fayyad, U., Haussler, D., and Stolorz, P., 1996, KDD for science data analysis: Issues and examples, in: *Proceedings of the Second International Conference of Knowledge Discovery and Data Mining*, AAAI Press, Menlo Park, CA, pp. 50–56.

Forbus, K.D., 1984, Qualitative process theory, *Artificial Intelligence* 24:85–168.

Glymour, C., Scheines, R., Spirtes, P., and Kelly, K., 1987, *Discovering Causal Structure: Artificial Intelligence, Philosophy of Science, and Statistical Modeling*, Academic Press, New York.

Karp, P.D., 1990, Hypothesis formation as design, in: *Computational Models of Scientific Discovery and Theory Formation*, J. Shrager and P. Langley, eds., Morgan Kaufmann, San Francisco, CA, pp. 275–317.

Kocabas, S., 1991, Conflict resolution as discovery in particle physics, *Machine Learning* 6: 277–309.

Kuhn, T.S., 1962, *The Structure of Scientific Revolutions*, University of Chicago Press, Chicago, IL.

Kulkarni, D. and Simon, H.A., 1990, Experimentation in machine discovery, in: *Computational Models of Scientific Discovery and Theory Formation*, J. Shrager and P. Langley, eds., Morgan Kaufmann, San Francisco, CA, pp. 255–274.

Langley, P., 1979, Rediscovering physics with BACON.3, in: *Proceedings of the Sixth International Joint Conference on Artificial Intelligence*, Morgan Kaufmann, San Francisco, CA, pp. 505–507.

Langley, P., 2000, The computational support of scientific discovery, *International Journal of Human-Computer Studies* 53:393–410.

Langley, P., Simon, H.A., Bradshaw, G.L., and Żytkow, J.M., 1987, *Scientific Discovery: Computational Explorations of the Creative Processes*, MIT Press, Cambridge, MA.

Lenat, D.B., 1977, Automated theory formation in mathematics, in: *Proceedings of the Fifth International Joint Conference on Artificial Intelligence*, Morgan Kaufmann, San Francisco, CA, pp. 833–842.

Lindsay, R.K., Buchanan, B.G., Feigenbaum, E.A., and Lederberg, J., 1980, *Applications of Artificial Intelligence for Organic Chemistry: The DENDRAL project*, McGraw-Hill, New York.

Mitchell, F., Sleeman, D., Duffy, J.A., Ingram, M.D., and Young, R.W., 1997, Optical basicity of metallurgical slags: A new computer-based system for data visualisation and analysis, *Ironmaking and Steelmaking* 24:306–320.

Newell, A., and Simon, H.A., 1956, The logic theory machine, *IRE Transactions on Information Theory* IT-2:61–79.

O'Rorke, P., Morris, S., and Schulenberg, D., 1990, Theory formation by abduction: A case study based on the chemical revolution, in: *Computational Models of Scientific Discovery and Theory Formation*, J. Shrager and P. Langley, eds., Morgan Kaufmann, San Francisco, CA, pp. 197–224.

Ourston, D. and Mooney, R., 1990, Changing the rules: a comprehensive approach to theory refinement, in: *Proceedings of the Eighth National Conference on Artificial Intelligence*, AAAI Press, Menlo Park, CA, pp. 815–820.

Polanyi, M., 1958, *Personal Knowledge: Towards a Post-Critical Philosophy*, University of Chicago Press, Chicago, IL.

Potter C.S. and Klooster, S.A., 1997, Global model estimates of carbon and nitrogen storage in litter and soil pools: Response to change in vegetation quality and biomass allocation, *Tellus* 49B:1–17.

Potter, C.S. and Klooster, S.A., 1998, Interannual variability in soil trace gas (CO_2, N_2O, NO) fluxes and analysis of controllers on regional to global scales, *Global Biogeochemical Cycles* 12:621–635.

Rajamoney, S., 1990, A computational approach to theory revision, in: *Computational Models of Scientific Discovery and Theory Formation*, J. Shrager and P. Langley, eds., Morgan Kaufmann, San Francisco, CA, pp. 225–254.

Rogers, S., Langley, P., and Wilson, C., 1999, Learning to predict lane occupancy using GPS and digital maps, in: *Proceedings of the Fifth International Conference on Knowledge Discovery and Data Mining*, ACM Press, San Diego, pp. 104–113.

Rose, D., and Langley, P., 1986, Chemical discovery as belief revision, in: *Machine Learning* 1:423–451.

Saito, K., Langley, P., Grenager, T., Potter, C., Torregrosa, A., and Klooster, S.A., 2001, Computational revision of quantitative scientific models, in: *Proceedings of the Fourth International Conference on Discovery Science*, Springer, Heidelberg, Germany, pp. 336–349.

Saito, K. and Nakano, R., 1997, Law discovery using neural networks, in: *Proceedings of the Fifteenth International Joint Conference on Artificial Intelligence*, Morgan Kaufmann, San Francisco, CA, pp. 1078–1083.

Simon, H.A., 1954, Spurious correlation: A causal interpretation, *Journal of the American Statistical Association* 49:467–479.

Simon, H.A., 1966, Scientific discovery and the psychology of human problem solving, in: *Mind and Cosmos: Essays in Contemporary Science and Philosophy*, R.G. Colodny, ed., University of Pittsburgh Press, Pittsburgh, PA.

Shrager, J., and Langley, P., eds., 1990, *Computational Models of Scientific Discovery and Theory Formation*, Morgan Kaufmann, San Francisco, CA.

Shrager, J., Langley, P., and Pohorille, A., 2002, Guiding revision of regulatory models with expression data, in: *Proceedings of the Pacific Symposium on Biocomputing*, Lihue, Hawaii, pp. 486–497.

Todorovski, L., and Džeroski, S., 2001, Theory revision in equation discovery, in: *Proceedings of the Fourth International Conference on Discovery Science*, Springer, Heidelberg, Germany, pp. 389–400.

Washio, T. and Motoda, H., 1998, Discovering admissible simultaneous equations of large scale systems, in: *Proceedings of the Fifteenth National Conference on Artificial Intelligence*, AAAI Press, Menlo Park, CA, pp. 189–196.

Żytkow, J.M. and Simon, H.A., 1986, A theory of historical discovery: The construction of componential models, *Machine Learning* 1:107–137.

Żytkow, J.M., Zhu, J., and Hussam, A., 1990, Automated discovery in a chemistry laboratory, in: *Proceedings of the Eighth National Conference on Artificial Intelligence*, AAAI Press, Menlo Park, CA, pp. 889–894.

ENCODING AND USING DOMAIN KNOWLEDGE ON POPULATION DYNAMICS FOR EQUATION DISCOVERY

Sašo Džeroski and Ljupčo Todorovski

Department of Intelligent Systems, Jožef Stefan Institute
Jamova 39, 1000 Ljubljana, Slovenia
saso.dzeroski@ijs.si, ljupco.todorovski@ijs.si

Abstract   This chapter is concerned with integrating knowledge-based modeling or modeling from first principles, with data-driven or automated modeling of dynamic systems. The approach presented here includes methods for equation discovery: Unlike mainstream system identification methods, which work under the assumption that the form of the equations is known, equation discovery systems explore a space of possible equation structures. We propose a formalism for representing knowledge about processes in population dynamics domains and a method to transform such knowledge into an operational form that could be used by equation discovery systems. We also describe the extensions of the equation discovery system LAGRAMGE necessary to incorporate this kind of knowledge in the process of equation discovery.

1.   Introduction

Most of the work in scientific discovery [Langley et al., 1987] is concerned with assisting the empirical approach to modeling of physical systems. Following this approach, the observed system is modeled on a trial-and-error basis to fit observed data. None or a very limited portion of the available domain (or background) knowledge about the observed system is used in the modeling process. This is especially the case in domains where a limited amount of knowledge is expressed in the form of mathematical laws, such as biology, medicine and other life sciences.

The empirical approach is in contrast to the theoretical approach to modeling, where the basic physical processes involved in the observed

L. Magnani, N.J. Nersessian, and C. Pizzi (eds.),
Logical and Computational Aspects of Model-Based Reasoning, 227–247.

system are first identified. A human expert then uses the domain knowledge about the identified processes to write down a proper structure of the model in the form of differential equations. Finally, the values of the constant parameters of these equations are fitted against the observed data using standard system identification methods [Ljung, 1993].

The focus of this chapter is on integrating background knowledge from the domain of use, supplied by a domain expert, into the process of automated modeling of population dynamics with equation discovery. The equation discovery system LAGRAMGE [Todorovski and Džeroski, 1997] has made initial steps toward this integration. It allows the user to define the space of possible equations using a context free grammar, written on the basis of the user background knowledge about the domain at hand. However, one can argue that it is difficult to encode a context free grammar from expert knowledge.

In this chapter, we propose a formalism for encoding population dynamics modeling knowledge that is more accessible to human experts. It allows an automated generation of a grammar for equation discovery. The generated grammar is context dependent (and not context free as in LAGRAMGE), so LAGRAMGE 2.0 was developed that, among other improvements, allows the use of context dependent constraints in the grammar specifying the space of possible equations.

The chapter is organized as follows. The basics of population dynamics modeling are introduced in section 2. The formalism for encoding the population dynamics modeling knowledge and the process of its transformation into a grammar for equation discovery are presented in section 2. section 3 gives a brief introduction to different approaches to the problem of equation discovery, especially differential equations. The focus is on methods for incorporating background knowledge from the domain of use in the process of equation discovery. The necessary improvements of LAGRAMGE are presented in section 4. The experimental evaluation of LAGRAMGE 2.0 is given in section 5. The last Section 6 summarizes the chapter and gives directions for further work.

2.   Population dynamics modeling

Population ecology studies the structure and dynamics of populations, where each population is a group of individuals of the same species inhabiting the same area. In this chapter, we consider modeling the dynamics of populations, especially the dynamics of change of their density. We consider models of predator-prey population dynamics, where the interaction between predator and prey is antagonistic in the sense that it causes increase of the predator population and decrease of the prey population. The models take the form of systems of differential equations [Murray, 1993].

2.1 A generalized Volterra-Lotka model of population dynamics

Consider a simple model based on two populations, foxes and rabbits. The latter are grazing on grass and the foxes are carnivores that hunt rabbits. We assume that rabbits are the only food of foxes, unlimited supply of grass is available to the rabbits and ignore seasonal changes. Under these assumptions, if the rabbit population is large, the fox population grows rapidly. However, this causes many rabbits to be eaten, thus diminishing the rabbit population to the point where the food for foxes is not sufficient. Consequently, the fox population decreases, which causes faster growth of the rabbit population.

The oscillatory behavior of the two population densities can be modeled by the Volterra-Lotka population dynamics model [Murray, 1993]. This model can be generalized to the following schema:

$$\dot{N} = growth\_rate(N) - feeds\_on(P, N) \tag{1}$$
$$\dot{P} = feeds\_on(P, N) - decay\_rate(P), \tag{2}$$

where N is the prey (rabbit) population density and P is the predator (fox) population density, while $\dot{N} = \frac{dN}{dt}$ and $\dot{P} = \frac{dP}{dt}$ denote the time derivatives or rates of change of the two populations.

The generalized schema above allows for relaxing the unrealistic assumptions made in the original Volterra-Lotka model. These assumptions include exponential growth of the prey population and unlimited predation capacity of the predator population. By relaxing these assumptions, as described below, we can build models of predator-prey population dynamics with different complexities.

The term $growth\_rate(N)$ defines the model of the prey population growth in absence of predation. Two models of single population growth are usually used [Murray, 1993]: (a) $growth\_rate(N) = aN$; and (b) $growth\_rate(N) = aN(1 - N/K)$. The first model (a) assumes that the population growth is exponential and unlimited. However, real-world environments typically have some carrying capacity for the population, which limits the density of the population. In such cases, logistic growth model (b) can be used as an alternative, where K is a constant, determining the carrying capacity of the environment.

The second assumption made in simple population models is that the predation rate is proportional to the densities of predator and prey

populations ($feeds\_on(N, P) = bPN$). As for population growth, this
means that the predation growth is exponential and unlimited. Again,
in some cases the predators have limited predation capacity. When the
prey population density is small, the predation rate is proportional to
it, but when the prey population becomes abundant, the predation capacity saturates to some limit value ($feeds\_on(N, P) = bPs(N)$). Three
alternative terms are often used to model the predator saturation response to the increase of the prey density: $(a)\, s(N) = AN/(N + B)$,
$(b)\, s(N) = AN^2/(N^2 + B)$ and $(c)\, s(N) = A(1 - e^{-BN})$, where A is the
limit value of the predation capacity saturation, and B is the constant,
which determines the saturation rate [Murray, 1993].

The modeling knowledge about population growth and saturation presented here can be very useful as background knowledge for automated
modeling of ecological systems with equation discovery.

2.2   Encoding of domain and modeling knowledge

The knowledge about modeling population dynamics can be divided
in two types. The first type is domain specific knowledge about the
populations and their role in the food-chain. In the Volterra-Lotka model
of population dynamics, this knowledge is represented by the single fact
that foxes feed on rabbits. This type of knowledge can be expected
from a biologist without any experience in mathematical modeling of
population dynamics.

The second type of knowledge is domain independent knowledge about
population dynamics modeling, of the kind presented above. This type of
knowledge is independent of the particular populations and food chain
in a given domain, but is specific to the area of population dynamics
modeling. It can be provided by, for example, a mathematician with
some population dynamics modeling experience.

When modeling a new system of populations, domain specific knowledge about the populations and food chains can be provided by a biologist familiar with the domain, but not necessarily with population
dynamics modeling. We can re-use the area specific, but domain independent modeling knowledge across different domains, although it has
been provided by somebody who is not necessarily familiar with the
ecosystem structure and the interactions involved in the new domain.
By combining the two types of knowledge, we can obtain a space of
possible population dynamics models (systems of differential equations)
specified by a grammar. Measured data about the behavior of the pop-

ulations can then be used to select among these models through the process of equation discovery.

Domain specific knowledge. We first provide a specification of the food-chain in the domain: for our example rabbits and foxes domain it is given in Table 1. We use three first-order predicates: the domain(domain_name) predicate is used to specify the name of the domain at hand; each population in the domain is specified using the predicate population(domain_name, population_name); finally, we use the predicate feeds_on(domain_name, predator_population, prey_population) to specify each predator-prey interaction between two populations. For now, only predator-prey interactions can be specified. However, the formalism can be easily generalized to allow other types of interactions between populations, such as parasitism, competitive exclusion and symbiosis [Murray, 1993].

```
domain(volterra_lotka).
population(volterra_lotka, fox).
population(volterra_lotka, rabbit).
feeds_on(volterra_lotka, fox, rabbit).
```

Table 1. Description of a simple Volterra-Lotka population dynamics domain consisting of two populations: foxes (predator) and rabbits (prey).

Note that by using the predicates population and feeds_on, the user is allowed to specify an arbitrary number of populations and predator-prey interactions between them.

Domain independent modeling knowledge. The second part of the modeling knowledge, which is domain independent, is given in Table 2. We use the predicate template(process_name, input_variable, process_model_template) to specify a set of alternative models for population dynamics processes like population growth and saturation. Note that the symbol const is used to specify a constant parameter, whose value has to be fitted against measured data.

A constraint of the form [L:U] can be assigned to each constant parameter specifying that the value v of the constant parameter should be within the interval $L \leq v \leq U$. Omitting the U (L) value in this constraint means that there is no upper (lower) bound on the constant parameter value. For example, the symbol const[0:] means that the constant parameter should be non-negative.

```
template(saturation, X, (X)).
template(saturation, X, (X / (X + const[0:]))).
template(saturation, X, (X * X / (X * X + const[0:]))).
template(saturation, X, (1 - exp(-const[0:] * X))).

template(growth, X, (const * X)).
template(growth, X, (const * X * (1 - X / const[0:]))).
```

Table 2. Templates with alternative sub-expressions used for modeling the processes of saturation and population growth.

More complex food chains (lattices).   In the Volterra-Lotka example, we are dealing with a very simple food chain with two populations and one predation (feeds_on) link. In general, a population dynamics domain will have many populations and the graph consisting of predation links will be a lattice. We can have a situation where one predator population can prey on several other populations.

For example, we can have three populations of foxes, rabbits and pheasants, where foxes can feed on both rabbits and pheasants as alternative food sources: This will be represented by the facts feeds_on(fox, rabbit) and feeds_on(fox, pheasant). The two predation process are happening in parallel and are combined additively, which means that there are two feeds_on terms in the differential equation for the (predator) population of foxes.

Similarly, we can have a situation where several predator populations prey on a single prey population. An example situation is represented by the facts feeds_on(fox, rabbit) and feeds_on(wolf, rabbit). Again, the two predation processes are happening in parallel and are combined additively, i.e., there are two feeds_on terms in the differential equation for the (prey) population of rabbits.

In contrast to alternative food sources, a population may depend on several food sources at the same time. For example, phytoplankton needs both phosphorus and nitrogen at the same time to achieve optimal growth. In our formalism for representing domain knowledge, this would be represented as feeds_on(phytoplankton, [nitrogen, phosphorus]). The processes of phytoplankton feeding on nitrogen and phosphorus combine multiplicatively, i.e., the feeds_on term for phytoplankton contains a product of the two terms for the processes of feeding on nitrogen and phosphorus. Considering the food chains as graphs, the above feeds_on fact would correspond to something like an AND node in AND/OR graphs [Bratko, 2001]. Alternative food sources for a predator population, on the other hand, would correspond to OR nodes.

Transforming the background knowledge into a grammar: A program. We have written a program in Prolog [Bratko, 2001] that transforms a given food chain with nodes and links as outlined above to a grammar that derives expressions for the differential equations. The Program is given in Appendix 6.

The top level productions of the grammar follow the generalized schema (1) and (2). For each prey population that is not a predator, a growth term is added on the right-hand of the equation. For all other populations, a decay term is added. In the equation for a predator population, a positive term is added for each outgoing feeds_on link. In the equation for a prey, a negative term is added for each incoming feeds_on link.

Productions are then added for each of the basic processes (growth, decay, predation) taking place in the domain. For each feeds_on(X, Y) link, a production specifies the form the associated term can take: the predator density X is multiplied by the term nutrient(Y). Productions for nutrient(Y) allow for different saturation terms, based on the templates specified in Table 2. For each feeds_on(X, [Y1, ..., Yn]) link, the product nutrient(Y1) * ... * nutrient(Yn) is used instead of the nutrient(Y) term above.

The above assumes that we know exactly what predation processes are taking place in our domain. A useful extension might be to consider the given food chain as the set of possible processes, not all of which need take place. In such a case, we can allow any of the feeds-on, growth, and decay terms go to zero, essentially specifying that the process in question is not taking place (or has negligible effect in the domain overall). Also, for feeds_on(X, [Y1, ..., Yn]) links, we can allow only a subset of the given set of n potential prey populations (nutrients) be relevant to the predator, and have appropriate productions for each of the non-empty subsets of [Y1, ..., Yn]. The Prolog program already implements these extensions, although our experiments with deriving grammars from domain knowledge and using these grammars for equation discovery did not use these extensions.

Transforming the background knowledge into a grammar: An example. Using the definitions of the background knowledge from Tables 1 and 2, a grammar for equation discovery can be automatically generated. The process of transformation of the knowledge into grammar is automated using the predator-prey model schema. The grammar for the example population dynamics domain, consisting of rabbits and foxes, is given in Table 3.

```
volterra_lotka ->
  time_deriv(rabbit) = growth(rabbit) - feeds_on(fox,rabbit);
  time_deriv(fox) = feeds_on(fox,rabbit) - const * fox

feeds_on(fox,rabbit) -> const * fox * nutrient(rabbit)

nutrient(rabbit) -> rabbit
nutrient(rabbit) -> rabbit/(rabbit + const[0:])
nutrient(rabbit) -> rabbit*rabbit/(rabbit*rabbit+const[0:])
nutrient(rabbit) -> 1-exp(-const[0:]*rabbit)

growth(rabbit) -> const*rabbit
growth(rabbit) -> const*rabbit*(1-rabbit/const[0:])
```

Table 3. A grammar for equation discovery constructed from the background knowledge in Tables 1 and 2.

The starting non-terminal symbol in the grammar `volterra_lotka` is used to generate the system of two differential equations of the population dynamics model, using the generalized schema (1) and (2). The growth of the prey population of rabbits in the absence of predation is modeled using the non-terminal symbol `growth(rabbit)` with two alternative productions, reflecting the two **template** predicates for growth from Table 2. The third non-terminal symbol `feeds_on(fox, rabbit)` models the predation of foxes on rabbits. The predation rate is always proportional to the density of the fox population and the non-terminal `nutrient(rabbit)` is used to introduce the model of predator response to the increase of rabbit population density (the productions reflect the templates from Table 2). The terminal symbols `fox` and `rabbit` represent the system variables (population densities).

Strictly speaking, the grammar in Table 3 is not context free. The production for the starting symbol `volterra_lotka` generates two `feeds_on(fox, rabbit)` symbols, one in the first and another one in the second equation. In context free grammar, these two non-terminal symbols can generate two different expressions. In population dynamics models, however, these two expressions have to be the same. The use of context dependent constraints can overcome this limitation of context free grammars.

3.   Equation discovery

Equation discovery is the area of machine learning that develops methods for automated discovery of quantitative laws, expressed in the form of equations, in collections of measured data [Langley et al., 1987]. It is related to the area of system identification. However, mainstream system identification methods work under the assumption that the structure of the model, i.e., the form of the equations, is known and are concerned with determining the values of the constant parameters in the model [Ljung, 1993]. Equation discovery systems, on the other hand, aim at identifying both an adequate structure of the equations and appropriate values of the constant parameters.

3.1 Background knowledge and language bias

Equation discovery systems search through the space of possible equation structures. Most of the equation discovery systems emulate the empirical approach to scientific discovery: different equation structures are generated and fitted against measured data. However, some of the possible equation structures may be inappropriate for modeling the observed system. For example, consider the case where the measured variables of the observed system are not dimensionless. In that case some algebraic combinations of the system variables, such as addition or subtraction of mass and energy, are not valid. Beyond this simple example, there are also more sophisticated inconsistencies of equation structures with some background knowledge from the domain of the observed system.

Different equation discovery systems explore different spaces of possible equations, or in other words they use different language biases. One possibility is to use some pre-defined (built-in) language bias that restricts the space of possible equation structures to some reasonably small class, such as polynomials or trigonometric functions, like in LA-GRANGE [Džeroski and Todorovski, 1995]. In this case, the user can only influence the space of possible equations in a limited way and cannot use domain specific knowledge in the process of equation discovery.

It is much better to use a declarative bias approach, where the user is allowed to influence or directly specify the space of possible equations. This approach provides users with a tool for incorporating their background knowledge about the domain at hand in the process of equation discovery. The use of background knowledge in the sense of a declarative language bias can avoid the problems of inconsistency of the discovered equations with the knowledge about the domain of the observed system, mentioned above.

Several equation discovery systems make use of domain specific knowledge. In equation discovery systems that are based on genetic programming, the user is allowed to specify a set of algebraic operators that can be used. A similar approach has been used in the EF [Zembowicz and Żytkow, 1992] equation discovery system. The equation discovery system SDS [Washio and Motoda, 1997] effectively uses scale-type information about the dimensions of the system variables and is capable of discovering complex equations from noisy data.

However, expert users can usually provide much more modeling knowledge about the domain at hand than merely enumerating the algebraic operators to be used or (the scale-type of) dimensions of the measured system variables. In order to incorporate this knowledge in the process of equation discovery, we should provide the user with a more sophisticated declarative bias formalism. In LAGRAMGE [Todorovski and Džeroski, 1997], the formalism of context free grammars has been used to specify the space of possible equations. Note here that context free grammars are a far more general and powerful mechanism for incorporating domain specific knowledge than the ones used in SDS [Washio and Motoda, 1997] and EF [Zembowicz and Żytkow, 1992].

The use of declarative bias in the form of a context free grammar was crucial for modeling the phytoplankton growth in Lake Glumsoe in Denmark from real-world sparse noisy measurements [Džeroski et al., 1999]. However, one can argue that it is difficult for the users of LAGRAMGE to express their knowledge about the domain in the form of a context free grammar. In this chapter, we present a formalism for encoding the domain knowledge at a higher, more user-friendly level, which can be automatically transformed to the operational form of grammars for equation discovery.

3.2   Discovery of differential equations

In this chapter, we consider the problem of modeling dynamic systems, i.e. systems that change their state over time. Differential equations are the most common tool for modeling dynamic systems. LAGRANGE was the first equation discovery system that extended the scope of equation discovery systems to ordinary differential equations [Džeroski and Todorovski, 1995]. The basic idea was to introduce the time derivatives of the systems variables through numerical differentiation and then search for algebraic equations. This simple approach has a major drawback: large errors are introduced by numerical differentiation.

The problem was partly resolved in the equation discovery system LAGRAMGE [Todorovski and Džeroski, 1997], where numerical integration is used instead of differentiation for the highest-order derivatives. However, LAGRAMGE is only capable of discovering one differential equation for a single user specified system variable at a time. In order to discover a system of simultaneous differential equations, LAGRAMGE has to be invoked several times, once for each system variable.

4. The equation discovery system LAGRAMGE 2.0

In order to use grammars like the one from Table 3 for equation discovery, we developed the equation discovery system LAGRAMGE 2.0, an improved version of LAGRAMGE 1.0 [Todorovski and Džeroski, 1997]. Improvements were made in three directions. First, the context dependent constraints have to be checked for each expression. Second, the grammar in Table 3 generates all model equations at once, therefore a system of simultaneous equations has to be discovered instead of discovering an equation for each system variable separately. Third, the constraints on the lower and upper bound of the values of the constant parameters have to be considered. The top-level algorithm of the LA-GRAMGE 2.0 exhaustive search procedure is presented in Table 4.

	procedure LagramgeSearch(V, D, G, V_d, b)
1	$Q \leftarrow \{\}$
2	$S \leftarrow$ enumerate all derivation trees in G
3	**foreach** T **in** S **do**
4	**if** CheckConstraints(T, G) **then**
5	T.error $\leftarrow 0.0$
6	**foreach** v **in** V_d
7	T.error $\leftarrow T$.error + Fit($\dot{v} = T.v$, D)
8	**endfor**
9	**endif**
10	$Q \leftarrow Q \cup T$
12	**endwhile**
11	return the b best parse trees in Q

Table 4. Outline of the search procedure of LAGRAMGE 2.0.

The search procedure of LAGRAMGE 2.0 takes as an input a set of variables V, a data set D with measured time behaviors of the variables in V, a (context dependent) grammar G, a set $V_d \subseteq V$ of dependent variables and a parameter b that determines the number of best models (systems of equations) returned as output of LAGRAMGE.

The search space of LAGRAMGE is the set of parse trees that can be derived with the user provided grammar G. (A parse tree represents a set of simultaneous equations, i.e., a system of differential equations.) The search space is ordered according to the height of the parse trees using the refinement operator defined in [Todorovski and Džeroski, 1997]. Starting with an empty parse tree, it can be repeatedly used to generate all parse trees.

4.1   Context dependent constraints

Each generated parse tree is first checked to see if it satisfies the context dependent constraints in G (line 4 of the algorithm in Table 4). The user is allowed to specify an arbitrary number of context dependent constraints for each production in the grammar. Examples of productions with context dependent constraints are presented in Table 5.

E -> A + B, B - A { A.1 == A.2; }
E -> A + B, B - A { A.1 == A.2; B.1 == B.2; }
E -> A * E { A <= E; }

Table 5. Examples of grammar productions with context dependent constraints.

In the first production, a single constraint A.1 == A.2 specifies that the first (A.1) and the second (A.2) occurrence of the symbol A on the right hand side of the production should generate the same subexpression. For example, the expression a1 + b1, b2 - a2 can not be derived using that production (A.1 -> a1 is different from A.2 -> a2), whereas the expression a1 + b1, b2 - a1 can. However, the latter expression can not be derived using the second production due to the second constraint B.1 == B.2. a1 + b1, b1 - a1 is an example of an expression that can be derived using both productions.

The third production illustrates the use of a context dependent constraint to avoid redundant generation of expressions that are equivalent due to the commutativity of multiplication. The context free production E -> A * E can generate both a * b and b * a. On the other hand, using the context dependent constraint A <= E (where the operator <= stands for lexicographic comparison), the second expression b * a can not be derived.

4.2   Simultaneous equations

In order to evaluate a system of simultaneous equations for the user provided set of dependent variables V_d, the sub-trees T_v of the generated tree T are identified for each dependent variable $v \in V_d$. Then the error of each equation of the form $\dot{v} = T.v$ is evaluated (using the Fit function in line 7 of the algorithm in Table 4), where $T.v$ denotes the expression derived by the sub-tree T_v. The errors of the equations for all dependent variables are added together to obtain the error of the whole parse tree T (lines 5-8).

4.3   Constraints on the values of the constant parameters

The function Fit(*equation*, *D*) is used to fit the values of the constant parameters of the equation to the given data set *D*. The discrepancy between the measured data *D* and the data obtained by simulating the equation is used to evaluate the error of the equation. In LAGRAMGE 1.0, the downhill simplex algorithm [Press et al., 1986] was used to fit the values of the constant parameters [Todorovski and Džeroski, 1997]. However, the downhill simplex algorithm can not impose any constraints on parameter values. Because of this, we replaced the downhill simplex algorithm with the nonlinear regression algorithm proposed in [Bunch et al., 1993]. The latter allows the use of simple constraints specifying the lower and upper bounds on the values of the constant parameters.

5.   Experiments

The goal of the experiments with LAGRAMGE, presented in this section, is to evaluate the effect of using the new type of background knowledge in the process of equation discovery. For that purpose, we compared the performance of LAGRAMGE 2.0 with the performance of LAGRAMGE 1.0 on the task of reconstructing different models of a simple aquatic ecosystem consisting of two populations of plankton (phytoplankton and zooplankton), as well as an inorganic nutrient (which we also treat as a population). The predator-prey food-chain for this ecosystem is given in Table 6.

```
domain(aquatic).
population(aquatic, nut).        inorganic(_,nut).
population(aquatic, phyto).      feeds_on(aquatic, phyto, nut).
population(aquatic, zoo).        feeds_on(aquatic, zoo, phyto).
```

Table 6.   Description of the aquatic environment population domain consisting of two populations (phytoplankton and zooplankton) and an inorganic nutrient (also treated as a population).

5.1   Experimental setup

Using the food-chain description along with the domain independent modeling templates from Table 2, the context dependent grammar presented in Table 7 has been built using the algorithm described in Section 2.2. The grammar in Table 7 generates thirty-two different models, i.e. systems of three simultaneous equations, which are used in the ex-

periments. The experimental evaluation of LAGRAMGE 2.0 consisted
of attempting to reconstruct each of these 32 models from simulated
data. Note that for inorganic nutrients we assume no external inflow
(no growth term); this is reflected in the equation for nut.

```
aquatic ->
  time_deriv(nut) = - feeds_on(phyto,nut);
  time_deriv(phyto) = growth(phyto) + feeds_on(phyto,nut)
                         - feeds_on(zoo,phyto);
  time_deriv(zoo) = const * zoo + feeds_on(zoo,phyto)

feeds_on(phyto,nut) -> const[0:]  * phyto * nutrient(nut)
feeds_on(zoo,phyto) -> const[0:]  * zoo * nutrient(phyto)

nutrient(nut) -> nut
nutrient(nut) -> nut/(nut+const[0:])
nutrient(nut) -> nut*nut/(nut*nut+const[0:])
nutrient(nut) -> 1-exp(-const[0:]*nut)

nutrient(phyto) -> phyto
nutrient(phyto) -> phyto/(phyto+const[0:])
nutrient(phyto) -> phyto*phyto/(phyto*phyto+const[0:])
nutrient(phyto) -> 1-exp(-const[0:]*phyto)

growth(phyto) -> const*phyto
growth(phyto) -> const*phyto*(1-phyto/const[0:])
```

Table 7. A grammar for equation discovery in the aquatic ecosystem domain constructed from the background knowledge in Tables 6 and 2.

Using the grammar, we generated all thirty-two different model structures. In order to obtain simulation models, the values of the constant
parameters have to be set. We used randomly generated values uniformly
distributed on the $[0, 1]$ interval. We simulated each of the thirty-two
obtained models from ten different randomly selected initial states (initial values of nut, phyto and zoo) for 100 time steps of 1. Thus, ten
different behaviors were obtained of each of the thirty-two models.

In order to test the robustness of the approach to noise in the data,
we added artificially generated random Gaussian noise to the behaviors.
The noise was added at three different relative noise levels: 1%, 5%
and 10%. The relative noise level of $l\%$ means that we multiplied the
original value x with $(1 + l * G/100)$ to obtain a noisy value, where G
is a normally distributed random variable with mean zero and standard
deviation one.

Two evaluation criteria were used for evaluating the performance of
the equation (re)discovery. First, the leave-one-out procedure was ap-

plied in order to estimate the error of discovered equations on test data, unseen during the discovery process. In each iteration of the leave-one-out procedure, nine out of ten behaviors were used to discover a system of differential equations with LAGRAMGE. The obtained differential equations were then simulated using the initial state of the remaining (test) behavior. The simulation error was measured as the sum of squared differences between the simulated behavior and the test behavior. Second, the structure of the best model discovered by LAGRAMGE, was matched against the structure of the original model equations. The structure of the equations is obtained by abstracting the values of the constant parameters in them and replacing them with the generic symbol const.

5.2   Experimental results

We compared the performance of (1) LAGRAMGE 2.0 using the context dependent grammar with the performance of (2) LAGRAMGE 2.0 using the grammar without constraints on the values of the constant parameters, and (3) LAGRAMGE 1.0 (where no context dependent constraints, and no constraints on the values of the constant parameters can be used). The results of the comparison are summarized in Tables 8 and 9.

Before we discuss the experimental results, we should note here that the use of context dependent constraints in the grammar reduces the space of possible models. The grammar in Table 7 generates thirty-two models when interpreted as a context dependent grammar. On the other hand, when interpreted as a context free grammar (by LAGRAMGE 1.0) this grammar generates 512 possible models. Therefore, using context dependent constraints reduces the search space of LAGRAMGE by factor of sixteen.

NOISE	AVERAGE TEST ERROR			STRUCTURE RECONSTR.		
LEVEL	L2.0	L2.0-NCC	L1.0	L2.0	L2.0-NCC	L1.0
0%	0.0031	0.0020	0.0006	29	28	5
1%	0.0083	*(1)	*(10)	8	6	0
5%	0.1490	*(6)	*(13)	3	5	1
10%	0.6187	*(6)	*(13)	4	5	2

Table 8.   Performance of LAGRAMGE 2.0 (L2.0), LAGRAMGE 2.0 without applying constraints on constant parameters (L2.0-NCC) and LAGRAMGE 1.0 (L1.0). Left hand side: average sum of squared errors on the test behavior. Right hand side: number of successfully reconstructed original model structures.

NOISE LEVEL	L2.0 vs. L2.0-NCC	L2.0 vs. L1.0
0%	10-13-9	26-4-2
1%	4-23-5	18-12-2
5%	7-13-12	16-10-6
10%	6-10-16	14-8-10

Table 9. Win-tie-loss counts for comparison of the test error of (L2.0) with the test errors of L2.0-NCC and L1.0 (see caption of Table 8).

In the left hand side of Table 8 the average leave-one-out testing error of the thirty-two (re)discovered models is given. Note that the symbol *(N) means that N out of thirty-two (re)discovered models could not be simulated (and therefore the average error could not be properly evaluated) due to the singularities, such as division by zero or unstable behavior of the discovered system of differential equations. Some of the singularities were caused by inappropriate values of the constant parameters (e.g., a negative saturation limit or carrying capacity). For this reason, LAGRAMGE 2.0 without applying constraints on the values of constant parameters often fails to discover a valid model. In models discovered by LAGRAMGE 1.0, some of the simulation failures are caused by inappropriate model structures, due to the lack of context dependent constraints.

These results show one important aspect of the noise robustness of LA-GRAMGE 2.0: at all noise levels it discovers models that can be simulated and have stable behaviors. This is due both to the context dependent constraints and the constraints on the values of the constant parameters. This is very important: in our earlier experiments on modeling phytoplankton growth in Lake Glumsoe [Džeroski et al., 1999], we had to manually filter out the models discovered by LAGRAMGE 1.0 which had inappropriate values of the constant parameters [Džeroski et al., 1999].

In Table 9, the win-tie-loss counts for the comparison of the test error is presented. We counted the number of wins, ties, and losses in the following manner. For each of the thirty-two experiments, we compared the simulation error e_1 of LAGRAMGE 2.0 with the simulation error e_2 of the other two algorithms (L2.0-NCC and L1.0 in Table 9). Comparisons where the relative difference of the simulation errors is less than 10% (i.e., $90\% < e_2/e_1 < 110\%$) are considered ties.

The comparison of simulation errors shows clear performance improvement of LAGRAMGE 2.0 over LAGRAMGE 1.0 for all noise levels. On the other hand, models discovered by LAGRAMGE 2.0 without applying the constraints on the values of the constant parameters better fit noisy data than the ones generated with LAGRAMGE 2.0. This observation shows that models with inappropriate values of the constant parameters (or inappropriate structure for models discovered by LAGRAMGE 1.0) can sometimes fit the observed data better. However, these models do not make sense from biological point of view.

6. Discussion

We have presented an approach that allows for the representation and use of knowledge about processes in population dynamics systems and data about the behavior(s) of such systems, when trying to learn models that describe observed population behavior(s) in an automated fashion.

The formalism for encoding knowledge about population dynamics, allows us to encode two types of knowledge. The first is domain specific knowledge about the predator-prey food chains in the domain and can be provided by a biologist without any experience in mathematical modeling. The second is modeling knowledge in the form of typical models or sub-models used for modeling different population dynamic processes, such as the growth of a population and the saturation of predation. The modeling knowledge is provided by a population modeling expert, not necessarily familiar with the domain at hand.

In both cases, we are dealing with high-level knowledge represented in first-order logic, which can be automatically transformed to the operational form of grammars used to guide the search for models in the process of equation discovery. This can be done for arbitrarily complex predator-prey models consisting of any number of populations and interactions between them. The proposed formalism can be easily extended with predicates for specifying other types of interactions between populations, such as parastism, competitive exclusion and symbiosis.

The grammars generated using the presented approach are context dependent and and generate complete models, i.e., systems of simultaneous (differential) equations. The equation discovery system LAGRAMGE 2.0 was developed that makes use of such grammars. Using context dependent constraints reduces the space of possible models as compared to using purely context free grammars (as in LAGRAMGE 1.0). Therefore, context dependent constraints improve the efficiency of LAGRAMGE.

Experimental evaluation of LAGRAMGE 2.0 shows that both context dependent constraints and constraints on the values of the constant parameters improves the noise robustness of LAGRAMGE in several ways. First, all the models (re)discovered by LAGRAMGE 2.0 at different noise levels can be simulated and generate stable behaviors. Second, all the models have clear interpretations from biological point of view. Finally,

LAGRAMGE 2.0 (re)discovers the original model structure more often than LAGRAMGE 1.0. The experimental results should be further confirmed with experiments on the real-world observational data. These include modeling phytoplankton growth in the Danish lake Glumsoe [Džeroski et al., 1999], predicting algae blooms in Lagoon of Venice [Džeroski et al., 1999] and modeling plankton population dynamics in the Japanese lake Kasumigaura [Whigham, 2000].

Our approach is similar to the ECOLOGIC approach [Robertson et al., 1991] in the sense that it allows for representing modeling knowledge and domain specific knowledge. However, in ECOLOGIC, the user has to select himself among the alternative models, whereas in our approach observational data is used to select among the alternatives. It is also related to process-based approaches to qualitative physics [Forbus, 1984]. We can think of the food-chain or domain specific part of the knowledge as describing processes qualitatively, whereas the modeling part together with the data introduces the quantitative component.

Bradley et al. [Bradley et al., 2001], combine artificial intelligence techniques with traditional engineering methods for system identification, in order to automate the modeling of dynamic systems. They use meta-domain knowledge which allows for a unified representation of modeling knowledge, valid across different engineering domains, such as the mechanical, electrical or hydraulic domain. Their knowledge representation formalism is component-based, i.e., it incorporates knowledge about sub-models for individual components that can appear in the systems and about how these sub-models are combined into a model of the whole system. On the other hand, our representation of domain knowledge is process-based.

Currently, the presented formalism focuses on representing domain knowledge for population dynamics modeling. However, the presented formalism can be extended so knowledge about dynamic processes from other areas can be encoded and used for equation discovery. The knowledge should include ontology of typical processes in the area (such as predator-prey interaction and population growth in population dynamics) and templates of models typically used for modeling these processes. Finally, knowledge from different areas can be organized in the form of libraries of background knowledge for equation discovery.

Appendix: The Prolog program for transforming the population dynamics domain knowledge into grammar form

```prolog
grammar(Domain) :-
  setof(Population,population(Domain,Population),Pops),
  write('Populations of '), write(Domain),
    write(' are '), write(Pops), nl,
  StartSymbol =.. [Domain|Pops],
  write('Start symbol is '), write(StartSymbol), nl,
  setof(feeds_on(P1,P2),feeds_on(Domain,P1,P2),Feeds),
  write('Food chain contains the links '), write(Feeds), nl,
  getTopProduction(Domain,Pops,Expressions,Feeds),
  RHSTLP =.. [expressions|Expressions],
  write('Top level production is '),
    write(StartSymbol), write(' -> '), nl,
    write(RHSTLP), nl,
  feeds_onProductions(Feeds),
  nutrientProductions,
  growthProductions.


getTopProduction(_,[],[],_). getTopProduction(Domain,[P|Pops],
                              [Expression|Es],Feeds) :-
  (inorganic(Domain,P), Start=0;
   \+ inorganic(Domain,P), prey(Domain,P),
      Start=growth(P), assert(growth(P));
   \+ inorganic(Domain,P), \+ prey(P,Feeds), Start=const*P),
  getTopP(Domain,P,Start,Expression,Feeds),
  getTopProduction(Domain,Pops,Es,Feeds).


getTopP(_,_,CE,CE,[]). getTopP(Domain,P,CE,NE,[X|Fs]) :-
  X =.. [feeds_on, P, Q],
  !,
  getTopP(Domain,P,CE + X,NE,Fs).
getTopP(Domain,P,CE,NE,[X|Fs]) :-
  X =.. [feeds_on, Q, L], (P=L ; member(P,L)),
  !,
  getTopP(Domain,P,CE - X,NE,Fs).
getTopP(Domain,P,CE,NE,[_|Fs]) :-
  getTopP(Domain,P,CE,NE,Fs).


prey(Domain,P) :- feeds_on(Domain,_,P). prey(Domain,P) :-
                  feeds_on(Domain,_,[L|List]),
member(P,[L|List]).


feeds_onProductions([]). feeds_onProductions([Link|Ls]) :-
  feeds_onProduction(Link),
  feeds_onProductions(Ls).
```

```prolog
feeds_onProduction(feeds_on(X,Y)) :-
  atomic(Y), !,
  write('Production '), write(feeds_on(X,Y)), write(' -> '),
  write(const*X), write('*'), write(nutrient(Y)), nl,
  (\+ nutrient(Y), assert(nutrient(Y)); true).

nutrientProductions(Y) :-
  template(saturation,Y,T),
  write(nutrient(Y)), write(' -> '), write(T), nl.

feeds_onProduction(feeds_on(X,List)) :-
  write('Production '), write(feeds_on(X,List)), write(' -> '),
  write(const*X), write('*'), write(nutrient_list(List)), nl,
  nutrient_list_production(List).

% This can easily be replaced with a set of productions that
% allow a subset of nutrient terms to appear in the product

nutrient_list_production(Y) :-
  write(nutrient_list(Y)), write(' -> '), wnl(Y).
wnl([]) :- write(1), nl. wnl([N|Ns]) :-
  write(nutrient(N)), write('*'),
  (\+ nutrient(N), assert(nutrient(N)); true),
  wnl(Ns).

nutrientProductions :-
  retract(nutrient(Y)), nutrientProductions(Y), fail.
nutrientProductions.

growthProductions :-
  retract(growth(X)), growthProductions(X), fail.
growthProductions.

growthProductions(Y) :-
  template(growth,Y,T),
  write(growth(Y)), write(' -> '), write(T), nl.

member(X,[X|_]). member(X,[_|L]) :- member(X,L).

once(X) :- X, !.
```

References

Bradley, E., Easley, M., and Stolle, R., 2001, Reasoning about nonlinear system identification, *Artificial Intelligence* 133:139–188.

Bratko, I., 2001, *Prolog Programming for Artificial Intelligence*, Addison-Wesley, Reading, MA, third edition.

Bunch, D.S., Gay, D.M., and Welsch, R.E., 1993, Algorithm 717; subroutines for maximum likelihood and quasi-likelihood estimation of parameters in nonlinear regression models, *ACM Transactions on Mathematical Software* 19:109–130.

Džeroski, S. and Todorovski, L., 1995, Discovering dynamics: From inductive logic programming to machine discovery, *Journal of Intelligent Information Systems* 4:89–108.

Džeroski, S., Todorovski, L., Bratko, I., Kompare, B., and Križman, V., 1999, Equation discovery with ecological applications, in: *Machine Learning Methods for Ecological Applications*, A.H. Fielding, ed., Kluwer Academic Publishers, Boston, MA, pp. 185–207.

Forbus, K.D., 1984, Qualitative process theory, *Artificial Intelligence* 24:85–168.

Langley, P., Simon, H.A., Bradshaw, G.L., and Żythow, J.M., 1987, *Scientific Discovery*, MIT Press, Cambridge, MA.

Ljung, L., 1993, Modelling of industrial systems, in: *Proceedings of Seventh International Symposium on Methodologies for Intelligent Systems*, Springer, Berlin, pp. 338–349.

Murray, J.D., 1993, *Mathematical Biology*, Springer, Berlin, second, corrected edition.

Press, W.H., Flannery, B.P., Teukolsky, S.A., and Vetterlin, W.T., 1986, *Numerical Recipes*, Cambridge University Press, Cambridge, MA.

Robertson, D., Bundy, A., Muetzelfield, R., Haggith, M., and Uschold, M., 1991, *Eco-logic: Logic-based Approaches to Ecological Modelling*, MIT Press, Cambridge, MA.

Todorovski, L. and Džeroski, S., 1997, Declarative bias in equation discovery, in: *Proceedings of the Fourteenth International Conference on Machine Learning*, Morgan Kaufmann, San Francisco, CA, pp. 376–384.

Washio, T. and Motoda, H., 1997, Discovering admissible models of complex systems based on scale-types and identity constraints, in: *Proceedings of the Fifteenth International Joint Conference on Artificial Intelligence*, volume 2, Morgan Kaufmann, San Francisco, CA, pp. 810–817.

Whigham, P.A., 2000, An inductive approach to ecological time series modelling by evolutionary computation, in: *Abstract Book of the Second International Conference on Applications of Machine Learning to Ecological Modelling*, Adelaide University, Adelaide, Australia, p. 35.

Zembowicz, R. and Żytkow, J.M., 1992, Discovery of equations: Experimental evaluation of convergence, in: *Proceedings of the Tenth National Conference on Artificial Intelligence*, Morgan Kaufmann, San Francisco, CA, pp. 70–75.

REASONING ABOUT MODELS OF NONLINEAR SYSTEMS

Reinhard Stolle

Palo Alto Research Center (PARC), 3333 Coyote Hill Road, Palo Alto, CA 94304, USA

stolle@parc.com

Matthew Easley

Rockwell Scientific, 444 High Street, Palo Alto, CA 94301-1671, USA

measley@rwsc.com

Elizabeth Bradley

Department of Computer Science, University of Colorado, Campus Box 430, Boulder, CO 80309-0430, USA

lizb@cs.colorado.edu

Abstract   An engineer's model of a physical system balances accuracy and parsimony: it is as simple as possible while still accounting for the dynamical behavior of the target system. PRET is a computer program that automatically builds such models. Its inputs are a set of observations of some subset of the outputs of a nonlinear system, and its output is an ordinary differential equation that models the internal dynamics of that system. Modeling problems like this have immense and complicated search spaces, and searching them is an imposing technical challenge. PRET exploits a spectrum of AI and engineering techniques to navigate efficiently through these spaces, using a special first-order logic system to decide which technique to use when and how to interpret the results. Its representations and reasoning tactics are designed both to support this flexibility and to leverage any domain knowledge that is available from the practicing engineers who are its target audience. This flexibility and power has let PRET construct accurate, minimal models of a wide variety of applications, ranging from textbook examples to real-world engineering problems.

L. Magnani, N.J. Nersessian, and C. Pizzi (eds.),
Logical and Computational Aspects of Model-Based Reasoning, 249–271.

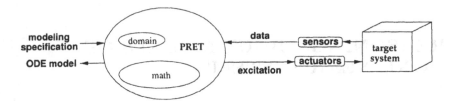

Figure 1.   PRET combines AI and formal engineering techniques to build ODE models of nonlinear dynamical systems. It builds models using domain-specific knowledge, tests them using an encoded ODE theory, and interacts directly and autonomously with target systems using sensors and actuators.

1. Reasoning about nonlinear system identification

Programs that reason about physical systems, about their behavior, and about their models are central to research in scientific discovery, automated modeling, and model-based reasoning. This paper describes the computer program PRET, an engineer's tool for automated modeling of nonlinear systems – also known as *nonlinear system identification*. Modeling the dynamics of a system is an essential first step in a variety of engineering problems. Faced with the task of designing a controller for a robot arm, for instance, a mechanical engineer performs a few simple experiments on the uncontrolled robot arm, observes the resulting behavior, makes some informed guesses about what model fragments could account for that behavior, and then combines those terms into a model and checks it against the physical system. This model then becomes the mathematical core of the controller. Accuracy is not the only requirement; for efficiency reasons, engineers work hard to construct minimal models – those that ignore unimportant details and capture only the behavior that is important for the task at hand. The subtlety of the reasoning skills involved in this process, together with the intricacy of the interplay between them, has led many of its practitioners to classify modeling as "intuitive" and "an art" [Morrison, 1991].

PRET formalizes these intuitions and automates a coherent and useful part of this art. Its inputs are a set of observations of the outputs of a target system, some optional hypotheses about the physics involved, and a set of tolerances within which a successful model must match the observations; its output is an ordinary differential equation (ODE) model of the internal dynamics of that system. See Figure 1 for a block diagram. PRET uses a small, powerful domain theory to build models and a larger, more-general mathematical theory to test them. It is designed to work in any domain that admits ODE models; adding a new domain is simply a matter of coding one or two simple domain rules. Its architecture wraps a layer of artificial intelligence (AI) techniques around a set of traditional formal engineering methods. Models are represented using a component-based modeling framework that accommodates different domains, adapts smoothly to varying amounts of domain knowledge, and allows expert users to create model-building frameworks for new application domains easily. An input-output modeling subsystem allows PRET to observe target systems actively, manipulating actuators and reading sensors to perform experiments whose results augment its knowledge in a manner that is useful to the modeling problem that it is trying to solve. The entire reasoning process is orchestrated by a special first-order logic inference system, which automatically chooses, invokes, and interprets the results of the techniques that are appropriate for each point in the model-building procedure. This combination of techniques lets PRET shift fluidly back and forth between domain-specific reasoning, general mathematics, and actual physical experiments in order to navigate efficiently through an exponential search space of possible models.

In general, system identification proceeds in two interleaved phases: first, *structural identification*, in which the form of the differential equation is determined, and then *parameter estimation*, in which values for the coefficients are obtained. In linear physical systems, structural identification and parameter estimation are fairly well understood. In *nonlinear* systems, however, both procedures are vastly more difficult – the type of material that is covered only in the last few pages of standard textbooks. Unlike system identification software used in the control theory community, PRET is not just an automated parameter estimator; rather, it uses sophisticated reasoning techniques to automate the structural phase of model building as well.

PRET's basic paradigm is "generate and test". It first uses its encoded domain theory – the upper ellipse in Figure 1 – to assemble combinations of user-specified and automatically generated ODE fragments into a candidate model. In a mechanics problem, for instance, the generate phase uses Newton's laws to combine force terms; in electronics, it uses Kirchhoff's laws to sum voltages in a loop or currents in a cutset. In order to test a candidate model, PRET performs a series of inferences about the model and the observations that the model is to match. This process is guided by two important assumptions: that abstract reasoning should be chosen over lower-level techniques, and that any model that cannot be proved wrong is right. PRET's inference engine uses an encoded mathematical theory (the lower ellipse in Figure 1) to search for contradictions in the sets of facts inferred from the model and from the observations. An ODE that is linear, for instance, cannot account for

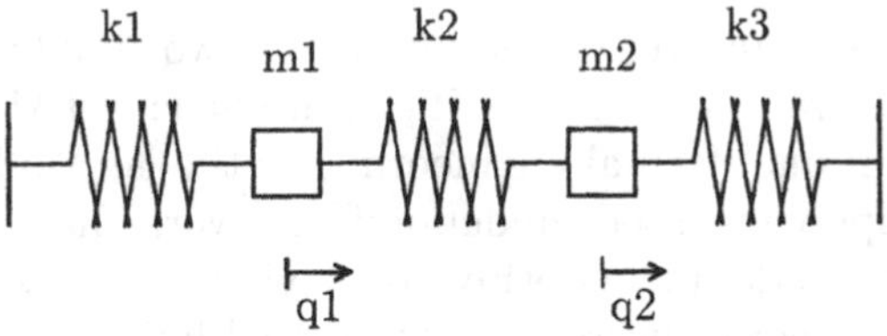

```
(find-model
  (domain mechanics)
  (state-variables (<q1> <point-coordinate>) (<q2> <point-coordinate>))
  (observations
    (autonomous)
    (oscillation <q1>)
    (oscillation <q2>)
    (numeric (<time> <q1> <q2>) ((0 .1 .1) (.1 .109 .110) ...)))
  (hypotheses
    (<force> (* k1 <q1>))
    (<force> (* k2 (- <q1> <q2>)))
    (<force> (* k3 <q2>))
    (<force> (* m1 (deriv (deriv <q1>))))
    (<force> (* m2 (deriv (deriv <q2>))))
    (<force> (* r1 (deriv <q1>)))
    (<force> (* r2 (square (deriv <q1>))))
    (<force> (* r3 (deriv <q2>)))
    (<force> (* r4 (square (deriv <q2>)))))
  (specifications
    (<q1> relative-resolution 1e-2 (-infinity infinity))
    (<time> absolute-resolution 1e-6 (0 120))))
```

Figure 2.   Modeling a simple spring/mass system. Angle brackets (e.g., `<time>`) identify state variables and other special keywords that play roles in PRET's use of its domain theory. The `teletype` font identifies terms that play roles in a user's interaction with PRET.

chaotic behavior; such a model should fail the test if the target system has been observed to be chaotic. Furthermore, establishing whether an ODE is linear is a matter of simple symbolic algebra, so the inference engine should not resort to a numerical integration to establish this contradiction. Like the domain theory, PRET's ODE theory is designed to be easily extended by an expert user.

To make these ideas more concrete, consider the spring/mass system shown at the top right of Figure 2.

To instruct PRET to build a model of this system, a user would enter the `find-model` call at the left of the figure. The `domain` statement instantiates the relevant domain theory; the next two lines inform PRET that the system has two point-coordinate state variables.[1]

---

[1] As described in [Bradley et al., 2001], PRET uses a variety of techniques to infer this kind of information from the target system itself; to keep this example simple, we bypass those facilities by giving it the information up front.

Observations are measured automatically by sensors and/or interpreted by the user; they may be symbolic or numeric and can take on a variety of formats and degrees of precision. For example, the first observation in Figure 2 informs PRET that the system to be modeled is autonomous.[2] An optional list of hypotheses about the physics involved – e.g., a set of ODE terms ("model fragments") that describe different kinds of friction – may be supplied as part of the find-model call; these may conflict and need not be mutually exclusive, whereas observations are always held to be true. Finally, specifications indicate the quantities of interest and their resolutions. It should be noted that this spring/mass example is representative neither of PRET's power nor of its intended applications. Linear systems of this type are very easy to model [Ljung, 1987]; no engineer would use a software tool to do generate-and-test and guided search on such an easy problem. We chose this simple system to make this presentation brief and clear.

To construct a model from the information in this find-model call, PRET uses the mechanics domain rule (point-sum <force> 0) from its knowledge base to combine hypotheses into an ODE. In the absence of any domain knowledge – omitted here, again, to keep this example short and clear – PRET simply selects the first hypothesis, producing the ODE $k_1q_1 = 0$. The model tester, implemented as a custom first-order logic inference engine [Stolle, 1998], uses a set of general rules about ODE properties to draw inferences from the model and from the observations. In this case, PRET uses its ODE theory to establish a contradiction between the model's order (the model's highest derivative) and its oscillatory behavior. The way PRET handles this first candidate model demonstrates the power of its abstract-reasoning-first approach: only a few steps of inexpensive qualitative reasoning suffice to let it quickly discard the model.

PRET tries all combinations of <force> hypotheses at single point coordinates, but all these models are ruled out for qualitative reasons. It then proceeds with ODE systems that consist of *two* force balances – one for each point coordinate. One example is

$$k_1q_1 + m_1\ddot{q}_1 = 0$$
$$m_2\ddot{q}_2 = 0$$

PRET cannot discard this model by purely qualitative means, so it invokes its nonlinear parameter estimation reasoner, which uses knowledge derived in the structural identification phase to guide the parameter

---

[2]That is, it does not explicitly depend on time.

estimation process (e.g., choosing good approximate initial values and thereby avoiding local minima in regression landscapes) [Bradley et al., 1998]. This parameter estimator finds no appropriate values for the coefficients k_1, m_1, and m_2, so this candidate model is also ruled out. This, however, is a far more expensive proposition than the simple symbolic contradiction proof for the one-term model above – roughly five minutes of CPU time, as compared to a fraction of a second – which is exactly why PRET's inference guidance system is set up to use the parameter estimator only as a last resort, after all of the more-abstract reasoning tools in its arsenal have failed to establish a contradiction.

After having discarded a variety of unsuccessful candidate models via similar procedures, PRET eventually tries the model

$$k_1 q_1 + k_2(q_1 - q_2) + m_1 \ddot{q}_1 = 0$$
$$k_3 q_2 + k_2(q_1 - q_2) + m_2 \ddot{q}_2 = 0$$

Again, it calls the parameter estimator, this time successfully. It then substitutes the returned parameter values for the coefficients and integrates the resulting ODE system with fourth-order Runge-Kutta, comparing the result to the numeric time-series observation. The difference between the integration and the observation stays within the specified resolution, so this candidate model is returned as the answer. If the list of user-supplied hypotheses is exhausted before a successful model is found, PRET generates hypotheses automatically using Taylor-series expansions on the state variables – the standard engineering fallback in this kind of situation. This simple solution actually has a far deeper and more important advantage as well: it confers black-box modeling capabilities on PRET. The implications of this are discussed in the following section.

The technical challenge of this model-building process is efficiency; the search space is huge – particularly if one resorts to Taylor expansions – and so PRET must choose promising model components, combine them intelligently into candidate models, and identify contradictions as quickly and simply as possible. In particular, PRET's *generate* phase must exploit all available domain-specific knowledge insofar as possible. A modeling domain that is too small may omit a key model; an overly general domain has a prohibitively large search space. By specifying the modeling domain, the user helps PRET identify what the possible or typical "ingredients" of the target system's ODE are likely to be, thereby narrowing down the search space of candidate models. This "grey-box" modeling approach differs from traditional black-box modeling, where the model must be inferred only from external observations of the target system's behavior. It is also more realistic, as described in more depth in section 3: the engineers who are PRET's target audience do not operate in a complete vacuum, and its ability to leverage the kinds of domain knowledge that such users typically bring to a modeling problem lets PRET tailor the search space to the problem at hand.

The key to our approach is to classify model and system behavior at the highest possible abstraction level. PRET knows about several different reasoning techniques that are appropriate in different situations and that operate at different abstraction levels and in different domains. Examples of such techniques are symbolic algebra, qualitative reasoning about symbolic model properties, phase-portrait analysis, and so on. See [Stolle, 1998; Stolle and Bradley, 1998] for a complete discussion. To effectively build and test models of nonlinear systems, PRET must determine which methods are appropriate to a given situation, invoke and coordinate them, and interpret their results.

Perhaps the most important of these techniques – and one that is unique in the AI/modeling literature – is *input-output modeling*, in which interacts directly and autonomously with its target systems, using sensors and actuators to perform experiments whose results are useful to the model-building process. Sensor data processing is reasonably straightforward to automate, as it is an inherently passive process: PRET's input-output module receives a data stream and then processes the information using available data analysis tools to yield abstract, qualitative observations about the system. Manipulating an actuator, however, is significantly harder; to do so, PRET must not only formulate a plan about how to control that actuactor, but also consider the set of possible states that are reachable with the available control input and also reason effectively about what new knowledge would be useful, given the current modeling problem. Space limitations preclude detailed discussion of these issues in this paper; please see the references listed below for coverage of:

1 control theory's *controllability/reachability* problem: given a system, an initial condition, and an actuator configuration, what state-space points are reachable with the available control input? (see [Easley and Bradley, 1999])

2 intelligent data analysis methods for distilling high-level qualitative information out of megabytes of detailed sensor data (see [Bradley and Easley, 1998; Easley, 2000])

3 reasoning techniques that allow model-building knowledge from one behavioral regime to be effectively – and appropriately – applied in others (see [Easley and Bradley, 1999])

2. Automated modeling and scientific discovery

In the AI literature, work on automatically finding a model for a given dynamic system falls under the rubrics of "reasoning about physical systems," "automated modeling", "machine learning", and "scientific discovery". Traditionally, the areas of automated modeling and scientific discovery have been perceived as fairly distinct disciplines, and unfortunately the interaction between these fields has been limited. In this section, we discuss some of these perceived differences; we also describe what these fields have in common. Finally, we show where PRET resides within the spectrum of automated modeling and scientific discovery.

In the field of Qualitative Reasoning (QR) [Weld and de Kleer, 1990], a model of a physical system is mainly used as a representation that allows an automated system to *reason* about the physical system [Forbus, 1984; Kuipers, 1993]. QR reasoners are usually concerned with the physical system's structure, function, or behavior. For example, qualitative simulation [Kuipers, 1986] builds a tree of qualitative descriptions of possible future evolutions of the system. Typically, the system's structural and functional properties are known, and the task of *modeling* consists of finding a formal representation of these properties that is most suitable to the intended reasoning process [Nayak, 1995]. Such models frequently highlight qualitative and abstract properties of the system so as to facilitate efficient qualitative inferences. Modeling of systems with known functional and structural properties is generally called *clear-box modeling*.

The goal of Scientific Discovery (e.g., [Langley, 1987]) and System Identification [Ljung, 1987] is to investigate physical systems whose structural, functional properties are not – or are only partially – known. Modeling a target system, then, is the process of inferring an intensional (and finite) description – a *model* – of the system from extensional (and possibly infinite) observations of its behavior. For example, a typical system identification task is to observe a driven pendulum's behavior over time and infer from these time series measurements an ordinary differential equation system that accounts for the observed behavior. This process is usually referred to as *black-box modeling*. It amounts to inverting simulation, which is the process of predicting a system's behavior over time, given the equations that govern the system's dynamics.

Whereas the desired model in a system identification task usually takes the form of differential equations, the field of scientific discovery comprises a wider range of tasks with a broader variety of possible models. According to [Langley, 2000], a scientific discovery program typically tries to discover regularities in a space of entities and concepts that has been designed by a human. Such regularities may take the form of qualitative laws, quantitative laws, process models, or structural models (which may even postulate unobserved entities). The discovery of process models amounts to explaining phenomena that involve change over time; it is the kind of scientific discovery that comes closest to system identification. Most of the scientific discovery literature (e.g., [Huang and Żytkow, 1997; Langley, 1987; Todorovski and Džeroski, 1997; Washio et al., 1999; Żytkow, 1999]) revolves around the discovery of natural laws. Predator-prey systems or planetary motion are prominent examples. System identification, on the other hand, is typically performed in an engineering context – building a controller for a robot arm, for example.

For the purposes of this paper, we distinguish between *theories* and *models*. Żytkow's terminology [Żytkow, 1999] views theories as analytical and models as synthetic products. We prefer to draw the distinction along the generality/specificity axis. A theory and a model are similar in the sense that both are derived from observations of target systems. However, theories aim at a more-general and more-comprehensive description of a wider range of observations. Constructing a theory includes the definition (or postulation) of relevant entities and quantities; the laws of the theory, then, express relationships that hold between these entities and quantities. The developer of a theory tries to achieve a tight correspondence between the postulated structural setup of the entities and the laws that describe the behavior of the entities. One of the major goals of theory development is an increased *understanding* of the observed phenomena: the quality of a theory depends not only on how accurately the theory accounts for the observations, but also on how well the theory connects to other theories, on whether it generalizes or concretizes previous theories, and on how widely it is applicable. Therefore, research in scientific discovery must address the question about whether the discovered theory accurately models the target system (e.g., nature), or whether it just happens to match the observations that were presented to the discovery program. Likewise, machine learning systems routinely use validation techniques (such as cross-validation) in order to ensure the "accuracy" of the learned model.

Engineering modeling is much less general and much more task-specific. A given *domain theory* sets up the space of possible quantities of interest.[3] A model, then, is a mathematical account of the behavioral relationships between these quantities. Some or all model fragments may

---

[3]By choosing a very specific domain theory and by setting up a specific space of possible model fragments, a human may actually convey substantial information about the target system

or may not correspond to a structural fragment of the modeled system. For example, one may recognize a particular term as a "friction term" or a "gravity term." Whether such correspondences exist, however, is of secondary concern. The primary concern is to accurately describe the *behavior* of the system within a faily limited context and with a specific task (e.g., controller design) in mind.

The distinction between this kind of modeling and theory development is not a clear one. Even though many engineering models are very specific compared to the scope of a scientific theory, engineers also broaden their exploration beyond a single system, in order, for example, to build a cruise control that works for most cars, not just a particular Audi on a warm day. Furthermore, one may argue that the distinction between "accounting for" and "explaining" an observation is arbitrary. For example, there may not be a big difference between saying "the resistor explains the dissipation of energy" and saying "this term (which corresponds to the resistor) accounts for that behavior (which corresponds to the dissipation of energy)." Nevertheless, it is important to note that explanation and understanding are stated *goals* of scientific discovery; in system identification, they are often merely *byproducts* of the modeling process. Furthermore, scientific discovery approaches often introduce higher-level concepts (e.g., the label "linear friction force" for the mathematical term $c\dot{x}$) in order to achieve structural coherence and consistency with background knowledge and/or other theories. Such higher-level concepts are useful in automated system identification as well, but PRET's approach is to restrict the search space in an effective, efficient grey-box modeling approach. Finally, in system identification, the engineering task at hand provides an objective measure as to when a model is "good enough" – as opposed to a scientific theory, which is always only a step toward a further theory: something that is more general, more widely applicable, more accurate, and/or expressed in more fundamental terms.

The long-term vision in scientific discovery is even more ambitious than the previous paragraph suggests. Rather than "just" manipulating existing concepts, quantities, and entities in order to construct theories, scientific discovery programs may even invent or construct new concepts or entities on the fly. Furthermore, Chapter 10 of [Langley, 1987] speculates about the automated "discovery of research problems, the invention of scientific instruments, and the discovery and application of good problem representations." Such tasks are clearly outside the scope of

---

to the automatic modeling program. As mentioned in Section 2, we call this compromise between clear- and black-box modeling *grey-box modeling*.

modeling from an engineering perspective. This difference in long-term vision between theory developers and model constructors has important consequences concerning the *parsimony* of the developed theory or constructed model. Both system identification and scientific discovery strive for a simple representation of the target system. However, from a scientific discovery viewpoint, parsimony must be achieved within the constraint that the theory be behaviorally accurate and structurally coherent with background knowledge and other theories, as described in the previous paragraphs. In system identification, however, parsimony is critically important. Modelers work hard to build abstract, minimal models that account for the observations. Typically, the desired model is the one that is just concrete enough to capture the behavior that is relevant for the task at hand.

PRET's techniques, which we present in the rest of this paper, resemble ideas from automated modeling, scientific discovery, and machine learning. Some of PRET's roots as an engineer's tool can be found in "the dynamicist's workbench" [Abelson, 1989; Abelson and Sussman, 1989]. Its representational scheme and its reasoning about candidate models build on a large body of work in automated model building and reasoning about physical systems (see, for example, [Addanki et al., 1991; Falkenhainer and Forbus, 1991; Forbus, 1984; Nayak, 1995]). In particular, our emphasis on qualitative reasoning and qualitative representations and their integration with numerical information and techniques falls largely into the category of qualitative physics. The project in this branch of the literature that is most closely related to PRET is the qualitative reasoning-based viscoelastic system modeling tool developed by Capelo et al. [Capelo et al., 1998], which also builds ODE models from time-series data. PRET is more general; it handles linear *and nonlinear* systems in a variety of domains using a richer set of model fragments that is designed to be adaptable. (Indeed, one of PRET's implemented modeling domains, `viscoelastics`, allows it to model the same problems as in [Capelo et al., 1998]).

PRET takes a strict engineering approach to the questions of accuracy and parsimony. Its goal is to find an ODE system that serves as a useful model of the target system *in the context of engineering tasks*, such as controller design. PRET's notion of "accuracy" is relative only to the given observations: it finds an ODE system that matches the observations to within the user-specified precision, and does not try to second-guess these specifications or the user's choice of observations. It is the user's power and responsibility to ensure that the set of observations and specifications presented to PRET reflect the engineering task at hand. It is, of course, possible to use PRET as a scientific discovery tool by supplying several sets of observations to it in separate runs and then unifying the results by hand. PRET can also be used to solve the kinds of cross-validation problems that arise in the machine learning literature: one would simply use it to perform several individual validation runs and then interpret the results.

The project in the scientific discovery/machine learning branch of the literature that is most closely related to PRET is LAGRANGE [Džeroski and Todorovski, 1995], which builds ODE models of completely observed systems by generating candidate models from the observed variables and testing them using qualitative constraints and linear regression techniques. PRET and LAGRANGE differ in several important ways. Whereas LAGRANGE's input is a complete set of time-series data for all system variables, PRET can handle *incomplete*, semi-numerical, or even purely qualitative input data. Both PRET and LAGRANGE extract qualitative constraints from the observed data and test the candidate models against them. However, PRET's user may submit additional constraints and qualitative observations that are formulated at various abstraction levels, and candidate models must meet these additional constraints. Moreover, LAGRANGE constructs candidate models by combining terms that are arbitrarily complex *products* of the observed state variables and their time derivatives; *sin* and *cos* operators are introduced for variables that are measured in radians. PRET allows for arbitrary terms to be suggested by the user in the form of hypotheses; if these are insufficient, it covers the complete space of all ODEs by resorting to Taylor series expansion. Furthermore, LAGRANGE's power is limited by its *linear* regression parameter estimator. PRET's performs *nonlinear* parameter estimation, which is vastly more computationally expensive. Finally, PRET autonomously generates additional constraints in a true input/output modeling approach, interacting directly with the target system via actuators and sensors. The program's architecture has been designed in such a way as to allow for additional modules to be plugged in, each of which further restricts the space of possible candidate models for a given target system and each of which tightens the tests that a successful candidate model must pass. Our research project's intended target applications in the long term are high-dimensional, nonlinear grey-box systems, hence the varied arsenal of techniques described in the rest of this paper.

3. Representations for model building

In the late 1950s and early 1960s, inspired by the realization that the principles underlying Newton's third law and Kirchhoff's current law were identical,[4] researchers began combining multi-port methods from a number of engineering fields into a generalized engineering domain with prototypical components [Paynter, 1961]. The basis of this *generalized physical networks* (GPN) paradigm is that the behavior of an ideal two-terminal element – a "component" – may be described by a mathematical relationship between two dependent variables: generalized flow and generalized effort, where *flow(t) * effort(t) = power(t)*. This pair of variables manifests differently in each domain: (*flow, effort*) is (*current, voltage*) in an electrical domain and (*force, velocity*) in a mechanical domain.

The GPN representation has three important advantages for model building. Firstly, its two-port nature makes it easy to incorporate sensors and actuators as integral parts of a model. For example, a current source often has an associated impedance that creates a loading effect on the rest of the circuit. With a network approach, these effects naturally become part of the model, just as they do in real systems. Secondly, GPNs bring out similarities between components and properties in different domains. Electrical resistors ($v = iR$) and mechanical dampers ($v = fB$) are physically analogous; both dissipate energy in a manner that is proportional to the operative state variable, and so both can be represented by a single GPN component that incorporates a *proportional* relationship between the flow and effort variables. Two other useful GPN components instantiate *integrating* and *differentiating* relationships; the representation also allows for flow and effort source components. Thirdly, the GPN representation makes it very easy to incorporate varying amounts and levels of information. This is closely related to its ability to capture behavioral analogs. Both of the networks in Figure 3, for example, can be modeled by a series proportional/integrating/differentiating GPN; knowledge that the system is electronic or mechanical would let one refine the model accordingly (to a series RLC circuit or damper-spring-mass system, respectively). The available domain knowledge, then, can be viewed as a lens that expands upon the internals of some GPN components, selectively sharpening the model *in appropriate and useful ways*. See [Easley and Bradley, 1999] for further details on this representation and how PRET uses it.

PRET currently incorporates five specific GPN-based modeling domains: `mechanics`, `viscoelastics`, `linear-electronics`, `linear-rotational`, and `linear-mechanics`. Domains are constructed by domain experts, stored in the domain-theory knowledge base, and instantiated

---

[4]Summation of {forces, currents} at a point is zero, respectively; both are manifestations of the conservation of energy.

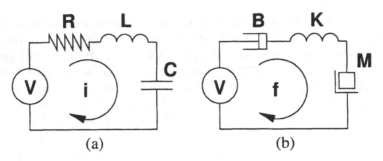

Figure 3.   Two systems that are described by the same GPN model: (a) a series
RLC circuit and (b) a series damper-spring-mass system. **V** is a voltage source in (a)
and a velocity source in (b).

by the domain line of the find-model call. Each consists of a set of
component primitives and a framework for connecting those components into a model. The basic linear-electronics domain, for example, was built by an electrical engineer; it comprises the components
{linear-resistor, linear-capacitor}, the standard parallel and series connectors, and some codified notions of model equivalence (e.g.,
Thévenin). Specification of state variables in different domains – type,
frames of reference, etc. – is a nontrivial design issue, as discussed in
[Easley, 2000]. Finally, these modeling domains are dynamic: if a domain does not contain a successful model, it automatically expands to include additional components and connections (also described in [Easley,
2000]).

If a user wants to apply PRET to a system that does not fall in an
existing domain, he or she can either build a new domain from scratch
– a matter of making a list of components and connectors – or use
one of PRET's *meta-domains*: general frameworks that arrange hypotheses into candidate models by relying on modeling techniques that tran-
scend individual application domains [Bradley et al., 2001; Easley, 2000].
The xmission-line meta-domain, for instance, generalizes the notion of
building models using an iterative pattern, similar to a standard model
of a transmission line, which is useful in modeling distributed parameter systems. The linear-plus meta-domain takes advantage of fundamental linear-systems properties that allow the linear and nonlinear
components to be treated separately under certain circumstances, which
dramatically reduces the model search space. Both can be used directly
or customized for a specific application domain. We chose this particular pair of meta-domains as a good initial set because they cover such a
wide variety of engineering domains. We are exploring other possibili-

ties, especially for the purposes of modeling competing systems (e.g., in economics or biology).

Choosing a modeling domain for a given problem is not trivial, but it is not a difficult task for the practicing engineers who are PRET's target audience. Such a user would first look through the existing domains to see if one matched his or her problem. If none were appropriate, he or she would choose a meta-domain based upon the general properties of the modeling task. If there is a close match between the physical system's components and the model's components (i.e., it is a lumped parameter system), then `linear-plus` is appropriate; `xmission-line` is better suited to modeling distributed parameter systems. There is significant overlap between the various domains and meta-domains; a linear electronic circuit can be modeled using the specific `linear-electronics` domain, the `xmission-line` meta-domain, or the `linear-plus` meta-domain. In all three cases, PRET will produce the same model, but the amount of effort involved will be very different. The advantage of the `linear-electronics` domain is its specialized, built-in knowledge about linear electrical circuits, and the effect of this knowledge is to focus the search. A capacitor in parallel with two resistors, for instance, is equivalent to a single resistor in parallel with that capacitor. The `linear-electronics` domain "knows" this, allowing it to avoid duplication of effort; the two meta-domains do not. Perhaps most important of all, their generality and overlap make the meta-domains particularly helpful if one does not know exactly what kind of system one is dealing with, which is not an uncommon situation in engineering practice.

Some of our ongoing work on model building representations involves extending PRET to application areas where the GPN conservation of energy constraint no longer holds. One interesting modeling approach, termed *Forrester System Dynamics* [Forrester, 1971], has met with success in diverse areas where no power relationship exists, such as biology, epidemiology, sociology, economics, and strategic management. Forrester models consist of a network of components that model the flow or accumulation of *other* important conserved quantities (e.g., pollutants in a river, ecological populations, or buyers in a market). This is a natural extension to our framework, and we are currently working out how to incorporate a more general notion of conservation into our current implementation, which would greatly expand the class of applications that PRET can handle. This extension should be fairly straightforward; indeed, LeFèvre [LeFèvre, 1997] showed that Forrester models use only capacitance, modulated flow sources, and generalized Kirchhoff's voltage and current laws.

The novelty and utility of this GPN-based modeling paradigm lie in its use of meta-level, two-terminal components for automated modeling of nonlinear systems. Previous work in the AI community using meta-level components similar to GPNs has typically been restricted to reasoning about causality [Top and Akkermans, 1991] and modeling hybrid systems [Mosterman and Biswas, 1997]. Amsterdam's work on automated model construction in multiple physical domains [Amsterdam, 1993] is an exception to this, but it is limited to linear systems of order two or less. Capelo et al. [Capelo et al., 1998], as described in the previous section, build ODE models of linear viscoelastic systems by evaluating time series data using qualitative reasoning techniques. Although these goals are similar to PRET's, the library of possible models is more restricted: it involves only two component types (linear springs and linear dashpots). PRET's GPN framework is much more general; not only does it use a large and rich set of meta-level components for automated model building of linear *and nonlinear* systems, but it also supports easy user customization of these components and domains [Easley, 2000].

4.   Orchestrating reasoning about models

As described in the previous section, PRET uses component-based representations, user hypotheses, and domain knowledge to generate candidate models of a given target system. In this section, we describe the reasoning framework in which PRET tests such a model against observations of the target system. PRET makes use of a variety of reasoning techniques at various abstraction levels during the course of this process, ranging from detailed numerical simulation to high-level symbolic reasoning. These modes and their interactions are described in Section 4.1. PRET's central task to quickly find inconsistencies between a candidate model and the target system. Section 4.2 describes the control techniques that allow it to do so by reasoning about which techniques are appropriate in which situations.

4.1   Reasoning modes

Qualitative Reasoning. Reasoning about abstract features of a physical system or a candidate model is typically faster than reasoning about their detailed properties. Because of this, PRET uses a "high-level first" strategy: it tries to rule out models by purely *qualitative* techniques [Forbus, 1996; Weld and de Kleer, 1990] before advancing to more-expensive semi-numerical or numerical techniques. For example, PRET's encoded ODE theory includes the qualitative rule that nonlinearity is a necessary condition for chaotic behavior. Using this rule and an observation that the target system is chaotic, PRET can discard any linear model without performing more-complex operations PRET's qualitative reasoning facilities are not only important for accelerating the search for inconsistencies between the physical system and the model; they also allow the user to express incomplete information.

Qualitative envisioning. After using its qualitative reasoning facilities to the fullest possible extent and before resorting to the numerical level, PRET attempts to establish contradictions by reasoning about the states of the physical system. PRET's qualitative envisioning module constrains the possible ranges of parameters in the candidate model. Currently, the qualitative states contain only sign information $(-, 0, +)$. For example, for the model $ax + by = 0$, the state $(x, y) = (+, +)$ constrains (a, b) to the possibilities $(+, -)$ or $(0, 0)$ or $(-, +)$.

Constraint reasoning. Often, information *between* the purely qualitative and the purely numeric levels is also available. If a linear system oscillates, for example, the imaginary parts of at least one pair of the roots of its model's characteristic polynomial must be nonzero. If the oscillation is damped, the real parts of those roots must also be negative. Thus, if the model $a\ddot{x} + b\dot{x} + cx = 0$ is to match an `damped-oscillation` observation, the coefficients must satisfy the inequalities $4ac > b^2$ and $b/a > 0$. PRET uses expression inference [Sussman and Steele, 1980] to merge and simplify such constraints.

Geometric reasoning. Other qualitative forms of information that are useful in reasoning about models are the geometry and topology of a system's behavior, as plotted in the time or frequency domain, state space, etc. A bend of a certain angle in the frequency response, for instance, indicates that the ODE has a root at that frequency, which implies algebraic inequalities on coefficients. In order to incorporate this type of reasoning, PRET processes numeric observations – e.g., sensor data – using MAPLE functions and simple phase-portrait analysis techniques, producing the type of abstract information that its inference engine can leverage to avoid expensive numerical checks. These methods are used primarily in the analysis of sensor data [Bradley and Easley, 1998].

Parameter estimation and numerical simulation. PRET's final check of any model requires a point-by-point comparison of a numerical integration of that ODE against all numerical observations of the target system. In order to integrate the ODE, however, PRET must first estimate values for any unknown coefficients. Parameter estimation is a complex nonlinear global optimization problem. PRET's nonlinear parameter estimation reasoner, described in [Bradley et al., 1998], solves this problem using a new, highly effective global optimization method that combines qualitative reasoning and local optimization techniques.

4.2   Control of reasoning

PRET's challenge in properly orchestrating the reasoning modes described in the previous section was to test models against observations using the cheapest possible reasoning mode and, at the same time, avoid duplication of effort. In order to accomplish this, the inference engine uses the following techniques.

Resolution theorem proving. The language in which observations and the ODE theory are expressed is that of generalized Horn clause intuitionistic logic [McCarty, 1988]. PRET's inference engine is a resolution-based theorem prover. For every candidate model, this prover combines basic facts about the target system, basic facts about the candidate model, and basic facts and rules from the ODE theory into one set of clauses, and then tries to derive `falsum` – which represents inconsistency – from that set. The special formula `falsum` may only appear as the head of a clause. Such clauses are often called *integrity constraints*; they express fundamental reasons for inconsistencies, e.g., that a system cannot be oscillating *and* non-oscillating at the same time. For a detailed discussion of PRET's logic system see [Hogan et al., 1998; Stolle, 1998; Stolle and Bradley, 1998].

Declarative meta level control. PRET uses declarative techniques not only for the representation of knowledge about dynamical systems and their models, but also for the representation of strategies that specify under which conditions the inference engine should focus its attention on particular pieces or types of knowledge. PRET provides meta-level language constructs that allow the implementer of the ODE theory to specify the *control strategy* that is to be used. The intuition behind PRET's declarative control constructs is, again, that the search should be guided toward a cheap and quick proof of a contradiction. For example, PRET's meta control theory prioritizes stability reasoning about the target system depending on whether the system is known to be linear.[5] For a discussion of PRET's meta control constructs, see [Beckstein et al., 1996; Hogan et al., 1998].

Reasoning at different abstraction levels. To every rule, the ODE theory implementer assigns a natural number, indicating its level of ab-

---

[5] If a system is known to be linear, its *overall* stability is easy to establish, whereas evaluating the stability of a nonlinear system is far more complicated and expensive.

straction. The inference engine uses less-abstract ODE rules only if the more-abstract rules are insufficient to prove a contradiction. This static abstraction level hierarchy facilitates strategies that cannot be expressed by the dynamic meta-level predicates alone: whereas the dynamic control rules impose an *order* on the subgoals and clauses of *one* particular (but complete) proof, the abstraction levels allow PRET to *omit* less-abstract parts of the ODE theory altogether. Since abstract reasoning usually involves less detail, this approach leads to short and quick proofs of the `falsum` whenever possible.

Storing and reusing intermediate results. In order to avoid duplication of effort, PRET stores formulae that have been expensive to derive and that are likely to be useful again later in the reasoning process. Engineering a framework that lets PRET store just the right type and amount of knowledge is a surprisingly tricky endeavor. On the one hand, remembering every formula that has ever been derived is too expensive. On the other hand, many intermediate results are very expensive to derive and would have to be rederived multiple times if they were not stored for reuse. PRET reuses previously derived knowledge in three ways. First, it remembers what it has found out about the physical system across all test phases of individual candidate models. The fact that a time series measured from the physical system contains a limit cycle, for example, can be reused across all candidate models. Second, every time the reasoning proceeds to a less-abstract level, PRET needs all information that has already been derived at the more-abstract level, so it stores this information rather than rederiving it.[6] Finally, many of the reasoning modes described in section 4.1 use knowledge that has been generated by previous inferences, which may in turn have triggered other reasoning modes. For instance, the PRET's parameter estimator relies heavily on qualitative knowledge derived during the structural identification phase in order to avoid local extrema in regression landscapes. To facilitate this, PRET gives these modules access to the set of formulae that have been derived so far.

In summary, PRET's control knowledge is expressed as a declarative meta theory, which makes the formulation of control knowledge convenient, understandable, and extensible. None of the reasoning techniques described in section 4.1 is new; expert engineers routinely use them when modeling dynamical systems, and versions of most have been used in at least one automated modeling tool. The set of techniques used by

---

[6]This requires the developer to declare a number of predicates as *relevant*, which causes all succeeding subgoals with this predicate to be stored for later reuse.

PRET's inference engine, the multimodal reasoning framework that integrates them, and the system architecture that lets PRET decide which one is appropriate in which situation, make the approach taken here novel and powerful.

5.   Conclusion

PRET is designed to produce the type of formal engineering models that a human expert would create – quickly and automatically. Unlike existing system identification tools, PRET is not just a fancy parameter estimator; rather, it uses sophisticated knowledge representation and reasoning techniques to automate the structural identification phase of model building as well. Unlike existing AI tools, PRET takes an *active* approach to the task of modeling complex, nonlinear systems, using techniques drawn from engineering, dynamical systems, and control theory to explore their behavior directly, via sensors and actuators. Unlike any existing software tools, PRET works with high-dimensional, nonlinear, grey-box systems: the kinds of hard problems with which engineers are faced on a daily basis.

Theoretically, PRET can model *any* system that admits an ordinary differential equation model – even in the most severe black-box situation, where it knows nothing whatsoever about that system. Because the `xmission-line` meta-domain allows PRET to use its lumped-element generalized physical network components to model spatially distributed systems, it can even model systems that technically require partial differential equations. In practice, however, the size of the associated search space and the available computer power limit PRET's range. The representation and reasoning tactics described in this paper mitigate this by intelligently streamlining the model-building and -testing processes. Incorporated input-output modeling techniques (described in [Easley, 2000; Easley and Bradley, 1999]) are also critical to PRET's success as knowledge learned from an earlier interaction with a target system may assist PRET in finding a model during a subsequent experiment. Thanks to these tactics, PRET has been able to successfully construct models of a dozen or so textbook problems (Rössler, Lorenz, simple pendulum, pendulum on a spring, etc.; see [Bradley et al., 1998; Bradley and Stolle, 1996]), as well as several interesting and difficult real-world examples, such as the well-aquifer, shock absorber, electrical transmission lines and driven pendulum in the previous section and a commercial radio-controlled car, which is covered in [Bradley et al., 1998]. These examples are representative of wide classes of dynamical systems, both linear and nonlinear. PRET's model of the radio-controlled car was particularly interesting; it not only fit the experimental data, but actually enabled the project analysts to identify what was wrong with their mental models of the system. Specifically, PRET's model matched the observations *but not their intuition*, and the disparities led them to understand the system dynamics better.

This anecdote brings out an important point: PRET is intended to be an engineer's tool, and that goal dictated a specific set of design choices. From an engineering standpoint, a successful model balances accuracy and parsimony. Accordingly, PRET's goal is not to infer physics that the user left implicit, but rather to construct the simplest model that matches the observed behavior to within the predefined `specifications`. Because evaluation criteria are always domain-specific, we believe that modeling tools should let their domain-expert users dictate them, and not simply build in an arbitrary set of thresholds and percentages. The notion of a *minimal* model that is tightly (some might say myopically) guided by its user's specifications represents a very different philosophy from traditional AI work in this area. Unlike some scientific discovery systems, PRET makes no attempt to exceed the range and resolution specifications that are prescribed by its user: a loose `specification` for a particular state variable, for instance, is taken as an explicit statement that an exact fit of that state variable is not important to the user, so PRET will not add terms to the ODE in order to model small fluctuations in that variable. Conversely, a single out-of-range data point will cause a candidate model to fail PRET's test. These are not unwelcome side effects of the finite resolution; they are intentional and useful by-products of the abstraction level of the modeling process. A single outlying data point may appear benign if one reasons only about variances and means, but engineers care deeply about such single-point failures (such as the temperature dependence of O-ring behavior in space shuttle boosters), and a tool designed to support such reasoning must reflect those constraints.

Acknowledgments

Apollo Hogan, Brian LaMacchia, Abbie O'Gallagher, Janet Rogers, Ray Spiteri, and Tom Wrensch contributed code and/or ideas to PRET. Parts of this paper are short versions of material previously published in the journal *Artificial Intelligence* [Bradley et al., 2001] and are republished here with kind permission of the publisher, Reed-Elsevier.

References

Abelson, H., Eisenberg, M., Halfant, M., Katzenelson, J., Sussman, G.J., and Yip, K., 1989, Intelligence in scientific computing, *Comm. ACM*.

Abelson, H. and Sussman, G.J., 1989, The Dynamicist's Workbench I: Automatic preparation of numerical experiments, in: *Symbolic Computation: Applications to Scientific Computing*, R. Grossman, ed., volume 5 of *Frontiers in Appl. Math.* SIAM, Philadelphia.

Addanki, S., Cremonini, R., and Penberthy, J.S., 1991, Graphs of models, *Artificial Intelligence* 51:145–177.

Amsterdam, J., 1993, Automated qualitative modeling of dynamic physical systems, Technical Report AI-TR-1412, MIT.

Beckstein, C., Stolle, R., and Tobermann, G., 1996, Meta-programming for generalized Horn clause logic, in: *META-96*, Bonn, Germany, pp. 27–42.

Bradley, E. and Easley, M., 1998, Reasoning about sensor data for automated system identification, *Intell. Data Analysis* 2(2):123–138.

Bradley, E., Easley, M., and Stolle, R., 2001, Reasoning about nonlinear system identification, *Artificial Intelligence*, in press.

Bradley, E., O'Gallagher, A., and Rogers, J., 1998, Global solutions for nonlinear systems using qualitative reasoning, *Annals Math. Artif. Intell.* 23:211–228.

Bradley, E. and Stolle, R., 1996, Automatic construction of accurate models of physical systems, *Annals Math. Artif. Intell.* 17:1–28.

Capelo, A., Ironi, L., and Tentoni, S., 1998, Automated mathematical modeling from experimental data: An application to material science, *IEEE Trans. Systems, Man and Cybernetics – C* 28:356–370.

Džeroski, S. and Todorovski., L., 1995, Discovering dynamics: From inductive logic programming to machine discovery, *J. Intell. Inf. Systems* 4:89–108.

Easley, M., 2000, *Automating Input-Output Modeling of Dynamic Physical Systems*, PhD thesis, University of Colorado at Boulder.

Easley, M. and Bradley, E., 1999, Generalized physical networks for automated model building, in: *IJCAI-99*, Stockholm.

Easley, M. and Bradley, E., 1999, Reasoning about input-output modeling of dynamical systems, in: *IDA-99*, volume 1642 of *LNCS*, Springer, Amsterdam, pp. 343–355.

Falkenhainer, B. and Forbus, K.D., 1991, Compositional modeling: Finding the right model for the job, *Artificial Intelligence* 51:95–143.

Forbus, K.D., 1984, Qualitative process theory, *Artificial Intelligence* 24:85–168.

Forbus, K.D., 1996, Qualitative reasoning, in: *CRC Computer Science and Engineering Handbook*, A.B. Tucker, Jr., ed., chapter 32, CRC Press, Boca Raton, FL, pp. 715–733.

Forrester, J., 1971, *World Dynamics*, Wright Allen Press, New York.

Hogan, A., Stolle, R., and Bradley, E., 1998, Putting declarative meta control to work, Technical Report CU-CS-856-98, University of Colorado at Boulder.

Huang, K.-M. and Żytkow, J.M., 1997, Discovering empirical equations from robot-collected data, in: *Foundations of Intelligent Systems (ISMIS-97)*, Z. Ras and A. Skowron, eds., volume 1325 of *LNCS*, Berlin, Springer, pp. 287–297.

Kuipers, B.J., 1986, Qualitative simulation, *Artificial Intelligence* 29(3):289–338.

Kuipers, B.J., 1993, Reasoning with qualitative models, *Artificial Intelligence* 59:125–132.

Langley, P., 2000, The computational support of scientific discovery, *Intl. J. Human-Computer Studies* 53:393–410.

Langley, P., Simon, H.A., Bradshaw, G.L., and Żytkow, J.M., eds., 1987, *Scientific Discovery: Computational Explorations of the Creative Processes*, MIT Press, Cambridge, MA.

LeFèvre, J., 1997, Reactive system dynamics: An extension of Forrester's system dynamics using bond graph-like notations, in: *Bond Graph Modeling and Simulations (ICBGM-97)*, Phoenix, AZ, pp. 149–155.

Ljung, L., ed., 1987, *System Identification; Theory for the User*, Prentice-Hall, Englewood Cliffs, N.J.

McCarty, L.T., 1988, Clausal intuitionistic logic I. Fixed-point semantics, *J. Logic Programming* 5:1–31.

Morrison, F., 1991, *The Art of Modeling Dynamic Systems*, Wiley, New York.

Mosterman, P.J. and Biswas, G., 1997, Formal specifications for hybrid dynamical systems, in: *IJCAI-97*, Nagoya, Japan, pp. 568–573.

Nayak, P.P., 1995, *Automated Modeling of Physical Systems*, volume 1003 of *LNCS*, Springer, Berlin.

Paynter, H., 1961, *Analysis and Design of Engineering Systems*, MIT Press, Cambridge, MA.

Stolle, R., 1998, *Integrated Multimodal Reasoning for Modeling of Physical Systems*, PhD thesis, University of Colorado, to appear in LNCS, Springer, Heidelberg.

Stolle, R. and Bradley, E., 1998, Multimodal reasoning for automatic model construction, in: *AAAI-98*, Madison, Wisconsin, pp. 181–188.

Sussman, G.J. and Steele, G.L., 1980, CONSTRAINTS – a language for expressing almost hierarchical descriptions, *Artificial Intelligence* 14:1–39.

Todorovski, L. and Džeroski, S., 1997, Declarative bias in equation discovery, in: *ICML-97*, Morgan Kaufmann, San Francisco, pp. 376–384.

Top, J. and Akkermans, H., 1991, Computational and physical causality, in: *IJCAI-91*.

Washio, T., Motoda, H., and Yuji, N., 1999, Discovering admissible model equations from observed data based on scale-types and identity constraints, in: *IJCAI-99*, pp. 772–779.

Weld, D.S. and de Kleer, J., eds., 1990, *Readings in Qualitative Reasoning About Physical Systems*, Morgan Kaufmann, San Mateo CA.

Żytkow, J.M., 1999, Model construction: elements of a computational mechanism, in: *Conference on Creativity*, Edinburgh, April.

MODEL-BASED DIAGNOSIS OF DYNAMIC SYSTEMS: SYSTEMATIC CONFLICT GENERATION

Bartłomiej Górny and Antoni Ligęza

Institute of Automatics AGH, University of Mining and Metallurgy

30-059 Kraków, al. Mickiewicza 30, Poland

bartlomiej.gorny@comarch.pl, ligeza@agh.edu.pl

Abstract   This paper presents a new and efficient solution for diagnosing dynamic systems within the framework of Reiter's theory. Reiter's theory allows for generation of potential diagnoses as minimal hitting sets of current conflicts. This paper analyzes the problems concerning systematic and efficient conflict generation in the case of continuous systems described with causal graphs and differential equations. The main idea of the paper is to introduce the notion of a *Potential Conflict Structure*, which is a local subgraph sufficient for definition of potential conflict. An outline of systematic, algorithmic approach to generation of a *complete set* of minimal conflicts is put forward. The paper includes also presentation of elements of an algebra of conflict sets. A prototype computer program for automatic generation of causal graphs from standard models of dynamic systems (Matlab-Simulink M-files) and conflict generation is also presented in brief.

1.   Introduction

Diagnostic reasoning is an activity oriented towards the detection of faulty behavior and its explanation, i.e. isolation of faulty components responsible for the observed misbehavior of the analyzed system. The main stages of a diagnostic procedure should include:

- Fault detection,
- Generation of potential diagnoses,
- Diagnoses verification,
- Repair.

L. Magnani, N.J. Nersessian, and C. Pizzi (eds.),
Logical and Computational Aspects of Model-Based Reasoning, 273-291.

Diagnosis of technical systems is thus a multi-stage procedure including searching and sophisticated reasoning. The inference methods applied in diagnostic systems contain deduction, abduction, hierarchical top-down fault location and many other non classical methods of reasoning, e.g. case-based reasoning, non-monotonic reasoning and different types of uncertain and imprecise reasoning (qualitative and quantitative). *Model-based* reasoning can be considered as search for system model consistent with observations, while consistency is checked with logical and algebraic methods. There are many approaches to diagnostics; some of the most popular are based on the use of:

- Diagnostic tests,

- Fault dictionaries,

- Rule-based (expert) systems,

- Decision trees,

- Decision tables,

- Case-based reasoning,

- Pattern recognition and neural nets,

- Model-based approach.

There is observed increasing interest in both research and applications of *model-based diagnosis* [Davis, 1984; Genesereth, 1984; de Kleer and Williams, 1987; Reiter, 1987; Davis and Hamscher, 1992; de Kleer et al., 1992; Bousson et al., 1994; Nicol et al., 1996; Milne and Nicol, 2000]. This is mainly due to the unquestionable advantages and well-founded theoretical backgrounds of this approach. As opposed to systems based on knowledge of experts and methods of learning from examples, model-based diagnosis allows for the elimination of the expert knowledge acquisition stage.

The *model-based diagnosis* is based on an explicit system model applied for diagnostic inference. Model-based diagnostic research has been used mostly within *logic based formalisms* applied to static binary-logic based systems. It is hard to find any work on application of consistency-based diagnosis to dynamic systems. An exception is the CA-EN system [Bousson et al., 1994]. CA-EN constitutes a part of the knowledge based gas turbine condition monitoring system Tiger [Nicol et al., 1996; Milne and Nicol, 2000]. The first installation of Tiger was in the Exxon Chemical plant in Scotland. Now Tiger is in continuous use on over 20 gas turbines across four continents [Milne and Nicol, 2000].

The diagnostic procedure applied in CA-EN is based on consistency-based approaches applied to a causal graph model, and, as such it falls into the Reiter's model-based diagnosis framework [Reiter, 1987]. In such diagnostic procedures the analysis is aimed at regaining consistency of the predicted model output with current observations by retracting some of the assumptions about the correct behavior of certain system components. The sets of elements suspected to contain at least one faulty component (i.e. the *conflict sets*) are identified by detecting inconsistency between the observed and the predicted behavior of certain output variables. The *conflict sets* generated by a diagnostic procedure are the basic products required to find diagnoses. However, one of the weakest points of the CA-EN diagnostic mechanism for dynamic systems is the lack of an efficient and complete conflict generation mechanism. The proposed conflict generation procedure is simplified; for more complex systems it is likely not to generate all the conflicts, and perhaps some sets which are not real conflicts can be generated.

A more complete diagnostic procedure following the consistency-based approach should perform the following tasks [Ligęza and Górny, 2000]:

- Detection of misbehavior; all the misbehaving variables should be found,

- Initial restriction of the search area to some smallest possible subsystem(s),

- Systematic conflict generation; all minimal conflicts should be found,

- Potential diagnoses generation (as minimal hitting sets),

- Verification of generated diagnoses.

This paper analyzes the problems concerning systematic conflict generation in the case of continuous systems described with causal graphs. The problem of conflict generation appears to be one of the most important problems in automated diagnosis of dynamic systems based on domain model of correct system behavior.

2. Consistency-based diagnosis: Reiter's Theory

Conflict sets [Reiter, 1987] (or *conflicts*, for short) are the sets of components of the system such that under the assumed model and the observed output *all* the components of any such set cannot be claimed to work correctly. Such sets of "suspected" elements are used then for potential diagnoses generation in the form of *hitting sets* for all the conflicts (i.e. a diagnosis is any set having nonempty intersection with any conflict set, and built from the elements of conflict sets only). This kind of diagnostic approach is based on the Reiter's theory [Reiter, 1987] of diagnosis from first principles, and DeKleer's work on diagnostic systems [de Kleer and Williams, 1987].

More formally, let $System := (SD, COMP, OBS)$, where SD denotes *system description* (a theory of system behavior, i.e. its model), $COMP$ are the *system components* (elements of interest which can become faulty), and OBS are the *observations* (knowledge about current state).

Let $AB(c_i)$ mean that component c_i is faulty; hence, $\neg AB(c_i)$ says that c_i is correct. If *failure* occurs then a theory formed by $SD \cup \{\neg AB(c_1), \ldots, \neg AB(c_n)\} \cup OBS$ is inconsistent.

A *diagnosis* is then any minimal set $\Delta \in COMP$ such that $SD \cup OBS \cup \{AB(c)|c \in \Delta\} \cup \{\neg AB(c)|c \in COMP - \Delta\}$ is consistent.

In other words, a diagnosis is determined by the smallest set of components with the following property: The assumption that each of these components is faulty (ABnormal), together with the assumption that all other components are behaving correctly (not ABnormal), is consistent with the system description and the observation.

A *conflict set* is any set $C = \{c_1, \ldots, c_k\} \subset COMP$, such that $SD \cup OBS \cup \{\neg AB(c_1), \ldots, \neg AB(c_k)\}$ is inconsistent.

In other words a conflict set is any set which has at least one faulty component.

A *hitting set* := any set $H \subseteq \bigcup_{C \in \mathbf{C}} C$, such that $H \cap C \neq \emptyset$ for any $C \in \mathbf{C}$.

We have the following theorem:

Theorem 1 *(Reiter):* $\Delta \subseteq COMP$ *is a diagnosis for* $(SD, COMP, OBS)$ *iff* Δ *is a minimal hitting set for the collection of conflict sets for* $(SD, COMP, OBS)$.

In other words the core idea of Reiter's theory is as follows:

Conflict sets describe sets of elements which all together cannot work correctly, so removing one element from each conflict set allows for restoring consistency. So the set of such elements, called a hitting set, is a potential diagnosis.

3.   Causal graph

Let $\mathbf{U}$ denote the set of unmeasured (unobserved) variables, $\mathbf{M}$ – the set of measured ones, and $\mathbf{I}$ the set of influences.

A *causal graph* indicates a set of variables represented by the graph nodes (system variables), and a set of *causal influences* defined with appropriate equations, and represented by the arcs of the causal graph.

Thus any causal graph is assumed to be a structure of the form **G** = (**U**, **M**, **I**). It is assumed that the equations defined by **I** are forward and backward calculable, i.e. having all input variables of one influence the output can be calculated, and having the all but one input variable values and the value of the output variable, the single undefined input variable can be calculated.

4. Graphical notation

For the sake of representing graphically various features concerning analyzed cases a simple extension of the causal graphs symbolics is proposed. Throughout the paper the following extension of the basic notation of [Bousson et al., 1994] will be used:

- A, B, ... – variables,

- $[U]$ – a not measurable variable,

- X^* – a variable observed to misbehave,

As usual, an arrow ($\longrightarrow$) means causality; also assumed to be described here with an equation (influence). Families of variables are to be denoted with boldface characters, e.g $\mathbf{X} = \{X_1, X_2, \ldots, X_k\}$. Influences (equations) are denoted by I; a component responsible for the correct work of I is to be denoted by c. Faulty components or influences will be also denoted c^* and I^*, respectively.

5. Problem formulation

A reasonable assumption is to start the diagnostic procedure by discovering that at least one of the observed (i.e. measurable and measured) variables misbehaves.

Our goal here is to determine *all minimal* conflict sets for the observed set misbehaves variables. These conflict sets may be used then for generation of diagnosesdiagnosis as hitting sets.

It seems reasonable to distinguish the following stages to be carried out in course of a systematic conflict generation procedure [Ligeza, 1996]:

1 *Domain restriction*: Restriction (possibly maximal) of the initial graph to a subgraph containing only the variables and components "involved" in the creation of the observed misbehavior; some details are given in [Ligeza, 1996].

2 *Strategy selection*: Establishing a strategy for conflict generation, e.g. "hot" starting points, order of generation, restrictions, etc.

3 *Efficient conflict generation*: Systematic conflict generation, usually from the "smallest" to the "biggest" ones with deleting non-minimal conflicts and efficient elimination of potential conflicts sets which are not real conflicts.

The conflict generation stage is a core one; it can be interleaved with diagnoses generation. Conflicts and diagnoses verification performed when the appropriate sets are available allows for efficiency improving.

6. Strategy for conflicts calculation and diagnoses generation

In order to consider the problems of strategy of conflicts generation, one should first answer the questions concerning preferences among diagnoses to be generated and sets of diagnoses viewed as "final solutions".

Considering preferences among diagnoses we must take into account the risk of basing the diagnostic procedure on incomplete set of conflicts, i.e. a case when not all the conflicts are calculated. The diagnoses calculated in such a case may be "false" or "partial".

Let us consider the problem of non-minimal conflicts, i.e. the case when we have non minimal conflicts instead of minimal ones. In such case diagnoses may be also "false" or "partial".

Finally, let us consider the problem of adding new conflict sets to an existing set of conflicts. If in the conflict set, which is already found there is one smaller than the one to be added, then adding the new one is not necessary; no new diagnoses will be generated.

The above considerations seem to justify the following assumptions concerning the strategy of conflict generation:

1 Conflicts should be generated in a systematic way; the procedure of generation conflict sets should assure that *all minimal* conflicts will be generated.

2 Conflicts should be generated from $i = 1$ towards $i = k$, where i is the number of components in a conflict set and k is the maximal number of components in the analyzed subgraph; this assures that more precise conflicts are generated first.

3 All conflicts comprising i elements should be calculated before ones comprising $i + 1$ elements; a conflict which is a superset of some previously generated conflict is not to be considered.

4 The procedure can be stopped when either no new conflicts can be generated, or for any new conflict to be generated a subset conflict has already been generated.

7. An approach to systematic conflict generation

From a purely mathematical point of view, in order to generate conflicts one must have more equations than variables; in our case the unmeasured variables are of interest. Thus if n denotes the number of equations (both for back- and for-propagation) defined for some subgraph, and m is the number of unmeasured variables involved in the computation, then the condition necessary for potential conflict generation is that $n \geq m + 1$. Further, for any such substructure there exists a potential possibility of generating no more than $\binom{n}{m+1}$ possible conflicts. This can be illustrated with Figure 1, where $n = 4$, $m = 1$, and we have $\binom{4}{2}$ potential conflicts.

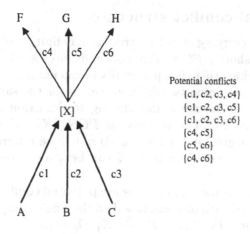

Potential conflicts
{c1, c2, c3, c4}
{c1, c2, c3, c5}
{c1, c2, c3, c6}
{c4, c5}
{c5, c6}
{c4, c6}

Figure 1. Examples – potential conflict structures selection.

Usually, there are less conflicts, since not all the structures described by n equations and containing m variables allow for calculation of a "full-size" conflict; examples include chains with pending unmeasured variables. Further, all the conflicts are only *potential* and must be verified. For further considerations only real conflicts should be taken into account.

In order to achieve better efficiency of conflict generation, the following, two-stage, transparent procedure of conflict generation is proposed:

- Identification of *potential conflict structures*, i.e. sets of influences assuring *necessary* conditions for conflict existence, and then

- Verification for any such structure and selected set of equations if a conflict really exists; this is to be done by an attempt at "solving" these equations.

Splitting the procedure of conflict generation into these two stages seems to be advantageous for the sake of transparency and systematic conflict generation. Moreover, identification of a conflict structure is equivalent to having the knowledge about its components. Thus exploration of non-minimal potential conflicts can be abandoned without performing the real calculations. Moreover, certain heuristics can be applied to reselect the potential conflict structures for further investigation, leaving a large part of them without performing costly mathematical calculations.

8.   Potential conflict structure

The key issue for carrying on is to introduce a definition of a *Potential Conflict Structure*, shortly *PCS*. This notion denotes a subgraph of the causal graphs, for which there is a possibility of calculating a conflict (always potential) via obtaining two different values for the same variable. A *PCS* comprising m variables and leading to detection of potential conflict at a variable X will be denoted as $PCS_m(X)$. The number of variables (both: measured and unmeasured) m will be referred to as the order of a conflict structure. Variable X can be a measured one or an unmeasured one.

Note that some of the most interesting are potential conflict structures having no unmeasured variables and based on the single component, they are always of the form $P_1, P_2, \ldots, P_j \xrightarrow{c} X$, where all the variables are measured; if such a structure provides a real conflict, then the conflict consists of one element c and in fact is a diagnosis or a partial diagnosis. In other words, component c is faulty and must be an element of any valid diagnosis; further the fault of c is a cause of the observed misbehavior of X. Therefore conflict structures without unmeasured variables should always be explored first (if existing).

Before the definition of *PCS* is formulated, it is necessary to put forward the following definition:

Definition 1 *A variable X is well-defined (defined, for short) iff:*

- *either it is a measured variable, or*

- *its value can be calculated on the base of some other variables, which are all well-defined.*

If there are two or more, e.g. k independent ways of calculating the value of X, then X is said to be k-defined.

For example variable B on Figure 2 is 3-defined (it is calculable from A, back-calculable from F and can be determined thought measurement).

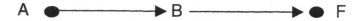

Figure 2. 3-defined variable B.

The independent ways of calculating the variable may consist in measuring the value and calculating it with different sets of equations. The calculation of a well-defined variable can be done no matter how; forward propagation is as good as backward one (at least from purely mathematical point of view). Now we can define a potential conflict structure on m variables.

Definition 2 *A* **Potential Conflict Structure** *for variable X defined on m variables is any subgraph of the causal graph, such that:*

- *it comprises exactly m variables (including X),*

- *all the variables are well-defined, and X is double-defined;*

- *for the PCS being defined on m variables it is necessary that all the values of the m variables are necessary for making X double-defined.*

Variable X will be called the head *of the PCS.*

Examples of PCS with different number of unmeasured variables are shown on Figure 3.

Note that any of the graphs represent m conflict structures, i.e. ones for different head variable (in fact all variables in PCS are double-defined); in these cases the PCS constitute the same graphs and the same conflicts will eventually be generated. Thus for any such PCS the calculation is to be performed only once, and the selection of the head variable is to be done arbitrarily, e.g. with respect to making the problem of equation solving easier.

Note that there is always a measured variable in PCS, so the algorithm of generation of all PCSs in the graph being the object of analysis may repeat calculation of PCSs for every measured variable in this graph. Of course repeated PCSs, i.e. ones different only with respect to the head variable should be abandoned.

Potential conflict structures can take arbitrary "shape" and it is, in general, difficult to say if some structure is a PCS at the first sight. The

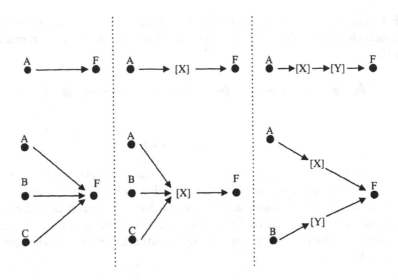

Figure 3.   Examples – potential conflict structures.

definition is in fact recursive, and so the algorithm for detection of *PCS* must be.

9.   An outline of algorithmic approach

To summarize, an outline of an algorithm for systematic conflict generation can be as follows:

1 Detect misbehavior of the system (the set of misbehaving variables **X***).

2 For any measured variable X sequentially detect all the *PCS*s.

3 Repeated the *PCS*s, i.e. ones different only with respect to the head variable should be abandoned (leaving exactly one of them for investigation); further, *PCS* leading to non-minimal conflicts can be abandoned before numerical investigation.

4 Stop the procedure when there are no more *PCS*s to be generated (or earlier, according to some heuristics, or when an appropriated diagnosis is generated).

After generation, if a system misbehaves, potential conflicts should be verified before calculating diagnoses. Non-real conflicts should be abandoned.

This algorithm may be changed and adapted to specific system, e.g. the procedure may calculate only PCSs without unmeasured variables and then the conflicts consisting of one element c are interpreted as a potential diagnoses and are verified. More complex conflicts can generated after this step if necessary.

10. Diagnoses calculation - elements of algebraic approach

Algorithm of calculating diagnoses from conflict sets given by Reiter [Reiter, 1987] require rather complicated data structures such as HS-tree and is quite difficult to implement. Now, a theorem, which allows calculate diagnoses for sum of conflicts when there are known diagnoses for those conflict sets will be proposed. This theorem may be used for instance in the case when diagnoses are generated simultaneously for some subsystems of one system; in such a case it can be used for combining together the separately generated sets of diagnoses. It can be also used for incremental generation of diagnoses for one set of conflict sets, and especially when one generates conflicts and diagnoses simultaneously (in such a case we interpreted each new conflict as a new set of conflicts, with one conflict of course and each element of this new conflict is a diagnosis.

Before the theorem is formulated, it is necessary to put forward the following definition:

Definition 3 *Let* $\mathbf{A}$ *is set of non empty sets. The reduced set* $\lfloor \mathbf{A} \rfloor$ *for* $\mathbf{A}$ *is the set containing this elements from* $\mathbf{A}$ *which are not supersets for other elements.*

Example 1 *Let* $\mathbf{A} = \{\{a, b\}, \{a, b, c\}\}$*. We have:* $\lfloor \mathbf{A} \rfloor = \{\{a, b\}\}$*. Let* $\mathbf{B} = \{\{a\}, \{a, b\}, \{a, b, c\}\}$*. Then* $\lfloor \mathbf{B} \rfloor = \{\{a\}\}$*. Finally, let* $\mathbf{C} = \{\{a\}, \{a, b\}, \{a, b, c\}, \{b, c\}, \{d, e\}\}$*. In this case* $\lfloor \mathbf{C} \rfloor = \{\{a\}, \{b, c\}, \{d, e\}\}$*.*

Now a special operator for combining diagnoses will be defined. The operator as its arguments takes the diagnoses for two different families of conflict sets.

Definition 4 *Let* $\mathbf{C}_i$ *denote sets of conflict sets,* $\mathbf{D}_i$ *sets of diagnoses calculable from* $\mathbf{C}_i$*, and* $\mathbf{H}_i$ *sets of all hitting sets for* $\mathbf{C}_i$*,* $i = 1, 2$*. Let* $D_1 \in \mathbf{D}_1$*,* $D_2 \in \mathbf{D}_2$*. The operator* $\oplus$ *is defined as follows:*

$$D_1 \oplus D_2 = \begin{cases} \{D_1, D_2\} & ; D_1 \in \mathbf{H}_2 \text{ and } D_2 \in \mathbf{H}_1 \\ \{D_1\} & ; D_1 \in \mathbf{H}_2 \text{ and } D_2 \notin \mathbf{H}_1 \\ \{D_2\} & ; D_1 \notin \mathbf{H}_2 \text{ and } D_2 \in \mathbf{H}_1 \\ \{D_1 \cup D_2\} & ; D_1 \notin \mathbf{H}_2 \text{ and } D_2 \notin \mathbf{H}_1 \end{cases}$$

Note that the result of operation $\oplus$ is a family of sets, which may contain one or two sets and each of these sets is a hitting set for $\mathbf{C}_i$, $i = 1, 2$.

Example 2 *Let us consider the following sets of conflict sets:*

$$\mathbf{C}_1 = \{\{a, \quad b, \quad c\},$$
$$\{a, \quad d\}\}$$

$$\mathbf{C}_2 = \{\{a, \quad c, \quad d\},$$
$$\{b, \quad e\}\}.$$

The sets of diagnoses that can be generated are respectively:
$\mathbf{D}_1 = \{\{a\}, \{b, d\}, \{c, d\}\}$, *and*
$\mathbf{D}_2 = \{\{a, b\}, \{a, e\}, \{b, c\}, \{c, e\}, \{b, d\}, \{d, e\}\}$.
We have:
$\{b, d\} \oplus \{a, b\} = \{\{b, d\}, \{a, b\}\}$,
$\{a\} \oplus \{a, b\} = \{\{a, b\}\}$,
$\{a\} \oplus \{b, c\} = \{\{a, b, c\}\}$.

Now let us define another operator, which constitutes a kind of extension of the previous one.

Definition 5 *Let $\mathbf{C}_i$ denote sets of conflict sets, and $\mathbf{D}_1 = \{D_1^1, D_1^2, \ldots, D_1^m\}$, $\mathbf{D}_2 = \{D_2^1, D_2^2, \ldots, D_2^n\}$ are the sets of diagnoses calculable from $\mathbf{C}_i$, $i = 1, 2$. We define operator $\bigoplus$ as follows:*

$$\mathbf{D}_1 \bigoplus \mathbf{D}_2 = \bigcup_{i=1, j=1}^{i=m, j=n} \{D_1^i \oplus D_2^j\}$$

In other words by using $\bigoplus$ one makes a union of results of operations with $\oplus$ for each diagnosis from $\mathbf{D}_1$ with each diagnosis from $\mathbf{D}_2$.

Finally the main theorem of this algebraic approach will be presented. The theorem is named a *Composition Theorem* since it allows for combining partial results (ones obtained separately or in turn) to the final set of diagnoses (proof in [Górny, 2001]):

Theorem 2 (Composition Theorem) *Let* $\mathbf{C}_i$ *denote sets of conflict sets and* $\mathbf{D}_i$ *sets of diagnoses calculable from* $\mathbf{C}_i$, $i = 1, 2, 3$. *If* $\mathbf{C}_3 = \mathbf{C}_1 \cup \mathbf{C}_2$ *then* $\mathbf{D}_3 = \lfloor \mathbf{D}_1 \oplus \mathbf{D}_2 \rfloor$.

Example 3 *Let*

$$\mathbf{C}_1 = \{\{a, \quad b, \quad c\},$$
$$\{a, \quad d\}\}$$

$$\mathbf{C}_2 = \{\{a, \quad c, \quad d\},$$
$$\{b, \quad e\}\}.$$

The sets of diagnoses are:
$\mathbf{D}_1 = \{\{a\}, \{b, d\}, \{c, d\}\}$, *and*
$\mathbf{D}_2 = \{\{a, b\}, \{a, e\}, \{b, c\}, \{c, e\}, \{b, d\}, \{d, e\}\}$. *Consider the combined set of conflict sets:*

$$\mathbf{C}_3 = \mathbf{C}_1 \cup \mathbf{C}_2 = \{\{a, \quad b, \quad c\},$$
$$\{a, \quad d\},$$
$$\{a, \quad c, \quad d\},$$
$$\{b, \quad e\}\}.$$

The set of diagnoses in this case is as follows:
$\mathbf{D}_3 = \{\{a, b\}, \{a, e\}, \{b, d\}, \{c, d, e\}\}$
Now let us calculate the combination of diagnoses as:

$$\mathbf{D}_1 \oplus \mathbf{D}_2 = \{\{a, b\}, \{a, e\}, \{a, b, c\}, \{a, c, e\},$$
$$\{b, d\}, \{a, d, e\}, \{b, c, d\}, \{c, d, e\}\}.$$

Let us reduce the combined set of diagnoses; we obtain:
$\lfloor \mathbf{D}_1 \oplus \mathbf{D}_2 \rfloor = \{\{a, b\}, \{a, e\}, \{b, d\}, \{c, d, e\}\}$.
It is easy to see that $\mathbf{D}_3 = \lfloor \mathbf{D}_1 \oplus \mathbf{D}_2 \rfloor$.

The Composition Theorem allows for calculating diagnoses for the sum of two families of conflict sets in the case where there are known diagnoses for each of these sets of conflicts. One does not need start generation of diagnoses from the beginning, i.e. without using the known diagnoses for each individual family of conflicts. Application of this theorem may therefore significantly increase efficiency of a procedure for calculation of diagnoses.

The Composition Theorem may be easily generalized to the following theorem (proof in [Górny, 2001]):

Theorem 3 (Generalized Composition Theorem) *Let* C_i *denote sets of conflict sets and* D_i *sets of diagnoses calculable from* C_i, $i = 1, 2, \ldots, n, n + 1$. *If*

$$C_{n+1} = C_1 \cup C_2 \cup \ldots \cup C_n$$

then

$$D_{n+1} = \lfloor\lfloor\lfloor D_1 \bigoplus D_2\rfloor \bigoplus D_3\rfloor \bigoplus \ldots\rfloor \bigoplus D_n\rfloor.$$

11.   An example

Let us consider the system shown in Figure 4. Diagnostics of such a multi-tank system has been an object of many publications.

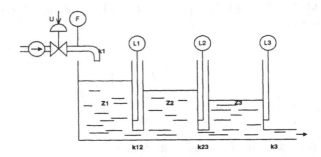

Figure 4.   Diagram of the three-tank system.

It consists of three interconnected tanks having inflow forced by a pump and controlled by a valve. The diagnostics are performed on the basis of the following signals: water levels L_1, L_2 and L_3 in the tanks, and a control signal U.

Behavior of the system is defined by the following non-linear differential equations (modeling the physical behavior of the system):

$$f(U) = F \tag{1}$$

$$A_1 \frac{dL_1}{dt} = F - F_{12} \tag{2}$$

$$A_2 \frac{dL_2}{dt} = F_{12} - F_{23} \tag{3}$$

$$A_3 \frac{dL_3}{dt} = F_{23} - F_3 \qquad (4)$$

where:
$F_{ij} = \alpha_{ij} C_{ij} \sqrt{2g(L_i - L_j)}$, $F_3 = \alpha_3 C_3 \sqrt{2gL_3}$ and A_i - cross-sectional area of the tanks, C_{jk} - cross-sectional areas of the flow channels.

The Matlab/Simulink model of this system is shown in Figure 5. Using Matlab/Simulink one can simulate the expected correct behavior of the system. If some calculated variables are different from the measured values, an inconsistency is observed and the diagnostic procedure should be activated.

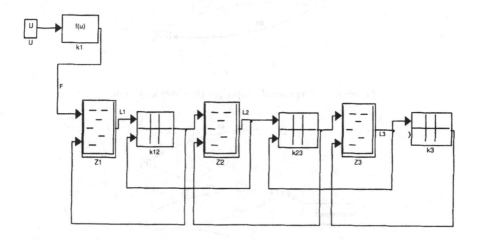

Figure 5.   Matlab/Simulink model of the three-tank system.

The causal graph for the example system is shown in Figure 6. The causal graph can be generated (automatically) from the Matlab/Simulink model of the system.

Assume, for example, that an inconsistency on variable L_1 has occurred. The correct value of L_1 depends on correct behavior of k_1, z_1 and k_{12} (Figure 7). Since L_1 is not correct, at least one of k_1, z_1 and k_{12} is faulty. Thus the set $\{k_1, z_1, k_{12}\}$ is a conflict for the misbehaving variable L_1. All supersets of $\{k_1, z_1, k_{12}\}$ are conflicts as well but they are not minimal.

A test program in C++ has been implemented. Its main applications are:

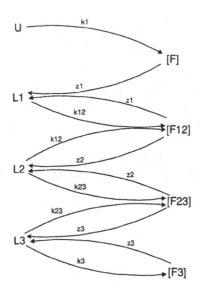

Figure 6.   The causal graph for three-tank system.

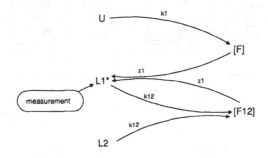

Figure 7.   The PCS for L_1.

- automatic generation of the causal graph from the Matlab/Simulink model,

- calculation of all minimal conflict sets for this graph.

The main window of the program and conflicts calculates by the program are shown in Figure 8.

The program is aimed at being a core module of a system for automatic conflict generation and diagnostic analysis for dynamic systems

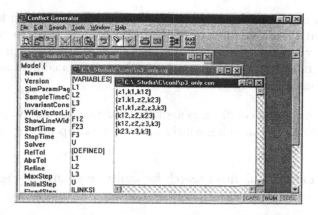

Figure 8. Generated conflicts by computer program.

specified with use of Matlab/Simulink models. Using these conflict sets (potential!): $\{z_1, k_{12}, k_1\}$, $\{k_{12}, z_2, k_{23}\}$, $\{z_1, z_2, k_1, k_{23}\}$, $\{k_{12}, z_2, z_3, k_3\}$, $\{z_1, z_2, k_1, z_3, k_3\}$ one can generate minimal diagnoses (again, potential!) as minimal hitting sets. Diagnoses are generated by separated module integrated with Matlab system. Example diagnoses are as follow: $\{z_1, k_{12}\}$, $\{k_{12}, z_2\}$, $\{k_{12}, k_{23}, k_3\}$.

12. Conclusions

In this paper an outline of an approach to systematic conflict generation is proposed. This approach is based on the use of double definition of variables; a variable calculated in at least two ways (double defined) may indicate existence of potential inconsistency.

The proposed approach seems to be a step towards building a systematic conflict generator for a class of static and dynamic modelsmodels, dynamic based on causal graphs. It should be a useful component of inconsistency-based diagnostic systems.

The following additional remarks may be useful in case building a diagnostic system:

- Conflict generation efficiency depends on graphs structure, initial area restriction, "calculability" of influences, etc.

- A small number of well-localized diagnoses can be obtained only if the unmeasured variables are relatively sparse.

- Heuristics and expert domain knowledge seems to be important, e.g. modes of faulty behavior of the components and their influence on the observed behavior of variables; an a priori knowledge about possible faults of components can be used to model the faulty behavior and thus skip certain possible faults if the modeled behavior is not observed. Further, certain conflict sets can be eliminated during generation.

- Generation of conflicts, generation of diagnoses and diagnoses verification done simultaneously may to speed up finding the correct diagnosis.

- Combination of direct search for faulty components [Ligeza et al., 1996] with procedures based on conflict generation may be useful.

To summarize, the presented approach allows for diagnosis of problems in a wide class of static and dynamic systems. Both single and multiple faults can be diagnosed. In order to be applied no expert knowledge, past experience or training sequences are necessary; the diagnostic procedure is solely based on the use of system model for simulation of its normal behavior.

In this paper elements of the algebraic approach to calculation of diagnoses in the form of minimal hitting sets is also presented.

References

Bousson, K., Trave-Massuyes, L., and Zimmer, L., 1994, Causal model-based diagnosis of dynamic systems, *LAAS Report No. 94231*.

Davis, R., 1984, Diagnostic reasoning based on structure and behavior, *Artificial Intelligence* 24:347–410.

Davis, R. and Hamscher, W., 1992, Model-based reasoning: Troubleshooting, in: *Readings in Model-Based Diagnosis*, W. Hamscher, L. Console, and J. de Kleer, J., eds., Morgan Kaufmann Publishers, San Mateo, CA, pp. 3–24.

de Kleer, J., Mackworth, A., and Reiter, R., 1992, Characterizing diagnoses and systems, *Artificial Intelligence* 56:197–222.

de Kleer, J. and Williams, B., 1987, Diagnosing multiple faults, *Artificial Intelligence* 32:97–130.

Genesereth, M., 1984, The use of design descriptions in automated diagnosis, *Artificial Intelligence* 24:411–436.

Górny, B., 2001, Consistency-Based Reasoning in Model-Based Diagnosis, Ph.D. Thesis, AGH Kraków.

Ligeza, A., 1996, A note on systematic conflict generation in CA-EN-type causal structures, *LAAS Report No. 96317*.

Ligeza, A., Fuster-Parra, P., and Aguilar-Martin, J., 1996, Causal abduction: Backward search on causal logical graphs as a model for diagnostic reasoning, *LAAS Report No. 96316*.

Ligeza, A. and Górny, B., 2000, Systematic conflict generation in model-based diagnosis, in: *Safeprocess'2000. 4th IFAC Symposium on Fault Detection Supervision and Safety for Technical Processes*, A.M. Edelmayer, ed., volume 2, Budapest, pp. 1103–1108

Milne, R. and Nicol, C., 2000, Tiger: Continuous diagnosis of gas turbines, in: *ECAI 2000. 14th European Conference on Artificial Intelligence*, W. Horn, ed., IOS Press, Berlin, pp. 711–715.

Nicol, C., Trave-Massuyes, L., and Quevedo, J., 1996, Tiger: Applying hybrid technology for industrial monitoring, *LAAS Report, No. 96039*.

Reiter, R., 1987, A theory of diagnosis from first principles, *Artificial Intelligence* 32:57–95.

MODELING THROUGH HUMAN-COMPUTER INTERACTIONS AND MATHEMATICAL DISCOURSE

Germana Menezes da Nóbrega
LIRMM, 161 Rue Ada, 34392 Montpellier, France
nobrega@lirmm.fr

Philippe Malbos
Laboratoire G.T.A., Université Montpellier II, Place Eugène Bataillon, 34095
Montpellier, France
malbos@math.univ-montp2.fr

Jean Sallantin
LIRMM, 161 Rue Ada, 34392 Montpellier, France
sallantin@lirmm.fr

Abstract   The project φ-calculus, or *philosophical calculus*, concerns the research on a rational framework that, in the context of interactive design, can be source and guarantee of the adequacy of a theory to its application context. The framework focus (i) on the *process* of theory construction by an Artificial Agent under the supervision of Human Agent(s), (ii) on the *knowledge* evolving through this process and (iii) on the *adequacy* of experiments with respect to knowledge represented in a given moment. The project is being developed in three different perspectives, namely, conceptual, formal, and experimental. Conceptual work consists on examining how experiments may point out the need for revising a theory and how to use experiments in order to perform revision. Formal work proposes formalisms to apprehend the dynamics on the process of theory construction, as well as building languages that may, from the one hand, facilitate interaction and, on the other hand, be adapted for computing. Experiments have been carried out in domains like e-commerce and Human Learning. In this paper, we focus on conceptual and formal aspects of the work on φ-calculus in its current state.

L. Magnani, N.J. Nersessian, and C. Pizzi (eds.),
Logical and Computational Aspects of Model-Based Reasoning, 293–311.
© 2002 Kluwer Academic Publishers. Printed in the Netherlands.

Introduction

As well-known, a good model is rarely obtained in one shot. Often, it is constructed progressively, by successive refinement, as long as a human modeler confronts the reality being modeled with the model being constructed. According to that thought, the successful evolution of a model is strongly dependent on the modeler's skill and experience telling him/her when and how to modify a model. Also, large and complex models are hard to develop and time-spenders, even if one or more experimented modeler(s) are involved. We claim that the modeling task could be better carried out if human intervention would be carefully considered in modeling methodologies, frameworks or environments.

Let us consider theory formation as the target of a modeling process. In such a context, a theory can be seen as a model which verifies an adequacy with experiments. Such an adequacy is judged over both the predictability and explanatory qualities of the theory. Indeed, theory is formulated in a formal language which should have, on the one hand, natural semantics in order to allow phenomena description, and, on the other hand, a formal semantics providing suitable computation properties. Moreover, theory should be linked - by an induction mechanism - to the facts, hypothesis, and particular situations (examples) describing the reality being modeled.

History has shown that, before being declared as universal, scientific knowledge evolves through a number of phases ranging from its appearance/discovery in individual experiences, through its submission to the criticisms of a working group, up to its definitive consolidation in the widespread scientific community. During its diffusion path, scientific knowledge is eventually submitted to refutations and even to adaptations, if it resists [Lakatos, 1976b; Popper, 1959].

In computational scientific discovery, a number of systems has successfully simulated important discoveries registered in the history of science - e.g. EURISKO [Lenat, 1983], BACON [Langley et al., 1981], GLAUBER [Langley et al., 1987], and BOOLE [Ledesma et al., 1997]. The conception of those systems has strongly contributed to provide the basis for a normative theory of discovery. By considering discovery modeling as a problem resolution process, as proposed by Newell et al. [Newell et al., 1962], a normative theory of discovery points out that theories rarely appear from random search since this would be impracticable due to the extension of search spaces. In such a context, a fundamental notion is that of *rationality*. For a scientist, such a notion consists of using the best possible way to limit a search space to manipulable proportions. Discovery systems, however, seemed to exhibit a bounded behavior concerning continuous evolution of theories they handled. According to Simon et al. [Simon et al., 1997], "Discovery systems which solve tasks cooperatively with a domain expert are likely to have an important role, because in any nontrivial domain, it will be virtually impossible to provide the system with a complete theory which is anyway constantly evolving". We share this idea. We think that, in the context of discovery, a system which does not account for interaction with its user will certainly find a border concerning its capacity to exhibit results that follow the evolution in the working domain. Therefore, instead of discovery systems, an alternative practice would be to build systems for aiding discovery, if one is interested in incorporating in a system the ability of apprehending expertise continuous evolution. In the literature, the advantages of human-computer collaboration in scientific discovery is evidenced by a number of AI systems, as reported in [Langley, 1998; Langley, 2000].

In our teams, such an approach has been largely applied, e.g. by Ferneda [Ferneda et al., 1992], Py [Py, 1992], and da Nóbrega [da Nóbrega, 1998]. In these works, the systems were designed as rational agents [Sallantin et al., 1991; Sallantin, 1997] to the extent that they are able to build theory under the supervision of an expert. Indeed, theories that rational agents are able to build in cooperation with an expert, exhibit properties required by a scientific theory, namely *predictability* and *explicability* in the context of experimental improvement.

Moreover, we suggest that the development of future aiding discovery systems might be improved if designers could dispose of some guidelines telling them what to observe when building such systems. In this direction, Jong and Rip [Jong and Rip, 1997] and Valdés-Pérez [Valdés-Pérez, 1999] highlight some informal guidelines. In this paper, we present a methodology called φ-calculus to support theory building under human supervision. φ-calculus is formally presented as a formal system.

1. Introducing φ-calculus

The scientists act rationally to build a scientific knowledge representation when they establish a correspondence [Carnap et al., 1929] between an empirical and a theoretical knowledge representation. The search for corroboration [Lakatos, 1976a] between empirical and theoretical knowledge gives the dynamic of scientific knowledge building. Practical uses of theoretical knowledge include explanation and prediction of empirical phenomena. So the classes of scientific predictions and explanations arising from the theory are more constrained than the candidate explanations and predictions expressed with the apparatus of an empirical domain. In other words, the capacity of creativity, of liberty and of adequacy to the reality, which belongs to an empirical language, is greater than those of the theoretical one. In opposite, the computational and human learnability of knowledge is only possible when there is an adequacy [Kuhn, 1982; Kuhn, 1983] between empirical and theoretical knowledge. In other terms, theory is efficient mainly to formulate and to explain an existing empirical and technological discovery [Butts, 1968].

1.1   Noise and silence in rational dialog

The main scientific research activity is a collective rational work to detect the errors coming from a theory when established by a correspondence between theoretical and practical knowledge. In this context, a rational process of scientific discovery attempts to minimize the noise and the silence arising from an error detection coming from the theory.

A noise is a message corresponding to a false error detection that inhibits the belief about the capacity of explanation of the theory. A noise is an error coming from a negative explanation of the theory. In other words, a noise is the rational consequence of a theoretical refutation making falsely impossible an empirical hypothesis.

A silence comes from the non detection of errors by a not enough constrained theory. Silence is then the result of a too much audacious conjecture [Popper, 1959] included in a theory. In other terms, a silence is the rational consequence of a conjecture which gives a theoretical correct prediction that is shown a posteriori not correct. So a silence is an absence of a message which has the rational consequence of a non detection of errors.

The non correspondence between a theory and a practice is then revealed by errors. The identification of the errors is a collective scientific goal. Up to this point, we have considered errors as coming from the theory and judged by experiments. Surely, some imprecisions are due to experiments, and in this case, the theory resists. On the other hand, overcoming an error that does not arise from experimental imprecision should be the result of the revision of the theory and of the improvement of the techniques that are used to observe, to detect and to formulate the experimental event. A truth assertion is a corrected error [Bachelard, 1932].

1.2 The paradigmatic status of a theory under revision

The theories under revision are paradigmatic when they are a mould to produce explanation and prediction. This mould is constituted with knowledge tools and procedures shared by a working group. This working group is the only one who understands completely the theory building framework and who is able to make the built theory available to a society.

When a theory is a common practice in a group [Kuhn, 1962], the theory appears to be dogmatic. The reasons are that theory predominates reality. Theory inhibits by itself any potential criticism and therefore refutation, because the lobbying of the theoretical working group having built the theory fights against all the propositions of revision. This is always the case for a well-established theory. Fortunately, this is less the case for a theory when it is built by a computer which has no social lobbying. This is the case in our application in LAW, that is briefly described in section 2.

During a process of theory revision, any theory may change but the paradigmatic [Toulmin, 1953] character is kept for new theories because these new good scientific theories are more specialized representations which are isolated with respect to any other specialized knowledge. As a consequence, they create a space in which they are in juxtaposition and not in superposition.

We notice that there is no paradigm of a general theory of science. So it does not exist a meta-theory of science which should be a super abstraction of science, but there exists practical protocols to detect and to correct errors that may arise when using a computational mathematical theory.

1.3 Prolegomena

Let us supposed some prolegomena:

- empirical and theoretical knowledge are both described in a formal system;

- theoretical knowledge is described in a logically consistent formal system;

- theoretical knowledge is produced by abduction from empirical knowledge;

- theoretical terms are the definitions of the empirical terms that are used to make inferences in empirical knowledge.

Concretely, the alphabet of theoretical knowledge consists of the definitions of the terms that are used to carry out inference about the description of real situations. In the simplest case, the definition of a term is a *type* and a misunderstanding about a term is an *error* about a type. We consider a *definition* what Carnap called a "quasi-definition", which is formulated by means of a transitive semantic graph. In a quasi-definition, the nodes are the terms of the empirical language and the edges are the inferences among them.

Under our assumption about the theoretical aspect, the theory is consistent. Therefore, the errors in the theory are revealed by the refutation [Hanson, 1961] that appears when the researcher looks for a corroboration between the theory and the experiments. For the practical aspect, the errors are the result of inadequacy [Kuhn, 1962] between an empirical statement and the empirical verification rules, when these latters had been obtained by induction [Fann, 1970] from the theoretical statements. So we assume:

- empirical errors reflect inadequacy between the empirical kind of statements and the verifications rules;

- empirical verification rules are obtained by induction from the theoretical kind of statements;

- theoretical errors are a lack of corroboration between the empirical and the theoretical kind of statements.

We assume the existence of formation rules and verification rules for the scientific statements. The meaning of a statement comes from the verification rules (Figure 1).

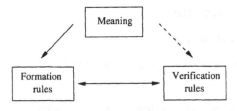

Figure 1.   Meaning of a statement in φ-calculus.

In the context of the formation of scientific theory, we have to define the terms, formulas and references that are used to express empirical and theoretical knowledge. We have rules to form and to verify Terms, Formulas and References:

- a Fact is a Term defined and verified by some verification forms called Law; a Hypothesis is a Formula defined and verified by some verification form called Postulate;

- errors arise as inadequacy between formation and verification of defined statements;

- the search for adequacy requires to induce new definition of the reference statements and to abduce new verification statements;

- there is no inadequacy between formation and verification for theoretical statements.

We summarize in Table 1 all the types of statement required for theory formation, according to what we have just presented.

Statement	Formation	Verification	Type of errors
practical	Defined statement	Defined statement	
Term	Fact	Law	Misunderstanding
Formula	Hypothesis	Postulate	Ambiguity
Reference	Example	Concept	Paradox
theoretical	Defined statement	Defined statement	
Term	Opinion	Principle	
Formula	Conjecture	Lemma	
Reference	Refutation	Proof	

Table 1. Required statements for theory formation in φ-calculus.

The rationality in dialog between a user and his/her artificial agent is related to the way to induce a set of good definitions for Terms, Formulas, and References used in practical knowledge. So we conserve the word but we modify the definition of the words in such a way that it is possible to carry out inference with the terms which respect to the definition of terms. In a philosophical viewpoint, the principle used to define makes possible the hypotheses. The rational process achieves when:

- *the Hypotheses are not refuted.* When the Hypotheses are not the object of correct refutations. In such a situation, the system has a good explanation for a Hypothesis: the refutation of the Hypothesis has not created a noisy message to the user.

- *The Examples are convincing.* The Examples are correct conjectures and a new statement which does not correspond to a correct conjecture is detected. So the system has a good generalization of the Examples and it sends a message in case of inappropriate generalization.

A rational process of theory building should assist the user to find a theory to produce statements that fit with empirical data. In our work, the notion of adequacy is used to estimate the quality of the prediction of a theory and to prepare adequate data for iterated steps of induction and abduction. In our current implementation, we use the Galois lattice theory [Duquenne, 1999] to give a mathematical formalism to the induce the definition of terms and an algebraic theory acting on these definitions [Liquière and Sallantin, 1998].

2.   An experiment in LAW

In 1994, one of us (js) have started a collaboration with the Lawyer's Company Fidal-KPMG grouping 1200 lawyers in France. Lawyer's daily activity consists of understanding, proving and comparing contracts. The issue for innovation, for them, is that laws, norms and events change continuously, so contracts have to be modified as well accordingly. Any (new) contract requires necessarily a compromise between the liberty of commitment between the parts and the security of the commitment with respect to foreseen events and possibly unforeseen events during the time the contract will bind the parts. In order to assist each of their lawyers in their activity, the Company has identified classes of contracts, and for each class has decided to offer lawyers a contract "template". The construction of these templates is assisted by a tool called *fid@ct*. This tool assists the iterative process of designing a template for each class of legal contracts, by explaining inadequacies, in order to generate a text - the template is a pattern of a class of instances of contract - judged coherent and ensuring freedom and security as described above.

The Company delegates the design of the template to a senior experimented lawyer and two or three junior novice lawyers familiar with the use of the design tool fid@ct. The juniors evaluate a set of pre-existing instance contracts in the class; extract terms such as the notions of franchise, duration, obligation and sanction; encode prescriptions from the senior in the form of logical constraints between terms; formulate a temporary template syntactically correct, therefore respecting literally the prescriptions of the senior. Their attitude, in doing this, is however to essay to challenge the senior by generating a template that respects the syntactic constraints but at the same time is, as much as possible, semantically absurd. The senior reacts to this aggression and modifies his/her theory. According to the senior reaction, juniors revise the set of terms and the set of prescriptions and iterate what described above.

The process converges in the average after 10 meetings of 2 hours separated by a week. During the first phase of iterations the revisions focus on the terms - between 800 and 1800, generally near 1200 - and constraints associated to their presence and absence in a formula. During the second phase the senior reaction to the juniors' insolent contract is about prescribing, i.e. fixing the context of use of a given formula. During the third phase the senior explains to the juniors why the contract they build is correct but atypical, therefore suggesting the juniors to introduce new prescriptions, fixing the reuse of this template.

During this social game, the legal team builds useful shared knowledge (the template, or contract pattern) by means of eliminating the errors of denomination coming from terms, the ambiguities of connotation coming from formulas and the paradoxes coming from examples. Three months of work are required in order to design a template for a class of contracts. Once that template is available, a trained lawyer may conceive and write a new contract in thirty minutes. 400 lawyers are now currently using the templates emerged from the application of the methodology described. Their clients estimate that the lawyers take a better account of their requirements. The Company has decided to depose a patent of both methodology and the computational tool fid@ct used to build templates.

3.   A formal system for φ-calculus

The methodology φ-calculus has been presented under the form of a formal system, which is only partially introduced in this section. By means of a *Description Language* a Human Agent (HA) may describe his/her experience to an Artificial Agent (AA), under the form of a number of *statements*. Then, in a *Constraint Language* the knowledge of the AA is represented. Such a knowledge may be generated out of the statements created by the HA, through a combination of machine learning mechanisms and human analysis. Finally, the notion of *Adequacy* is formalized as the way by which the HA's statements and the AA's knowledge are confronted. The detection of *Inadequacy*, resulting from such a confrontation may trigger the revision process.

In the paper we present the Description Language. The presentation of such a part of our formal system is illustrated by a scenario that handles a popular toy domain: the "ontology" of an arc. The domain used in the scenario is the simplest we have found: we have borrowed it from Minsky's discussion about (human) concept learning [Minsky, 1986]. The HA is then supposed to be an "expert" in the blocks world who is interested to build a "theory" of arcs. The AA supporting the scenario is Web-Contract-Praal (WCP), which is a particular implementation of φ-calculus' features.

3.1   The description language

In this section, we describe both the terms and the operations of the component Φ_D of the formal system, that is proposed as the language through which the HA can describe his/her experience to the AA.

Example 3.1 *In order to introduce our scenario, let us suppose that the "expert" on the blocks world disposes of the examples of Figure 2 as the "experience" that he/she intends to describe to WCP.*

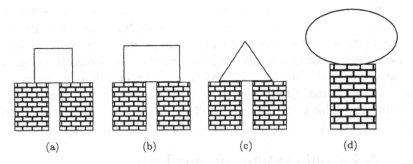

      (a)   (b)   (c)   (d)

Figure 2.   Available examples to build a "theory of arcs".

Terms.   Let us assume a set T of constants, called L_0-terms, coming from the natural language of the HA, called initial language and noted L_0. The set T is provided with a structure induced by the nature of L_0. In the present case, we are considering a language provided with natural semantics. Additionally, we provide T with a partial order relation $\leq$ in order to account for a hierarchical structure over the elements of the language L_0 represented in T.

Example 3.2 *Within our scenario, let us suppose that the "expert" proposes as shown in Figure 3 the organization of the terms naming the objects of Figure 2.*

It is important to notice that the organizing criterion given above is nothing but illustrative, just to enable us to describe the evolution of our scenario. In the general case, organizing criteria should be established by the HA according to his/her own criteria of organizing terms in the working domain.

φ-Facts, φ-Formulas, and φ-Examples.   The set T is provided with a set of evaluation functions

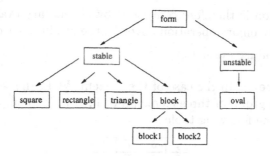

Figure 3. A fragment of the blocks world: hierarchy of terms naming the building blocks of Figure 2.

$$e : T \to \Sigma,$$

where Σ is the set of the possible states that the HA may assign to a term. In the present case, we are considering Σ a set of three elements, allowing the HA to declare a term as *present, absent,* or *not selected.* For T and Σ fixed, we note Σ^T the set of evaluations over T with value in Σ. We note

$$T_{ev} = \{(t, e(t)) : t \in T \text{ and } e \in \Sigma^T\},$$

the set of all evaluated terms. We call a φ-*Fact* a single evaluated term $(t, e(t))$, noted t_e. Now, being e $\in \Sigma^T$, we note

$$Te = \{t_e : t \in T\},$$

the terms evaluated by e. One can interpret T_{ev} as the disjoint reunion $T_{ev} = \coprod_{e \in \Sigma^T} T_e$. Each order relation $\leq$ over T induces an order relation over T_e, noted $\leq_e$. We note $\mathcal{G}(T)$ (resp. $\mathcal{G}_\Sigma(T)$) the set of graphs (resp. evaluated graphs) which set of nodes is T.

We call an *elementary evaluation* over T an evaluation that modifies the state of only one term. The elementary evaluation e_t modifying the state of t is defined by:

$$e_t(s) = \begin{cases} 0 & \text{if } s \neq t, \\ \in \{\top, \bot\} & \text{otherwise.} \end{cases}$$

Composition of evaluations.   As one might notice, Σ may be structured in different ways, and the evaluations verify different properties according to the structure of T. Every binary relation . over Σ induces over Σ^T a binary relation $\star$ defined by: if $e, e' : T \longrightarrow (\Sigma, .)$ are two evaluations, then:

$$(e \star e')(t) := e(t) . e'(t).$$

Every evaluation is therefore composed by elementary evaluations $e = \underset{t \in T}{\star}\, e_t$. For every binary operation . over Σ, the product $\star$ is commutative over the elementary evaluations.

Polarized tree.   In the case of the system WCP, the set Σ of states is defined as a group of three elements $0, \top, \bot$ which composition law . is defined by the following table:

.	0	$\top$	$\bot$
0	0	$\top$	$\bot$
$\top$	$\top$	0	$\top$
$\bot$	$\bot$	$\bot$	0

An evaluation $o : T \rightarrow \Sigma_{wc}$ is then said neuter if $o(t) = 0$ for every $t \in T$. For every set T, $\star$ provides the set of evaluations Σ^T with a neutral element o group structure. We provide Σ_{wc} with the partial order relation Σ_{wc} defined by:

$$\bot \leq_{wc} 0 \leq_{wc} \top.$$

A polarized term is a term provided with an elementary evaluation over Σ_{wc}. It is a term which presence $\top$ or absence $\bot$ has been explicitated by the Human Agent, or yet a term which has not been selected. In this latter case, the term has an evaluation equal to 0.

Polarization.   An evaluation e over Σ_{wc} is said coherent if it preserves the order relation:

if $t \leq t'$ in T, then $e(t) \leq_{wc} e(t')$.

Otherwise it is said incoherent. Coherent evaluations over T, called *polarizations*, correspond to the ordered set morphisms from T in Σ_{wc}.

Propagation rule for an elementary evaluation.   We define a propagation rule over the evaluated tree in order to construct the single polarization generated by an elementary evaluation. Let T be the set of terms, $t \in T$ and e_t an elementary evaluation with values in Σ. We define a propagation rule $\mathcal{R}_{e_t} : \mathcal{G}_\Sigma(T) \rightarrow \mathcal{G}_\Sigma(T)$ by stressing: $\mathcal{R}_{e_t}(T_{e_t}) = T_{\mathcal{R}_{e_t}}$ where $T_{\mathcal{R}_{e_t}}$ is the ordered set T provided with the polarization:

$$T_{\mathcal{R}_{e_t}}(s) = \begin{cases} \top & \text{if } t \leq s \text{ and } e_t(t) = \top, \\ \bot & \text{if } t \geq s \text{ and } e_t(t) = \bot. \end{cases}$$

Often, it is not possible to expand the propagation rule to a composition of elementary evaluations. In fact, the composition of two polarizations obtained by propagation of two elementary evaluations is not always a polarization.

At the beginning of this section, we have stated that the Description Language being proposed would allow a HA to build statements. The first of these kinds statements has been presented as a φ-Fact. Now let us consider that the HA intends to link different polarized terms in a single statement, or yet to assign an expression to a pre-defined term. These two actions correspond respectively to the operations of *conjunction* and *classification*. They are interpreted in φ-calculus by the operations of *Formulation* and *Exemplification*. In what follows, we give a formal description of these operations as well as the relations that they verify.

Formulation. The operation of formulation consists in grouping together in a single expression a set of terms evaluated accordingly in a single evaluation. Let $e \in \Sigma^T$ be an evaluation. The expression

$$t_e \otimes t'_e \otimes \ldots \otimes t_e^{(n)}$$

designates the conjunction of evaluated terms t_e, t'_e, ..., $t_e^{(n)}$. The result of such an operation of Formulation is called a φ-*Formula*.

Example 3.3 *Considering the object of Figure 2(a), a φ-Formula to represent such an object would be:*

$$(square, \top) \otimes (block1, \top) \otimes (block2, \top) \otimes (arc, \top)$$

We note F_e the set of all Formulas built by means of the operation of formulation over the terms evaluated by e:

$$F_e = \{ \underset{i \in I}{\otimes} t_e^i , \ t^i \in T \}.$$

For every evaluation e with value in Σ_{wc}, there exists a *Maximal φ-Formula*

$$\overline{f_e} = \underset{t_i \in T, \ e(t_i) \neq 0}{\otimes} t_e^i$$

including all the evaluated terms that may be obtained by means of the propagation rule, starting from a given φ-Formula. A set of *Minimal φ-Formulas*, which are composed of a single evaluated term, may be obtained starting from a given φ-Formula.

Definition 3.4 *The operation of formulation is associative, commutative, and idempotent. For every evaluation e with value in Σ, and for* *every term* $t, t', t'' \in T$, *we have:*

$$(Ass) \quad (t_e \otimes t'_e) \otimes t''_e = t_e \otimes (t'_e \otimes t''_e),$$
$$(Com) \quad t_e \otimes t'_e = t'_e \otimes t_e,$$
$$(Ide) \quad t_e \otimes t_e = t_e.$$

A φ-Formula is said *irreducible* if it cannot be reduced by means of *Ass, Com* ou *Ide*.

Exemplification.   The other basical operation is the Exemplification. Let us consider $e \in \Sigma^T$, f_e a φ-Formula, and $t \in T$ a term, then

$$\varphi_d t. f_e$$

designates the φ-Formula f_e assigned to the term t.
$\varphi_d t. f_e$ is called a φ-*Example*.

Example 3.5 *Considering yet the object of Figure 2(a), to classify this object as an example of "arc" (which is a term of the set T considered in the scenario) would lead us to the following φ-Example:*

$$\varphi_d arc.(square, \top) \otimes (block1, \top) \otimes (block2, \top) \otimes (arc, \top)$$

Now let $\varphi_d t. f_e$ be a φ-Example. If the term t appears in the φ-Formula f_e, then it is said to be *bound* in $\varphi_d t. f_e$, otherwise it is said to be a *free* term.

We distinguish then three kinds of φ-Examples:

- A *Neutral φ-Example* is a φ-Example $\varphi_d t. f_e$ such that the term t is free.

- A *Positive φ-Example* is a φ-Example $\varphi_d t. f_e$ such that the term t is bound, and that $e(t) = \top$ in f_e.

- A *Negative φ-Example* is a φ-Example $\varphi_d t. f_e$ such that the term t is bound, and that $e(t) = \bot$ in f_e.

Example 3.6 *The φ-Example shown in 3.5 is a Positive φ-Example since the term "arc" is bound and $e(arc) = \top$ in the φ-Formula.*

The interface of the system WCP is shown in Figure 4, allowing the HA to build each kind of statement of the Description Language just presented, i.e.: φ-Facts, φ-Formulas, and φ-Examples.

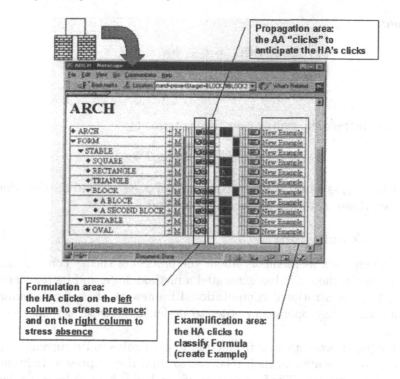

Figure 4. Interface of the system WCP allowing the HA to build statements.

Definition 3.7 *The operation φ_d verifies: if $t, t' \in T$ and $M, N \in \Phi_D$ then:*

$$(Dist1) \quad \varphi_d t.(M \otimes N) = (\varphi_d t.M) \otimes (\varphi_d t.N),$$
$$(Ide) \quad \varphi_d t.(\varphi_d t.M) = (\varphi_d t.M),$$
$$(Dist2) \quad \varphi_d t.(\varphi_d t'.M) = (\varphi_d t.M) \otimes (\varphi_d t'.M),$$
$$(\xi) \quad M = N \quad \Rightarrow \quad \varphi_d t.M = \varphi_d t.N.$$

The idempotence axiom (Ide) stresses the idea that the exemplification thus defined is the abstraction ultime that can be performed by the HA.

The theory Φ_D.

Definition 3.8 *Let T be a set of terms from L_0 and Σ^T the set of evaluations over T with value in Σ, and let us consider T_{ev} the set of evaluated terms, and t_e an evaluated term. The set of φ_d-terms, noted Φ_D is constructed over T through the operations of formulation and exemplifi-*

cation:

$$(\Phi_D 1) \quad t_e \in T_{ev} \Rightarrow t_e \in \Phi_D,$$
$$(\Phi_D 2) \quad M, N \in \Phi_D \Rightarrow M \otimes N \in \Phi_D,$$
$$(\Phi_D 3) \quad M \in \Phi_D, t \in T \Rightarrow \varphi_d t.M \in \Phi_D.$$

We can abstract the syntax over the φ_d-terms by:

$$\Phi_D ::= T_{ev} | \Phi_D \otimes \Phi_D | \varphi_d T \Phi_D.$$

We define Φ_D as a formal theory composed of φ_d-terms and the equations between these φ_d-terms, as defined in 3.4 and 3.7.

4.   Conclusion

φ-calculus is a meta-theory about the process of theory construction. As any useful theory, it has generated a methodology, and its associated tools, forming a rational computational framework that assists human agents when they operate in order to design and revise a theory on a domain.

This paper presents an introduction to φ-calculus in its current state; i.e. the components and their relations to what they represent in human theory construction. The foundations of φ-calculus have to be recognized in Computation more than in Mathematics and in Philosophy; even if a coherent mathematical formalization of φ-calculus is under development and the systematization seems on the one side to open new perspectives for a better definition of rationality and, on the other, to stimulate new arguments for the current philosophical debate.

The contribution of φ-calculus to Computing is in its integrated vision of knowledge and the process of knowledge construction in any domain; an issue that may give to Computing the rang of Natural Scientific discipline. As the process of Theory Construction in Nature coincides with the history of humans, and is typically a societal process of exchanges and transformations of Information, one may say that φ-calculus has emerged from observations of the Natural phenomena consisting of societal interactions for building theories.

The main practical application consolidated to date, is the one in the domain of legal contracts. The experiment of the methodology in this domain is at the origin of the tools, which supports contractual negotiations. Ongoing experimental activities, supported by contracts with the EU Intelligent System Technology, develop applications in the context of e-business. Also, experiments in the domain of Distance Learning are foreseen, as a Web-served Learning Environment is currently being implemented. Such a work lies on the theoretical basis of an interpretation of Lawyers' tool and methodology as a Learning Environment [da Nóbrega et al., 2001].

The contribution to Philosophy consists in offering a set of formal concepts based on Computation, that may orient the philosophical debate traditionally in search of formal foundations from mathematical formalisms. Insofar φ-calculus not only adopts a computational viewpoint about human processes, but is also mathematically well defined, it may be considered by philosophers to have a solid foundation for representing human theory formation. As φ-calculus adopts Rational Belief and Adequacy in the context of human interactive knowledge building as assumptions, it will be important to hear from philosophers not simply if they agree with the assumptions, but also if, under these assumptions, the consequences drawn may reflect better their intuitions and intellectual experiences.

Acknowledgment

Part of this work was supported by the European IST projects MK-BEEM and SMARTEC on e-Commerce.

References

Bachelard, G., 1932, *L'intuition de l'instant*, Stock, Delamain et Boutelleau, Paris.

Butts, R.E., ed., 1968, *William Whewell's Theory of Scientific Method*, University of Pittsburg Press, Pittsburg.

Carnap, R., Neurath, O., Hahn, H., 1929, Überwindung der metaphysik durch logische analyse der sprache, in: *Wissenschaftliche Weltauffassung. Der Wiener Kreis*, A. Wolf, Wien.

da Nóbrega, G.M., 1998, Especificação formal de um sistema de apoio à descoberta, Master's thesis, Universidade Federal da Paraíba (UFPB), Campina Grande, Paraíba.

da Nóbrega, G.M., Cerri, S.A., and Sallantin, J., 2001, DIAL: serendipitous DIAlectic Learning, in: *IEEE International Conference on Advanced Learning Technologies - ICALT'2001*, T. Okamoto, R. Hartley, Kinshuk, and J.P. Klus, eds., IEEE Computer Society, Madison, WI, pp. 109–110.

Duquenne, V., 1999, Latticial structure in data analysis, *Theoretical Computer Science* 217:407–436.

Fann, K.T., 1970, *Peirce's Theory of Abduction*, The Hague, Martinus Nijhoff.

Ferneda, E., Py, M., Reitz, P., and Sallantin, J., 1992, L'agent rationnel SAID: une application en géométrie, in: *Proceedings of the First European Colloquium on Cognitive Science*, Orsay, pp. 175–192.

Hanson, N.R., 1970, Is there a logic of scientific discovery?, in: *Readings in the Philosophy of Science*, B.A. Brody, ed., Prentice-Hall, Englewood Cliffs, NJ, pp. 620–633, originally published in: *Current Issues in Philosophy of Science*, H. Feigl and G. Maxwell, eds., Holt, Rinehart and Winston, New York, 1961, pp. 20–35.

Jong, H. and Rip, A., 1997, The computer revolution in science: Steps towards the realization of computer-supported discovery environments, *Artificial Intelligence* 91:225–256.

Kuhn, T.S., 1962, *The Structure of Scientific Revolutions*, The University of Chicago Press, Chicago.

Kuhn, T.S., 1982, Commensurability, comparability, communicability, in: *Philosophy of science association*, Pasqhith and T. Nickles, eds., volume 2, PSA, East Lansing, Michigan, pp. 669–688.

Kuhn, T.S., 1983, Rationality and theory choice, *The Journal of Philosophy* 80:563–579.

Lakatos, I., 1976a, Falsification and the methodology of scientific research programmes, in: *Criticism and the Growth of Knowledge*, I. Lakatos and A. Musgrave, eds., Cambridge University Press, Cambridge, pp. 91–196.

Lakatos, I., 1976b, *Proofs and Refutations: The Logic of Mathematical Discovery*, Cambridge University Press, Cambridge.

Langley, P., 1998, The computer-aided discovery of scientific knowledge, in: *Discovery Science First International Conference, DS'98, Fukuoka, Japan, December 14–16, 1998, Proceedings*, S. Arikawa and H. Motoda, eds., Springer, Berlin, pp. 25–39.

Langley, P., 2000, The computational support to scientific discovery, *Int. J. of Human-Computer Stud.* 53:393–410.

Langley, P., Bradshaw, G.L., and Simon, H.A., 1981, Bacon 5: The discovery of conservation laws, in: *IJCAI-81*, pp. 121–126.

Langley, P., Simon, H.A., Bradshaw, G.L., and Zytkow, J.M., 1987, *Scientific Discovery: Computational Explorations of the Creative Processes*, The MIT Press, Cambridge, MA.

Ledesma, L., Pérez, A., Borrajo, D., and Laita, L. M., 1997, A computational approach to George Boole's discovery of mathematical logic, *Artificial Intelligence* 91:281–307.

Lenat, D.B., 1983, Theory formation by heuristic search - the nature of heuristics ii: Background and examples, *Artificial Intelligence* 21:1–2.

Liquière, M. and Sallantin, J., 1998, Structural machine learning with Galois lattice and graphs, in: *5th International Conference on Machine Learning*, Madison, WI, pp. 305–313.

Minsky, M., 1986, *The Society of Mind*, Simon and Schuster, New York.

Newell, A., Shaw, J.C., and Simon, H.A., 1962, The process of creative thinking, in: *Contemporary Approaches To Creative Thinking*, H.E. Gruber, G. Terrel, and M. Wertheimer, eds., Atherton Press, New York.

Popper, K., 1959, *The Logic of Scientific Discovery*, Hutchinson, London.

Py, M., 1992, Un Agent Rationnel pour Raisonner par Analogie, PhD thesis, Université de Sciences et Techniques du Languedoc - Montpellier II, Montpellier, France.

Sallantin, J., 1997, *Les agents intelligents : essai sur la rationalité des calculs*, Hermès, Paris.

Sallantin, J., Quinqueton, J., Barboux, C., and Aubert, J.-P., 1991, Théories semi-empiriques: éléments de formalisation, *Revue d'Intelligence Artificielle* 5(1):69–92.

Simon, H., Valdés-Pérez, R.E., and Sleeman, D.H., 1997, Scientific discovery and simplicity of method, *Artificial Intelligence* 91:177–181.

Toulmin, S., 1953, *The Philosophy of Science*, Hutchinson University Library, London.

Valdés-Pérez, R.E., 1999, Principles of human computer collaboration for knowledge discovery in science, *Artificial Intelligence* 107(2):335–346.

COMBINING STRATEGY AND SUB-MODELS FOR THE OBJECTIFIED COMMUNICATION OF RESEARCH PROGRAMS

Ekkehard Finkeissen

University of Heidelberg, Institut for Medizical Biometry und Informatics,
Department of Medical Informatics, Im Neuenheimer Feld 400, D-69120 Heidelberg
ekkehard_finkeissen@med.uni-heidelberg.de

Abstract   Modeling is necessary for any kind of objectified planning and prognosis. A main problem of setting up and utilizing a generalized (objective) decision model is its demarcation to the correlated individual decision. The properties of the objective model can be generalized as they have to be of general nature. The differences as well as the interaction between objective and (complementary) individual decisions can be investigated by subtracting the objective part from the whole decision problem. That is why the structure of objective models has to be investigated more in detail. In the last century well known scientists have stated their demands on objectified models. From these demands the fundamental structure of an objective decision model will be derived. The article does not aim at formulating general (objective) truth about reality but at providing a general tool for building *objective models* – without being able to avoid uncertainty with respect to real life. From this point of view objective models can have a strongly normative nature. The article concentrates on the product rather than the process of scientific discovery. Many of the statements given in this article may sound trivial or at least well known. But combined as a structure it can be seen as a general model of objective models (meta-model), which can found a theory of formal decision systems as a formal representation of both the meta-model an its instantiations is in prospect. As the instantiations can found domain ontologies, the proposed meta-model can be seen as a meta-ontology.

L. Magnani, N.J. Nersessian, and C. Pizzi (eds.),
Logical and Computational Aspects of Model-Based Reasoning, 313–330.
© 2002 Kluwer Academic Publishers. Printed in the Netherlands.

1.   Scientific objectification of decision-making

Epistemology is the study of the nature, origin, and limits of human knowledge (compare [Britannica, 1998]). But the author takes the stance that – if knowledge is supposed to objectify nature – before knowing how to acquire knowledge the possibilities of *formulation and storage of an objective model (knowledge structure)* has to be clarified. Only on this basis the properties of an objective model can be derived. Moreover it can be sketched how to acquire knowledge and transfer it into the respective knowledge structure.

This article does not try to carry on with the discussion of [Bunge, 1996] about who are the enemies of science and academic learning – teaching that there are no objective and universal truths or who "smuggle in" fuzzy concepts, wild conjectures, or even ideology as scientific findings. Rather than analyzing the *truth of reality* the possible *types of representation of truth* will be addressed in the following.

If we assume that modeling is necessary for the formulation of truth, the concept of models has to be analyzed in the beginning. [Amigoni and Somalvico, 2001] describe a model as follows:

> A model is: i) *finite*: only a finite content of knowledge is drawn from the phenomenon, intended as a potentially infinite source of information, and is inserted in the model; ii) *objective*: the model is so neat that everyone has the same understanding of the truth that is embedded in the model itself; iii) *experienceable*: the model can be used in order to predict the happening of a new phenomenon.
>
> We think of a model as the result of a modeling activity which consists in adopting a given *formalism* and in shaping within it a given *form*.
>
> A model has the following three properties: i) a model is *perfect* within itself, because the built up form is precisely described since the adopted formalism (such as mathematics, logic, etc.) provides exact shape definitions; ii) a model is *imperfect* in knowing a phenomenon of reality, because the abduction is a creative process that expresses the knowledge embedded within the model, describing only some of the elements which contribute to the perception of the phenomenon; it is clear that the abduction cannot produce a model rich enough to describe the whole phenomenon; iii) a model is *perfectible*, since it can be indefinitely substituted with a better (in the sense of less imperfect) model. The new model is the result of a new abduction (namely, of a pulse of creativity) which takes into account additional elements of the real phenomenon. The process can be iterated in order to obtain a sequence of models, each one better than the previous one, but worse than the subsequent one. However, when the phenomenon to be modeled is very complex, various models of such phenomenon are not necessarily arranged into a hierarchical (monotonous) order, but into a heterarchical (non-monotonous) order (in this case the alternative models are called *paradigms*).

These definitions do not include individual models like the human brain. But since we want to describe objective models these attributes are supposed to apply.

A practical example of objective models will be given: The purpose of a knowledge-based system (KBS) in medicine is not the implementation of human creativity into machines, but the transfer of knowledge (acknowledged decision structures) into the daily routine for both researchers and practitioners (also compare [Amigoni and Somalvico, 2001]).

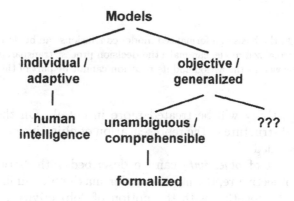

Figure 1. In contrast to individual models, objective models are supposed to unambiguously represent comprehensible knowledge structures.

In other words, a perfect model is supposed to describe the transition from the initial to the intended state unambiguously (compare Figure 2). Thus, a KBS should perform as an *unambiguous communication platform* (compare [Abernethy and Altman, 1999; Domingue and Motta, 1999]) – e.g. between researchers and practitioners. Hence, the decision-making should not be calculated in a mysterious "black-box"-system but be made transparent for all participants of the decision process.

It will be shown below that all kinds of perfect sub-models (e.g. a mathematical simulations) have to be embedded into a decision model to result in *one* comprehensive model – instead of fragmented partial models. So, before building a knowledge-based system the general properties of a *perfect decision-model* and the correlation to its *sub-models* are to be analyzed. This task will be the main point of this article.

Before discussing aspects of the representation of perfect decision models, the demands of well-known scientists with reference to the con-

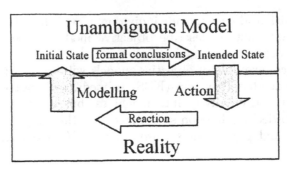

Figure 2.   On the basis of a formalized model conclusions can be derived unambiguously. This formalized model can make the decision process transparent. The process of modeling as well as the action and its reaction cannot be part of the formal model.

cept of objectivity will be treated more in detail. On their basis the fundamental structure of the representation of objective decision models can be concluded.

The concept of *objectivity* can be described with "strict matter-of-facts" or "objective representation under maximum elimination of subjectivity". So, together with the notion of "objectivity", an objective representation and an elimination of subjectivity has to be discussed more in detail.

There are so many definitions related to *objectivity* that only very few can be consulted representatively: Popper attributes a certain absoluteness to the notion of knowledge: "(1) There is knowledge in the subjective sense, which consist of dispositions and expectations. (2) But there is also knowledge in the objective sense, human knowledge, which consist of linguistically formulated expectations submitted to critical discussion" [Popper, 1972, p. 66]. This type he relates to problems, theories and arguments as such and emphasizes that no recognizing subject is necessary for this kind of knowledge. Lakatos underlines this point of view: he claims that the objective scientific value of a theory is independent from the human mind, namely the consciousness, which creates or recognizes it [Lakatos, 1978]. Heisenberg formulates a much more rigorous demand for a "closed" theory: he asks for its inner freedom of contradiction. According to him, it has to be possible to specify a notion derived from experience by definitions and axioms. The specified relations should be fixed in a way that mathematycal symbol can be assigned to the notion to result in a contradiction-free system. He emphasizes that we can never be sure in advance how far we can get with this kind of theory, which, as a consequence, limits the application range of theories [Heisenberg, 1971].

Heisenberg's demand for a formal freedom of contradiction can be correlated to the concept of causality: For Hessen an event can be predicted if it is causal, for Max Planck an event is causal if it can be predicted (both citations according to [Lukowski, 1966]). If the word causality is interpreted narrowly we also speak of a determinism meaning that fixed laws exist, predicting the future state of a system from its present state unambiguously (compare: [Heisenberg, 1971]). Therefore, the concept of causality is directly linked with the concept of prognosis.

On the other hand Russell claims that a thought is false if it cannot be integrated contradiction-free into the wholeness of our opinions. He concludes that true statements have to be a part of an enclosed system called "truth" [Russell, 1912].

According to Simon decisions can be programmed if they occur repeatable and unchangingly. Decisions cannot be programmed if they are innovative, complex and hardly structured [Simon, 1977]. As a generalized decision model has to support repetitions, its inner decision structure must be representable as a program.

According to these statements an objective model is of perfect type as described above. So, the following considerations will concentrate on the development of a model of objective decision models where both the meta-model and its instantiation can generally be programmed. First, the structure of a *decision model* will be deduced. Later its relation to *other types of models* will be derived.

It will not be treated, which concrete real-life problems can be represented with an objective decision model or how precise an objective decision model can be with respect to reality as no objective criterion can be found outside of the objective decision model for assessment. The process of building an instantiation of the meta-model for the representation of a real-life domain (also compare [Langley, 2000]) is also not the main point of the following discussion. For that reason techniques of acquiring and organizing coherent knowledge structure will not be considered.

2. Strategy: Superior structure of discrete decisions

To get an exhaustive and thus enclosed viewpoint onto formal decision models, the relation between the leading strategy and its adapted submodels has to be analyzed.

In business-economics the concept of a *strategy* is used in combination with decision-making for leading an undertaking. Here, several representative numbers are set up for the derivation of a strategy and the respective action guideline for the board of directors (compare [Petkoff, 1998]). The concept of a strategy is defined by Schneeweiß as a "sequence of planning decisions" [Schneeweiß, 1991].

From the point of view of computer science a *strategy* takes control over the order of the execution of partial tasks for a complex problem (compare: "strategic layer" [Breuker et al., 1994; Schäffer, 1996; Wielinga et al., 1992]).

Both definitions – in business-economics and computer science – have in common that the strategy consists of the superior decisions. In both viewpoints the notion of a strategy has a strongly subjective nuance as a precise demarcation to the adapted sub-models is missing.

Therefore, the following discussion tries to sketch a more precise concept of a strategy. On this basis the interaction between a strategy and its sub-models can be derived. Before this interaction can be analyzed, the role of a strategy in the context of the whole planning has to be defined.

2.1   Planning as an anticipation of decisions

From the point of view of economics planning is always related to an action. A model is needed for the prediction of the outcome and the assessment of the resulting states. Here, *planning* in its general meaning stands for the anticipation of decisions [Mag, 1993]. It can help raising the "outcome quality". Thus, the result of the actions can be estimated before actually acting. Therefore, during planning the notion of quality plays an important role: it describes the faultlessness of a solution or process. For Juran and Haux higher quality means: the reduction of errors, the reduction of reparation and loss, the reduction of customers dissatisfaction (patient dissatisfaction), the reduction of inspections and tests, the improvement of use and capacity with reference to financial expenses (compare [Juran, 1999] or [Haux, 1997]).

For the formulation of a general "model of procedure" a *"planning of planning"* (meta-planning) is necessary. If alternative actions are provided during the application of this *meta-plan* (planning), a decision is necessary from the viewpoint of the "ex-ante-design". Herein, decision means the choice of a possible action out of several alternatives which cannot be realized simultaneously (compare: "field of decisions" in [Schneeweiß, 1991]). Therefore, a decision during planning is – in contrast to e.g. a legal decision – not retrospective but prospective (compare Juran's roadmap to quality management, in [Juran, 1999]).

In the following the description of the various aspects of result quality will not be the main point. The further analysis will concentrate on guaranteeing the structural and process quality of the planning itself. The author assumes that an unambiguous comprehension of the planning cannot directly influence the result quality. But analyzing the decision structure and its process is of use for the scientific discussion and therefore for the quality management. So, in the following the demand is set up to comprehend the structure of an objective planning model unambiguously. Therefore, an objective formulation of problem solutions should be strived as a basis for both communication and investigation of any scientific problem solution.

Starting from the strategy as the superior decision level a top-down approach for an objective decision model as well as the fundamental boundaries of this kind of model have to be set up.

2.2   Fundamental properties of a decision-strategy

As observed above a strategy should be comprehensible unambiguously. To be able to make a choice within the discrete solution spectrum, decisions have to have a finite amount of solution alternatives.[1] We will start with the smallest unit of decision-making.

Each binary decision can be represented by a bit. The bit is well known to be the smallest unit of information [Shannon and Weaver, 1963]. Accordingly, the binary decision has to be the smallest unit of decision.[2] If a decision is interpreted as being a classification of the solution space, the binary decision accordingly can be called the smallest unit of classification.

From a formal point of view each sequence of binary sub-decisions can be joined to one discrete decision with more than two decision alternatives[3] (compare Figure 3). This encapsulation can be called "information hiding".

On the other hand, a decision with more than two decision alternatives can be divided in different ways:

$$A \vee B \vee C \vee D = (A \vee B) \vee (C \vee D) = (A \vee B \vee C) \vee D$$

---

[1]This demand distinguishes from other notions of strategy which presume a meta-modeling *either* in mathematical *or* logical manner (compare [Petkoff, 1998]).
[2]This fact does not mean that its reasoning has to be trivial.
[3]For each type of representation the justifications of the decisions will be different.

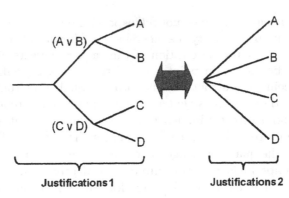

Figure 3.   Each decision can be analyzed into a sequence of binary decision. On the other hand a decision sequence can be summarized to one single decision. From a formal point of view these types of representation are equivalent. But the justifications for the single decision steps will vary.

These different types of representation for the same decision problem cannot be distinguished from a formal point of view as there are no formal criteria (within the formal model!) giving preference to one of these types of representation.[4]

But the justifications for the single decisions will vary between these types of representation as the context is different. An example for such a kind of decision-sequence can be given on the basis of a classification of a scientific research program (compare Figure 4). Herein, several competing and cooperating models, called paradigms, can be integrated (compare [Minsky, 1986]).

Figure 4.   A research program can be classified within its field of application. Vice versa decisions have to be made for the application of these models. Other kinds of classification are possible, but here an agreement is necessary to guarantee the same basis for reasoning.

Within a research program the decision alternatives can again be divided into distinct levels of decision-making. E.g. the gap between two

---

[4]Whether the minimal resulting tree (minimum number of branches and leaves) serves as the best representation of the real life problem cannot be defined a priori.

teeth can be described by a mathematical model. If the gap is larger than x then *do plan replacement*. Here, the output of the gap model serves as an input to the *replacement model*.

Like this one sub-model delivers information for the decision about the choice of the following sub-model and provides the necessary information for further planning.

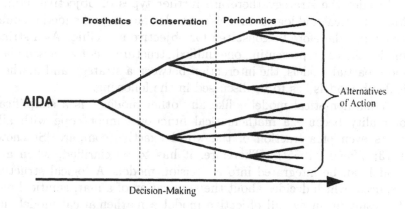

Figure 5. The sample decision sequence can be depicted as a decision tree. The decision alternatives represent a discrete, finite solution space.

Vice versa decisions are necessary for the location of both the application and the extension of a research program. In other words *classification* and *decision* have complementary character but depend on the same underlying decision structure.[5]

As a result, the choice of a sub-model can be called the *strategy*. The underlying decision-structure (including the research classification and the decision for the respective sub-model) can be called *strategy classification* or *paradigm* (scientific viewpoint, compare [Lakatos, 1978]). Here, only the *choice* of a sub-model can been objectified, but not necessarily its motivation.

---

[5]One can imagine other sequences for the classification of the AIDA project than shown in Figure 4. But such an alternative classification can be classified within the super-structure – as long as it is of scientific nature. Like this, an intuitive decision has to be made either to use the scientific model or an alternative one. In other words both models can be structured within a collective (super-posed) "root" where the decision process starts. This does not alter the fact that a classification is possible and necessary.

But it has not been clarified yet, whether the respective sub-model itself is formulated unambiguously. An according analysis of objective sub-models will be made in the following.

3. Embedding of sub-models into the decision-structure

Besides the strategy there are further types of objective modeling. Mathematical and logical models[6] can also be comprehended unambiguously and therefore be consulted for objective modeling. As a stringent model aims at representing one unified structure – not an assemblage of loose partial models, the interaction between a strategy and mathematical sub-models has to be discussed in the following.

A mathematical model – like any other model – is a simplification of reality because a mathematical function cannot cope with all aspects even of a fraction of the world exactly (compare [Stachowiak, 1973; Wiener, 1965]). Therefore, it has to be clarified, when a sub-model can be integrated into a decision model. A logical structure is required, which decides about the application of a mathematical model. To result in an overall objective model a mathematical model should be integrated into a decision model to objectify its application. In other words, a causal model can support the confinement of the problem space and assigns a domain to the mathematical sub-model.

As the appropriateness of a mathematical function has to be verified by humans building the decision structure, only a discrete amount of recognized mathematical models can be available.[7] Accordingly, only a finite amount of decisions about the application of a mathematical model is possible and necessary respectively. Thus, it can be expected to get a finite decision model about the application of mathematical models.

In the following the integration of different types of mathematical models into the strategic decision structure has to be clarified. In these considerations only mathematical models with continuous and such with discrete solution space will be distinguished.

3.1   Sub-models with continuous solutions space

A decision describing the application of a mathematical model with a continuous solution space can be objectified on the basis of a formal de-

---

[6]E.g.: Mechanical models and biometric models can be a combination of logical and mathematical nature. Here a similar structure to *if x is larger than y then follows z* can be found.

[7]The mathematical model might be specified by variables, but here the function will still remain the same.

cision model. But the integration of this kind of model into the *middle* of a decision structure is not possible as its solutions space is incompatible with the continuation of the decision structure.

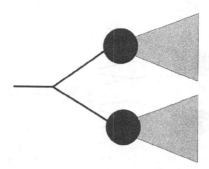

Figure 6. A mathematical model (circle) with a continuous solution spectrum cannot be bound *into* a decision structure (lines) but only at its end as a its solution spectrum is incompatible with the discrete decision structure. But the continuous solution spectrum (function) can help at the end of a decision structure to carry out an action (e.g. guide a medical operation).

An example for such a mathematical sub-model can be used for the inquiry of the tooth gap.

As far as no further measures are taken to generate a discrete number of solution alternatives from the solution spectrum the decision methodology has to terminate at this point.

3.2 Mathematical models with discrete solution space

If the continuous solution spectrum of the mathematical model is classified, a discrete solution spectrum results.[8] A decision about the application of a mathematical model with a discrete solution space is still possible and necessary. But as this kind of mathematical model provides a discrete amount of solution alternatives, its integration into the *middle* of a decision structure is possible without further measures. Each of its discrete solutions makes further alternatives in the decision structure possible.

---

[8]E.g. the solution space of a biometric model can be partitioned by a "confidence interval" of 95%. Here, the solution space is divided into two or three sections respectively.

From a different perspective the decision structure can be drawn including the alternatives from the mathematical model. From this point of view the mathematical model only assists the specific decisions (compare Figure 7).

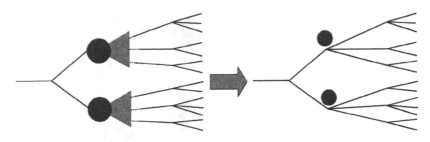

Figure 7. By classifying the solution space of the mathematical function the solution can be divided into discrete sections. From the point of view of the referring decision the mathematical function acts decision-supporting.

4. Imbedding of intra-individual preferences into the decision model

Not all aspects of a problem can be modeled formally.[9] Up to now is has not been clarified how these "intra-individual preferences" can be added to the formal decision model. On one hand they cannot be bound to the formal decision model formally as they underline the influences and interpretations of individuals and therefore do not satisfy the criteria of objectivity. On the other hand they are necessary for the application of a decision methodology as they can support decisions which cannot be predefined by mathematical functions.

But they can show the relation of the formal decision model to reality and thus guide the individual sub-decision. This type of guidance will be called "decision motivation" in the following. As a decision has a discrete solution spectrum – as shown above – only a discrete amount of different motivations is required for each sub-decision of the formal model.

One could hit upon the idea to build a formal model for a stringent motivation of a decision to get a strictly logical deduction of partial decisions. But this kind of sub-model could directly be integrated into

---

[9]E.g. the decision about the appropriateness of the decision model in an individual case.

the formal decision model because of its direct correlation. With other words it would lead to a simple extension of the strategy. Like this the problem of explaining the relation between the formal decision model and reality is not solved. Like this it can be made plausible that motivations are necessary which cannot be bound to the formal model formally.

An arbitrary formulation of motivations does not make sense as it could be interpreted arbitrarily. But even if a motivation cannot be included formally, it can be deposited explicitly.[10] So, it can be reproduced at any time. It can be stored on the basis of a formal convention – e.g. in a specific file format. This is correlated with a technical convention of storage and transmission and not the formalization of the interpretation within the decision methodology. Therefore, the formalization has only an indirect connection to the process of decision-making as it does not capture semantics.

The motivation can also initiate an action.[11] This action is not pre-defined precisely but necessary anyway.

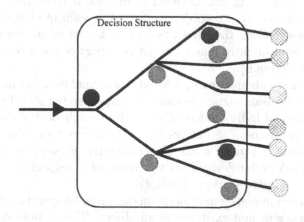

Figure 8. The relation of each partial decision to reality can be achieved on the basis of formally bound models (black) or non formally bound models (grey). Each of these partial sub-models can also initiate an action. The models at the end of the decision methodology on the other hand can *only* describe actions (formal sub-models = cross-hatched, non-formal motivations = checked).

---

[10]Text, Pictures, Sound, etc.

[11]E.g.: If the patient has not been disinfected yet, the nurse has to wash the wound.

5.   Generating individual problem solutions with the decision model

After the general structure of the decision model has been described, the individual decisions have to be derived on its basis. The user has to be able to get concrete action alternatives by the application of the decision model. In addition he needs decision motivations to be able to explain the single decisions and to find the optimal solution. The conscious decision can also be concluded at the end of a decision sequence that no action is necessary or useful. Though several repetitions of statements they will be shown from the perspective of the applying specialist.[12]

5.1   Comprehending a decision structure

Before a decision structure can be consulted for the solution of an individual decision problem, a principle decision about its application has to be made. As this decision cannot be derived from the decision model itself, it cannot be objective from the point of view of the decision methodology. Even the correlation of each part of the formal model to reality cannot be proven as it cannot be a part of the formal model itself (compare Figure 6).

If a decision about the application of a formal decision model has been made, the decision structure can be comprehended step by step for the specification of individual solutions. Here, each partial decision makes a choice from the alternatives from the current decision step. Each partial decision can be made either intuitively (supported by decision motivations) or be derived on the basis of a respective mathematical function (compare sections 3 and 4).

In decision models binary decision can often be found as they represent the existence or non-existence of an object. The ex-post decision about these two alternatives of the binary decision can be seen as the smallest unit of ex-post decisions. In general it is also called "diagnosis". The simplest type of decision therefore is a single diagnosis.

5.2   Application of sub-models

It has not been clarified yet, where the information comes from, which is inserted into the mathematical function for the derivation of the decision and the respective action. This kind of information cannot origin from the formal model itself and therefore has to be deduced from an ob-

---

[12]In the previous sections it has been shown from the perspective of the constructing expert.

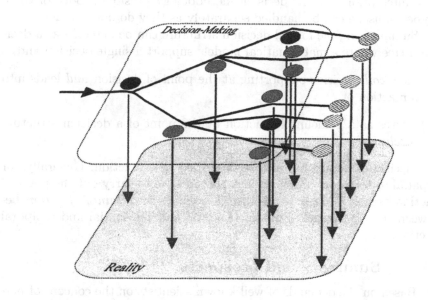

Figure 9. Each component of the objective decision methodology (black lines) has its specific relation to reality which can be illustrated by appropriate explanations (gray arrows). These explanations can be deposited formally, but do not have a formal relation with the formal relation.

servation. In other words an ex-post link of an observation with a scale is necessary. This procedure is usually called "measurement". In addition a binary decision about the success of a measurement is necessary and assigns the measured value to the decision model.

5.3   Guiding an action by a sub-model

Generally, a mathematical model can be support both, an action and the preparation of a decision (if there is a classification, which makes the integration into the decision methodology possible). Both of these types of usage can be handled separately as they do not interact.

Summarizing the formal decision structure can be depicted as a decision tree where a mathematical models support a single decision and

a) acts *both* decision-supporting at the point of decision *and* leads into an action or

b) serves as a guideline for action at the point of a decision structure *exclusively.*

A mechanical action e.g. can be described by a function. Generally, for spatial or temporal problems it is not possible to carry out the planned action exactly.[13] In other words, there is an *uncertainty of action* between the mathematical model of action and the spatial and temporal action.

6.   Summary and prospects

Based on the demands of well known scientists on the concept of perfect models and objectivity a general structure of an objective decision model has been derived. As the objective model has to be comprehensible and therefore represented formally, conclusions can be drawn for the fundamental structure of its meta-model. The relation of an objective model to reality cannot be objectified as it cannot be a part of the conclusive model itself.

It could be shown that a decision structure can be supported by mathematical sub-models and decision motivations. Here, the analysis concentrated on the product of a formal decision model.

Several aspects of objective models have to be emphasized: a) The process of acquiring the formal decision model has not been described in this article. b) A concrete formal decision model not necessarily has to utilize all possibilities of the meta-modeling for decision-making as

---

[13]In technical term we usually talk about "tolerance" e.g. of a piece of work.

described above. In other words, simplified versions can be built *only* consisting of e.g. a single mathematical model[14] or merely a formal decision structure.[15]

The instantiations of the developed meta-structure can serve as a domain ontology (also compare [Barley et al., 1997; Fernández et al., 1997]) and a communication platform (also compare [Arpfrez et al., 1998]) between scientists (also compare [Amigoni et al., 2000]). With appropriate tools managing these domain models and their multidimensional classification it should be possible to identify both redundancies and lacks in the world-wide research. A computer system seems to be most appropriate for the representation of such a decision-model and its sub-models as all of its parts can be represented on the basis of formal logic and mathematics (also compare [Langley, 2000]). Together they can form what Minsky called an agency of agents [Minsky, 1986].

Acknowledgment

The author would like to thank Falk Schubert, Stefan Skonetzki, Thomas Wetter and Francesco Amigoni for their comments on and Sophia Wille for the revision of the draft of this paper.

References

Abernethy, N. and Altman, R., 1999, SOPHIA: Delivering ontologies and knowledge bases over the web from a simple RDBMS, in: *Stanford Report SMI-1999-0773*, Stanford University, Palo Alto.

Amigoni, F., Schiaffonati, V., and Somalvico, M., 2000, A multilevel architecture of creative dynamic agency, *Foundations of Science* 5:157–184.

Amigoni, F., Somalvico, M., 2001, Dynamic agency: Models for creative production and technology applications, in: *Communications Through Virtual Technology: Identity, Community, and Technology in Internet Age*, G. Riva and F. Davide, eds., IOS Press, Amsterdam, pp. 167–192.

Arpfrez, J., Gómez-Pérez, A., and Lozano, A., 1998, (ONTO)^2 Agent: An ontology-based WWW broker to select ontologies, in: *ECAI-98 - Workshop on Applications of Ontologies and Problem Solving Methods. The 13th European Conference on Artificial Intelligence*, Brighton, UK, pp. 16–24.

Barley M, Clark, P., and Williamson, K., 1997, The neutral representation project, in: *Ontological Engineering - AAAI-97*, pp. 1–8.

Breuker, J., van de Velde, W., et al. *CommonKads Library for Expertise Modeling, Reusable Problem Solving Components*, vol. 21, Oxford, IOS Press, Amsterdam.

Britannica, 1998, classic edition, 15th revised edition, Chicago.

---

[14] These types of modeling can mostly be found in natural sciences.

[15] Most of the formal humanities concentrate on the construction of decision models. This becomes obvious in the law studies.

Bunge, M., 1996, In praise of intolerance to charlatanism in academia, in: *The Flight from Science and Change*, P.R. Gross, N. Levitt, and M.W. Lewis, eds., New York Academy of Sciences, New York, pp. 96–115.

Domingue, J. and Motta, E., 1999, A knowledge-based news server supporting ontology-driven story enrichment and knowledge retrival, in: *11th European Workshop on Knowledge Acquisition, Modelling and Management (EKAW'99)*, pp. 103–120.

Fernández, M., Gómez-Pérez, A., and Juristo, N., 1997, From ontological arts towards ontological engineering, in: *Ontological Engineering - AAAI-97*, pp. 33–40.

Haux, R., 1997, Aims and tasks of medical informatics, *Int. J. Med. Inform.* 44:9-20.

Heisenberg, W., 1971, *Schritte über Grenzen*, Piper, München (translated in English by P. Heath, *Across the Frontiers*, Harper & Row, New York, 1974).

Juran, J.M, et al., 1999, *Juran's Quality Handbook*, 5th ed., McGraw Hill, New York.

Lakatos, I., 1978, Science and pseudoscience, in: Worall J, Curie, G., and I. Lakatos, eds., *Philosophical Papers*, vol. 01, Cambridge University Press, Cambridge.

Langley, P., 2000, The computational support of scientific discovery, in: *Human-Computer Studies* 53:393-410.

Lukowski, A., 1966, *Philosphie des Arzttums*, Dt. Ärzte-Verlag, Köln.

Mag, W., 1993, *Planung und Unsicherheit*, 5th ed., Shaffer-Poeschel, Stuttgart.

Minsky, M., 1986, *The Society of Mind*, Simon & Schuster, New York.

Petkoff, B., 1998, *Wissensmanagement*, Addison-Wesley, Bonn.

Popper, K.R., 1972, *Objective Knowledge: An Evolutionary Approach*, Clarendon Press, Oxford.

Russell, B., 1912, *Problems of Philosophy*, Oxford University Press, London.

Schäffer, H., 1996, *Die Repräsentation von Managementwissen: Die Modellierung von Wissen für das Management von Expertensystemprojekten gemäßdem KADS-Ansatz*, Lang, Frankfurt am Main.

Schneeweiß, C., ed., 1991, *Planung 1 - Systemanalytische und entscheidungstheoretische Grundlagen*, vol. 1, Springer, Heidelberg.

Shannon, C.E. and Weaver, W., 1963, *Mathematical Theory of Communication*, University of Illinois Press, Urbana, IL.

Simon, H.A., *The New Science of Management Decision*, Prentice-Hall, Englewood Cliffs, NJ.

Stachowiak, H., 1973, *Allgemeine Modelltheorie*, Springer, Wien.

Wielinga, B., Schreiber, A.T., and Breuker, J.A., 1992, KADS: A modeling approach to knowledge engineering. Knowledge Acquisition - Special issue "The KADS approach to knowledge engineering" 1992,4.

Wiener, N., 1965, *Cybernetics or Control and Communication in the animal*, MIT Press, Düsseldorf.

Subject Index

Author Index

APPLIED LOGIC SERIES

1. D. Walton: *Fallacies Arising from Ambiguity.* 1996   ISBN 0-7923-4100-7
2. H. Wansing (ed.): *Proof Theory of Modal Logic.* 1996 ISBN 0-7923-4120-1
3. F. Baader and K.U. Schulz (eds.): *Frontiers of Combining Systems.* First International Workshop, Munich, March 1996. 1996   ISBN 0-7923-4271-2
4. M. Marx and Y. Venema: *Multi-Dimensional Modal Logic.* 1996
ISBN 0-7923-4345-X
5. S. Akama (ed.): *Logic, Language and Computation.* 1997
ISBN 0-7923-4376-X
6. J. Goubault-Larrecq and I. Mackie: *Proof Theory and Automated Deduction.* 1997   ISBN 0-7923-4593-2
7. M. de Rijke (ed.): *Advances in Intensional Logic.* 1997 ISBN 0-7923-4711-0
8. W. Bibel and P.H. Schmitt (eds.): *Automated Deduction - A Basis for Applications.* Volume I. Foundations - Calculi and Methods. 1998
ISBN 0-7923-5129-0
9. W. Bibel and P.H. Schmitt (eds.): *Automated Deduction - A Basis for Applications.* Volume II. Systems and Implementation Techniques. 1998
ISBN 0-7923-5130-4
10. W. Bibel and P.H. Schmitt (eds.): *Automated Deduction - A Basis for Applications.* Volume III. Applications. 1998   ISBN 0-7923-5131-2
(Set vols. I-III: ISBN 0-7923-5132-0)
11. S.O. Hansson: *A Textbook of Belief Dynamics.* Theory Change and Database Updating. 1999   Hb: ISBN 0-7923-5324-2; Pb: ISBN 0-7923-5327-7
Solutions to exercises. 1999.   Pb: ISBN 0-7923-5328-5
Set: (Hb): ISBN 0-7923-5326-9; (Pb): ISBN 0-7923-5329-3
12. R. Pareschi and B. Fronhöfer (eds.): *Dynamic Worlds from the Frame Problem to Knowledge Management.* 1999   ISBN 0-7923-5535-0
13. D.M. Gabbay and H. Wansing (eds.): *What is Negation?* 1999
ISBN 0-7923-5569-5
14. M. Wooldridge and A. Rao (eds.): *Foundations of Rational Agency.* 1999
ISBN 0-7923-5601-2
15. D. Dubois, H. Prade and E.P. Klement (eds.): *Fuzzy Sets, Logics and Reasoning about Knowledge.* 1999   ISBN 0-7923-5911-1
16. H. Barringer, M. Fisher, D. Gabbay and G. Gough (eds.): *Advances in Temporal Logic.* 2000   ISBN 0-7923-6149-0
17. D. Basin, M.D. Agostino, D.M. Gabbay, S. Matthews and L. Viganò (eds.): *Labelled Deduction.* 2000   ISBN 0-7923-6237-3
18. P.A. Flach and A.C. Kakas (eds.): *Abduction and Induction.* Essays on their Relation and Integration. 2000   ISBN 0-7923-6250-0
19. S. Hölldobler (ed.): *Intellectics and Computational Logic.* Papers in Honor of Wolfgang Bibel. 2000   ISBN 0-7923-6261-6

APPLIED LOGIC SERIES

20. P. Bonzon, M. Cavalcanti and Rolf Nossum (eds.): *Formal Aspects of Context.* 2000
ISBN 0-7923-6350-7

21. D.M. Gabbay and N. Olivetti: *Goal-Directed Proof Theory.* 2000
ISBN 0-7923-6473-2

22. M.-A. Williams and H. Rott (eds.): *Frontiers in Belief Revision.* 2001
ISBN 0-7923-7021-X

23. E. Morscher and A. Hieke (eds.): *New Essays in Free Logic.* In Honour of Karel Lambert. 2001
ISBN 1-4020-0216-5

24. D. Corfield and J. Williamson (eds.): *Foundations of Bayesianism.* 2001
ISBN 1-4020-0223-8

25. L. Magnani, N.J. Nersessian and C. Pizzi (eds.): *Logical and Computational Aspects of Model-Based Reasoning.* 2002
Hb: ISBN 1-4020-0712-4; Pb: ISBN 1-4020-0791-4

26. D.J. Pym: *The Semantics and Proof Theory of the Logic of Bunched Implications.* 2002
ISBN 1-4020-0745-0

27. P.B. Andrews: *An Introduction to Mathematical Logic and Type Theory: To Truth Through Proof.* Second edition. 2002
ISBN 1-4020-0763-9

KLUWER ACADEMIC PUBLISHERS – DORDRECHT / BOSTON / LONDON